DOMESTIC GAS INSTALLATION PRACTICE
(GAS SERVICE TECHNOLOGY VOLUME 2)

This book is the second of a series of three manuals devoted to the theory and practice of gas service. The manuals – designated volumes 1, 2 and 3 – were first produced in 1978, 1979 and 1980 respectively.

This second volume – originally entitled "Domestic Installation and Servicing Practice" – was first updated in 1992 and since then great changes have been taking place within the gas industry. A major change has been the replacement of the Health and Safety Commission's Approved Code of Practice (ACOP) with the Nationally Accredited Certification Scheme for Individual Gas Fitting Operatives (ACS). Certification under this scheme must conform with European Standard EN 45013. All Certification bodies must operate to the satisfaction of the United Kingdom Accreditation Service (UKAS). This scheme is now in place for domestic natural gas operatives and will cover all areas including LPG and non-domestic gas work by the end of 1999. It sets out the levels for both new entrants into the gas industry and also for experienced operatives. Most new operatives will pass through an NVQ course, existing operatives will have to be assessment tested under the ACS scheme, and hold a certificate of competence in the type of gas work they carry out.

The further update of volume 2 therefore came at an important time. It was carried out and completed in 1995 by Eric Glennon, a former customer service training officer, and Frank Saxon, a former service trainer, both with British Gas. This revised update – retitled "Domestic Gas Installation Practice" – has been carried out by Frank Saxon in 1999. The original format has been retained for the revision along with the names of the authors of the original chapters. Where necessary standards, legislation, etc., have been amended and additional materials have been added to cover recent developments in the industry.

Domestic Gas Installation Practice

(Gas Service Technology Volume 2)

Revised update by
Frank Saxon

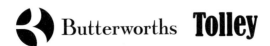 Butterworths **Tolley**

This title first published by
Benn Technical Books

First Edition 1979
Second impression 1980
Third impression 1982
Fourth (revised) impression 1986
Fifth impression 1987
Sixth impression 1988
Seventh impression 1989
Second Edition 1992
Reprinted 1993
Third Edition 1995
Revised update October 1999

Published by Butterworths Tolley
Tolley House, 2 Addiscombe Road, Croydon CR9 5AF

© Reed Elsevier (UK) Ltd 1999

Typeset by Letterpart Ltd, Reigate, Surrey

Printed in Great Britain by
Bath Press, Bath

All references and standards quoted in this manual were correct at the
time of publication. Readers should, however, satisfy themselves
that all such references and standards can still be substantiated. Whilst
care has been taken to ensure that information in this manual is up to date
and correct neither the individual contributors, the editor, nor the
publishers can accept responsibility for any errors or omissions.

ISBN 0 75450 553-7

Contents

Preface

In carrying out the second update of this title during 1994/5, its editors had encountered much change in the gas industry. During the research for this latest update it soon became apparent that considerable development is still taking place. Besides changes in standards and Building Regulations, the Gas Safety (Installation and Use) Regulations have been amended and the Health and Safety Commission's Approved Code of Practice (ACOP) has been replaced with the Nationally Accredited Certification Scheme for Individual Gas Fitting Operatives (ACS) in an attempt to standardise the assessment process across the country. Certification under this scheme must conform with European Standard EN 45013. All Certification bodies must operate to the satisfaction of the United Kingdom Accreditation Service (UKAS). The knowledge and experience required for success under the new ACS assessment system is much greater than that required under the old ACOP standard and this will increase the demand for training.

The editor feels that much, if not all, of the information needed by students studying for these qualifications and also by existing operatives, who are required to be assessment tested, will be found in this and the two other volumes of Gas Service Technology. To prevent the subject matter from becoming outdated, it has continued to be dealt with in a general rather than a particular way. However, in some instances reference has had to be made to specific appliances or equipment in order to cover some of the detail.

The reader should continue to consult current legislation, standards and manufacturers instructions, etc.

Wherever possible reference has been made to statutory obligations, particularly where they relate to safety. The Gas Safety (Installation and Use) Regulations 1998 should be studied in conjunction with the manual to ensure compliance with their requirements.

The installation and servicing methods and procedures described in this volume are believed to be accepted good practice at the time of writing. However, the editor and contributing authors cannot accept responsibility for any problems arising from the use of this information.

I would like to thank the technical services staff of the former British Gas (North Western) for the help they have given during the update of this volume and the Technology Department of Blackburn College for allowing the use of their premises and resources.

My thanks also to Mr A Pawsey for allowing the use of illustrations from his splendid magazine *Aspect*. Also a thank you to all those appliance and equipment manufacturers who supplied photographs, technical drawings and technical information.

Finally, I acknowledge Michael Webb, editor at Butterworths Tolley, and Chris Leggett of typesetters Letterpart, for all their constructive help in putting together this update.

Frank Saxon
September 1999

CHAPTER 1

Pipework Installation

Chapter 1 is based on an original draft by A. Davis

Introduction

This chapter is concerned with the basic principles of jointing and bending pipes and with the identification of the fittings and fixings associated with their installation.

The pipes, jointing materials and fittings are all the subjects of British Standard Specifications which regulate their composition, size, and construction. Reference is made throughout the manual to the requirements of these Standards.

There appears to be no satisfactory explanation why the words 'pipe' and 'tube' should be used for particular materials. Generally 'pipe' is applied to cast iron or steel, that is, the ferrous metals, while 'tube' is used for brass and copper. Lead is always 'pipe', and anything flexible is 'tubing'. British Standards seem to endorse these definitions although at some points they refer to 'copper pipe' and to 'steel pipes and tubes'.

This confusion has resulted in the two words being used synonymously. However, once pipes or tubes have been installed, they immediately become 'installation pipes' or 'pipework'.

Copper Tube

Tubes are produced by drawing, from de-oxidised, non-arsenical copper, 99.9% pure. They are made in two grades, both of which are specified in BS EN 1057: 1996.

Table 'X'

This tube is used for internal installations. It is supplied in half-hard condition in straight lengths of 6 metres. Smaller cut lengths can be obtained if required.

1

Table 'Y'

This is a heavier gauge tube used for external or buried installations. It is supplied in an annealed condition and is available in coils of 10, 25 or 50 metre lengths. Where liable to corrosion it should have a factory-bonded plastic sheath. In half-hard condition it is available in straight lengths up to 6 metres.

Since 1971, both Table 'X' and Table 'Y' have been in metric sizes based on the *outside diameter* of the tube. They use fittings specified in BS 864: 1983, Part 2 which may be compression or capillary soldered.

The original specifications of Imperial sizes were BS 659: 1967 for the equivalent of Table 'X' tube and BS 1386: 1957 for the equivalent of Table 'Y'. The sizes were based on the *nominal bore* of the tube. Fittings were to BS 864: 1972, Part 1. (All now withdrawn.)

Because of the differences between the outside diameter of metric and Imperial sized tubes, it has been necessary to use special adaptors when extending from an old imperial installation in metric tube. Table 1 shows the sizes and the adaptors required up to 54 mm.

Table 'Z' tube is used for waste or sanitation, but not for gas installations. It is a hard drawn, thin wall tube, not suitable for bending. The tube is available in sizes from 6 to 159 mm and the wall thickness ranges from 0.5 to 1.5 mm.

TABLE 1 Sizes of Copper Tube (up to 54 mm (2 in))

Imperial BS 659 Nominal bore (in)	Metric BS EN 1057: 1996 Outside diameter (mm)	Nominal wall thickness (mm)		Adaptors required	
		Table 'X'	Table 'Y'	Capillary joints	Compression joints
1/8	6	0.6	0.8	YES	YES
3/16	8	0.6	0.8	YES	YES
1/4	8	0.6	0.8	YES	YES
–	10	0.6	0.8	NO IMPERIAL EQUIVALENT	
3/8	12	0.6	0.8	YES	NO
1/2	15	0.7	1.0	YES	NO
–	18	0.8	1.0	NON-PREFERRED SIZE	
3/4	22	0.9	1.2	YES	YES
1	28	0.9	1.2	YES	NO
1 1/4	35	1.2	1.5	YES	YES
1 1/2	42	1.2	1.5	YES	YES
2	54	1.2	2.0	YES	NO

Steel Tube

Steel tube is manufactured to conform to BS 1387: 1985 (1990) and is technically equivalent to the standards issued by the International

Organisation for Standardisation (ISO). The nominal size of tube is identified by the letters DN followed by a number. It is a round number loosely related to a metric dimension and it is used on all fittings and components in a piping system. Screw threads used on steel tubes are to BS 21: 1985 Pipe threads for tube fittings where pressure-tight joints are made on the threads (metric dimensions). It is technically equivalent to the ISO standards. The thread size designation is based on the old Imperial sizes, without using the word 'inch' or its abbreviation.

Steel tube is made in three thicknesses or weights, each indicated by a colour band around the tube, Table 2. For each nominal size (DN), the outside diameter is the same for all three weights, so the same pipe thread may be used, Table 3.

TABLE 2 Thickness Series (Weights) of Steel Tube

Weight	Colour of band	Application
Light	Brown	Not used for gas or water supplies
Medium	Blue	used for internal gas installations and low pressure water supplies.
Heavy	Red	For gas service pipes, buried or external gas supplies and high pressure water supplies

TABLE 3 Dimensions of Steel Tubes (up to DN 50 (50 mm))

Nominal size (DN)	Designation of thread (BS 21)	Approx. outside diameter	Medium weight		Heavy weight	
			Thickness	Mass	Thickness	Mass
		mm	mm	kg/m	mm	kg/m
8	1/4	13.6	2.3	0.64	2.9	0.77
10	3/8	17.1	2.3	0.84	2.9	1.03
15	1/2	21.4	2.6	1.22	3.2	1.45
20	3/4	26.9	2.6	1.57	3.2	1.88
25	1	33.8	3.2	2.43	4.0	2.96
32	1 1/4	42.5	3.2	3.13	4.0	3.83
40	1 1/2	48.4	3.2	3.61	4.0	4.42
50	2	60.3	3.6	5.10	4.5	6.26

Steel tube is now very rarely used for domestic installations or service pipes, it has been replaced by other materials such as copper and polyethylene.

Because the metrication of steel tube has resulted in a change of name but not a change of diameter or thread, no adaptors are necessary when extending from existing supplies.

Plain steel pipe should be used only for gas supplies or closed water circuits, that is, central heating. All other steel water supplies must be coated externally and internally in accordance with BS 534: 1990 (Water by-laws 51 and 52).

Lead Pipe

Although used extensively in the past for gas and water installations, lead pipe is now prohibited by the Gas Safety (Installation and Use) Regulations 1998 (Part B Section 5(2a)) and Water by-law 9, (see Chapter 11, Vol. 1).

Methods of extending existing lead pipe installations will be dealt with later in this chapter.

Brass Tube

In the past, brass tube, either chrome plated or oxidized, was used extensively in the industry for making the final connection to fires and water heaters. The most common sizes were $\frac{1}{2}$ in (12.7 mm) and $\frac{3}{8}$ in (9 mm) outside diameter. They were threaded with BS 21 $\frac{1}{4}$ and $\frac{1}{8}$ threads, sometimes a non-standard thread of 26 T.P.I. (threads per inch) was used. Special miniature stocks and dies were used together with a rubber lined hand vice.

Brass tube has been superseded by copper tube which is much easier to manipulate and joint.

Stainless Steel Tube

This tube is manufactured to BS 4127: 1994. It has the same outside diameter as BS 2871 Table 'X' copper tube and can be manipulated and jointed in the same way. However, BS 6891: 1998, recommends that it should only be jointed with type A (non-manipulative) compression fittings complying with BS 864: Part 2 and not by soldering.

TABLE 4 Dimensions of Light Gauge Stainless Steel Tubes

Size of tube (outside diameter)	Nominal wall thickness
mm	mm
6	0.6
8	0.6
10	0.6
12	0.6
15	0.6
18	0.7
22	0.7
28	0.8
35	1.0
42	1.1

Bundy Tubing

This is a type of mild steel tubing marketed by Bundy (Telford) Ltd. It is used in the gas industry for gas supplies on appliances, for example, the oven supply on a gas cooker. The tubing is also used for hydraulic brake lines, refrigeration and conveying oil.

Bundy tube is made from mild steel strip which has been copper plated on both sides. The strip is wrapped twice around itself, lengthwise, to form a double walled tube, Fig. 1(a). It is then heated in a reducing atmosphere to produce a brazed joint where the copper-covered surfaces are in contact around the whole circumference.

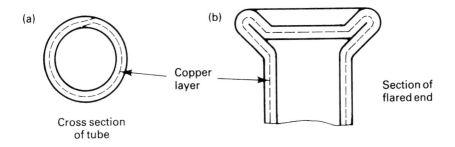

Fig. 1   Bundy tubing

The tubing is produced in 3 and 4.5 metre lengths and is made with the following outside diameters: 3.18, 4.76, 6.35 and 7.94 mm. The wall thickness for all sizes is 0.71 mm.

It may be bent, coiled, flanged, or swaged and may be jointed by soldering, brazing or compression fittings. Figure 1(b) shows an end flared, by the manufacturer, to form a compression joint.

Because of its copper plating, the tube resists rusting and may be mistaken for copper tube. It can be tinned or electro-plated and the makers can supply special lengths bent to any desired shape and fitted with any desired connection.

Plastic Pipe

Polyethylene

Polyethylene (PE) pipes are used extensively for mains and services in the gas and water industries (see Chapter 12, Vol. 1). The following information is permanently and legibly marked along the wall of each length of PE pipe used for distributing natural gas.

  (i)   Identity of manufacturer
  (ii) The letters PE
  (iii) Identification of the base polymer (This can be in the form of a code, e.g. X)
  (iv) Pipe diameter in millimetres
  (v)   The word METRIC or the letters MM
  (vi) The standard dimension ratio preceded by the letters SDR (This ratio is determined by the calculation)

$$\frac{\text{Specified minimum outside diameter}}{\text{Minimum specified wall thickness}}$$

  (vii) The letters BGC/PS/PL2 (if Phase 2 approval has been given by quality assurance)
  (viii) Identification of the shift production line and date of manufacture
  (ix)   Coiled pipe only – a sequential number, increasing at 1 metre intervals along the pipe between 000 to 999, or 0000 to 9999.

Examples of the outside diameters and wall thicknesses of some pipes used for gas distribution are given in Table 5.

TABLE 5 Outside Diameter and Wall thickness – PE pipes

| Pipe size | Outside diameter | | Wall thickness | | | |
| | | | SDR 11 | | SDR 17.6 | |
	min	max	min	max	min	max
mm	mm	mm	mm	mm	mm	mm
25	25	25.3	2.3	2.6		
50	50	50.4	4.6	5.2	2.9	3.3
63	63	63.4	5.8	6.5	3.6	4.1
75	75	75.5	6.9	7.8	4.3	4.9

In addition to their use for mains and services, PE pipes are now being used quite extensively in commercial and industrial situations, as installation pipes when they can be buried underground. PE pipes are adversely affected by sunlight and are only used below ground unless they are protected by a fibreglass cover or some other protective material.

Polyvinyl Chloride (PVC)

Ordinary PVC pipes are suitable only for use on cold water supplies. They become soft at high temperatures and brittle at very low temperatures. Because of their high coefficient of expansion (6 times that of steel pipe) provision for expansion must be made on long pipe runs.

Unplasticized PVC pipe, conforming to BS 3505: 1986 Unplasticized polyvinyl chloride (PVC-U) pressure pipes for cold potable water, is a thin-walled rigid pipe used for water supplies. It is supplied in nominal sizes from ³/₈ to 24 (17 to 610 mm outside diameter) and in 6 or 9 metre lengths. The pipes are classified for maximum sustained working pressure as follows:

(a) 9 bar (class C);
(b) 12 bar (class D);
(c) 15 bar (class E).

Table 6 gives details of small sizes of class E pipes.

Impact modified, or mPVC was used for gas distribution pipes and fittings, it was not suitable for internal gas installations.

TABLE 6 Pipe Dimensions for PVC-U 15 bar (Class E) Pipes

Nominal size	Mean outside diameter	Mean wall thickness
mm	mm	mm
³/₈	17.15	1.7
¹/₂	21.35	1.9
³/₄	26.75	2.2
1	33.55	2.5

Other Plastic Pipes

Thermoplastic pipes and fittings are available for use in domestic hot and cold water services and heating systems. The specification for the installation of these pipes and fittings is found in BS 5955: Part 8: 1995. Plastic pipework (thermoplastic materials) specification for the installation of thermoplastic pipes and fittings for use in domestic hot and cold water services and heating systems.

BS 7291 Thermoplastic pipes and associated fittings for hot and cold water for domestic purposes and heating installations in buildings is in four parts. Part 1: 1990 (1995) gives general requirements. Part 2: 1990 (1995) gives the specification for polybutylene (PB) pipes and associated fittings. Part 3: 1990 the specification for crosslinked polyethylene (PE-X) pipes and fittings and Part 4: 1990 (1995) Specification for chlorinated polyvinyl chloride (PVC-C) pipes and fittings and solvent cement.

Flexible hoses, for appliances that need to be pulled out for cleaning purposes (cookers, clothes driers, etc.) and burning first or second family gases, are made to BS 669: Part 1: 1989. These can be made from butyl, nylon or other plastics which do not absorb hydrocarbons. They can be all plastic or supported by a flexible metal tube or helical metal coil or wire.

Armoured flexible tube is used on commercial and industrial equipment and is covered in Vol. 3.

Flexible hoses for appliances burning third family gases are manufactured to BS 3212 and are made from neoprene. BS EN 549: 1995 covers the specification for rubber type materials for use with first, second and third family gas flexible hoses.

Polythene Pipe

This is a thick walled pliable pipe which was used only for above ground water services, it was manufactured to BS 1972: 1967.

The standard has now been withdrawn and partially replaced by BS 6572: 1985 Blue polyethylene pipes up to nominal size 63 for below ground use for potable water, and BS 6730: 1986 Black polyethylene pipes up to nominal size 63 for above ground use for potable water.

Black polyethylene pipe is not used below ground, to avoid confusion with buried power cables with black plastic sheathing.

Colour Coding of Plastic Pipes

Distribution pipes for carrying natural gas underground are coloured yellow. Distribution pipes, up to 75 mm external diameter, for carrying potable water underground are coloured blue.

Jointing

Joints between sections of pipe and pipe fittings or appliances must be:

- mechanically strong
- completely gas or water tight
- free from internal obstruction to flow of fluid
- neat and unobtrusive.

There are a number of different forms of joints which may be categorised as follows:

1.   Joints which may be easily disconnected and reconnected:

- screwed joints
- flanged joints
- compression joints and unions
- plug-in connections.

2.   Joints which can be disconnected and remade:

- capillary soldered joints
- blown joints.

3.   Joints which cannot be disconnected and remade:

- welded joints
- brazed joints
- chemical solvent joints.

Joints in this third category are seldom used on internal domestic gas installation pipes.

Screwed Joints

The pipe thread in use for tubes and fittings where pressure-tight joints are made on the threads is the British Standard pipe thread to BS 21: 1985, which is based on ISO 7/1 1982 specification. This thread is a 'Whitworth' thread form with an included angle of 55°, Fig. 2. The BS Whitworth thread is used in engineering on nuts and bolts and is named after Sir Joseph Whitworth who designed it as a standard thread form for British engineers.

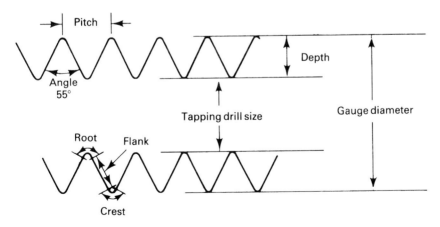

Fig. 2   Terms associated with threads

BS 21 pipe threads are indicated by the letter 'R'. Threads may be internal or external and either parallel or taper. A parallel thread has the same gauge diameter all along the thread. The taper thread has a taper of 1 in 16, that is, the diameter decreases by 1 mm in 16 mm length.

The type of thread is indicated by subscript letters:

c = internal taper
p = internal parallel
L = external parallel, longscrew.

The letter 'R' without a subscript indicates an external taper thread.

For example, the threads for 15 mm (½ in) pipe are:

$R\frac{1}{2}$ = external taper
$R_c\frac{1}{2}$ = internal taper
$R_L\frac{1}{2}$ = external parallel, longscrew
$R_p\frac{1}{2}$ = internal parallel.

The terms associated with screw threads are given in Fig. 2.

After the threads have been screwed up hand tight, a wrenching or tightening allowance of about $1\frac{1}{2}$ turns is provided for and damage to the fittings will occur if this is exceeded.

In order to judge the correct length of pipe to cut off, so that it will fit between two fittings, it is necessary to know how much thread will be screwed into the fittings. Table 7 gives details of the overall lengths of threads and their pitch, which is the distance the pipe enters the fittings for each complete turn.

TABLE 7 Sizes and Lengths of Pipe Threads

Nominal bore	Thread size	Pitch	Approximate length of thread (mm)	
			Screwed into fitting	Total length cut
mm		mm		
6	⅛	0.907	5	9
8	¼	1.337	7	11
10	⅜	1.337	9	13
15	½	1.814	11	16
20	¾	1.814	14	19
25	1	2.309	17	22
32	1¼	2.309	20	25
40	1½	2.309	20	25
50	2	2.309	23	29

For practical purposes, the pitch of pipe threads are approximately:

$R\frac{1}{8}$ = 1.0 mm
$R\frac{1}{4}$ and $R\frac{3}{8}$ = 1.5 mm

$R\frac{1}{2}$ and $R\frac{3}{4}$ = 2.0 mm
R1 and over   = 2.5 mm.

Pipe fittings, screwed to BS 21, have R_c (internal taper) threads. The couplers or sockets supplied on the pipe are normally screwed with R_p (internal parallel) threads. The steel pipe itself is supplied with R (external taper) threads.

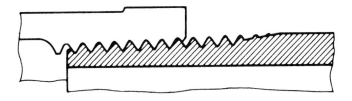

Fig. 3   Section through taper-to-taper screwed joint

An R pipe thread screwed into a R_c fitting thread makes a perfect joint, Fig. 3. All the threads should engage fully.

An R pipe thread screwed into a R_p fitting can also make an acceptable joint, Fig. 4. About half the threads are fully engaged but there is a risk of overtightening which would distort the fitting.

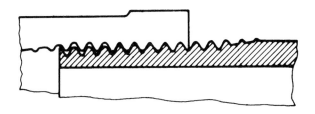

Fig. 4   Section through taper-to-parallel screwed joint

An R_L (external longscrew parallel) thread and an R_c fitting (internal taper) cannot make a satisfactory joint. A correctly sized male thread may not even enter the fitting. If it does, because of manufacturing tolerances, only one thread at most will engage.

An R_L pipe thread screwed into an R_p coupler thread can make an acceptable joint with the aid of a backnut, Fig. 5. This shows a longscrew with a coupler and a backnut.

The longscrew is used to connect pipe runs together. It consists of a short length of steel pipe with a taper thread at one end and a long parallel thread at the other. An R_c coupler and a backnut are screwed on to the R_L thread. To make the joint, the coupler is run off the parallel thread and on to the pipe thread. Because there is no seal between the coupler and the longscrew threads a soft hemp washer or 'grummet' is covered with jointing compound and compressed between the coupler and the backnut. These joints are only suitable for low pressure supplies.

Pipe threads where pressure-tight joints are not made on the threads should comply with BS 2779: 1986. They are also of Whitworth form and the BS follows the ISO 228/1 and 228/2 recommendations. These threads are parallel fastening threads in thread size designations $^1/_{16}$ to 6.

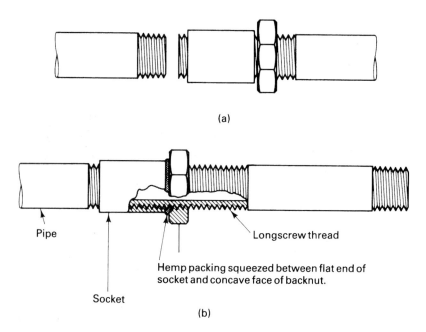

(a)

Pipe

Longscrew thread

Hemp packing squeezed between flat end of socket and concave face of backnut.

Socket

(b)

Fig. 5   Longscrew with coupler and backnut; (a) joint ready for connection; (b) joint made

The threads are used for fastening purposes, for example, the mechanical assembly of the component parts of cocks, valves and fittings. They are not suitable for pressure-tight seals. The difference between BS 21 and BS 2779 threads is in the tolerances permitted in manufacture. The BS 2779 threads are made with two classes of tolerance on the external threads. These are:

- class A – corresponding to a 'medium' tight fit
- class B – corresponding to a 'free' fit.

Fig. 12   Method of installing flexible tube to a cooker

Soft Soldered Joints

Solder has been described in Vol. 1, Chapter 12. Soldering can only be carried out effectively if the metal surfaces are properly prepared, suitably fluxed, heated to the correct temperature and joined by an appropriate solder. The essential points to be noted are:

1. Surfaces

All metals acquire a coating of oxide when exposed to the atmosphere. This must be cleaned off by scraping, filing or rubbing with clean steel wool. The finished surface must be bright, dry and free from grease. Do not touch the surface after you have cleaned it.

2. Flux

Immediately the surfaces have been cleaned they should be coated with the appropriate flux. This stops the metal from becoming oxidized again.

3. Heat

Apply heat from the blowlamp or torch evenly around the joint. If solder has to be applied, heat the joint rather than the solder.

4. Solder

Select the correct grade of solder for the particular joint. Where joints have exposed clean surfaces, rub the solder over the surface so that, as it melts, it forms a coating on the metal. This is called 'tinning' the metal. When all the surface is tinned, more solder may be added to complete the joint.

5. Final Cleaning

The completed joint should be wiped, while still warm, to remove any traces of surplus flux. Some fluxes are acidic and will corrode the pipe if allowed to remain.

Blown Joints

'Blown' joints are so called because they were originally made using a methylated spirit lamp and blowing air through the flame from a tube with a small nozzle. This produced a small aerated flame, the fierceness of which could be regulated by varying the strength of the blown air. The lamp was known as a 'mouth lamp'. It was superseded by the propane torch.

The blown joint was used to joint gas-weight lead pipe and to connect lead pipe to meter and connecting unions or to copper or brass pipes. Although lead pipes can no longer be used for gas installations – Regulation 5(2)(a), The Gas Safety (Installation and Use) Regulations 1998 – it might still be necessary to carry out some emergency repair or connect a copper supply to an existing lead installation pipe. Figs. 13 and 14 show a 22 mm copper pipe connected to a 20 mm lead gas pipe. Fig. 15 shows copper pipe branched into an existing lead gas pipe.

The procedure is as follows:

- square the end of the lead pipe with a rasp and drive in a turn pin to make a cup for the lining or other piece of copper pipe

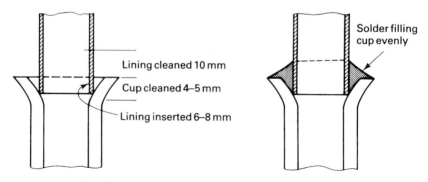

Lining cleaned 10 mm
Cup cleaned 4–5 mm
Lining inserted 6–8 mm

Solder filling cup evenly

Fig. 13   Blown joint, preparation   *Fig. 14   Blown joint, completed*

- clean the top edge and the inside of the cup, with a knife, to a depth not greater than the bottom of the lining or copper pipe
- clean the bottom of the lining or the copper pipe to a height of approximately 10 – 15 mm above the top of the cup using either a file for the lining, or steel wool for the copper pipe
- coat both parts with flux and assemble for soldering
- apply heat to the joint using a small nozzle on the torch and directing the flame at the brass lining or copper pipe. Continue until the solder will just melt when held against the lining or copper pipe
- keeping the solder just clear of the flame, tin the lining or copper pipe and then the edge of the cup
- melt the surplus solder on the lining or copper pipe so that it runs into the cup and add any more necessary to make the joint
- wipe off any surplus flux after the solder has set while the pipe is still hot
- take care to cool the joint before handling.

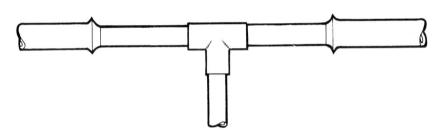

Fig. 15   Method of branching a copper pipe into an existing lead gas pipe

The traditional method of jointing or branching into water-weight lead pipes was by the plumbers' wiped joint. The prohibition of lead from potable water supplies – by-law 9, Water Supply by-laws – has meant that the wiped joint is no longer used. As with gas-weight lead it might be necessary to repair or branch into an existing water-weight lead supply pipe. Mechanical couplings have been developed, Fig. 16, that can be used to introduce an acceptable copper or plastic pipe into the installations.

Capillary Soldered Joints

The capillary soldered joint is used to connect light gauge copper tubes to BS EN 1057: 1996. The joints themselves are specified in BS 864: 1983, Part 2.

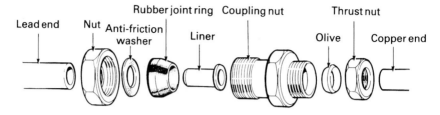

Fig. 16   Method of joining a copper or PE pipe to an existing water-weight lead pipe

A short length of the tube is inserted into the fitting leaving a gap between the outside of the tube and the inside of the fitting of between 0.02 mm and 0.2 mm. Molten solder is drawn into this space by capillary attraction (Vol. 1, Chapter 5).

There are two types of capillary fitting

- end feed – solder is applied to the mouth of the fitting, Fig. 17
- solder ring – the correct amount of solder is contained in an annular ring inside the fitting, Fig. 18.

Grades A and K solders are recommended for these joints in gas installations, grades B and F are sometimes used but they require the joints to be heated up by about an extra 20° C. A lead-free solder must be used for joints on potable water supply pipes (Vol. 1, Chapter 12).

The flux should be of a type recommended by the manufacturers of the fittings.

Procedure for making the joint:

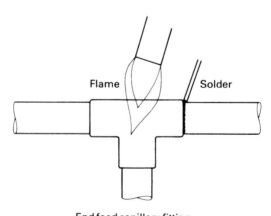

End feed capillary fitting

Fig. 17   End feed (capillary soldered)

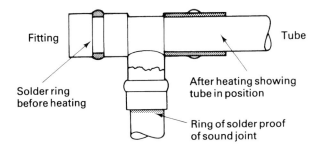

Fig. 18   Solder ring (capillary soldered)

- file the end of the tube square and remove any internal or external burrs. Do not file the surface of the tube
- clean the inside of the fitting and about 20 mm of the tube with steel wool or special wire brushes. Do not use emery cloth or sand paper, it leaves a deposit on the surface
- flux both surfaces, lightly
- assemble the joint. It may be made in any plane but the tube must fit in squarely and be fully inserted into the fitting
- protect any adjacent surface with a fireproof mat
- heat the fitting evenly, applying solder in the case of an end feed type, until a complete ring of solder is seen at the mouth of the fitting. Do not add more solder to a solder ring type
- wipe off any surplus solder or flux.

Capillary soldered joints may be disconnected by re-heating and may be re-made with added solder. However, excessive heating can cause the solder to alloy with copper and if both fitting and tube are of copper, it will be very difficult to disconnect them. For this reason it is advisable to use the lower melting point solders, A and K.

Copper tube may be jointed by means of a socket formed on the pipe itself. This is done by means of a socket-forming tool (Vol. 1, Chapter 13), which is a steel drift, driven into the end of the pipe. The socket so formed is prepared and soldered in the same way as an end-feed fitting.

Compression Joints

Compression fittings conform to BS 864: Part 2, 1983 for light gauge copper and stainless steel pipe and Part 5, 1990 for plastic tubes. They are of two types:

- type A, non-manipulative, Fig. 19
- type B, manipulative, Fig. 20.

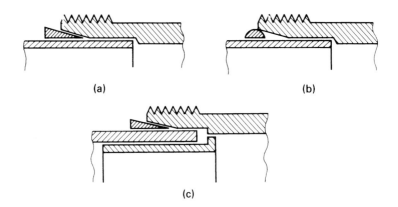

(a)   (b)

(c)

Fig. 19   Type A, non-manipulative compression joint

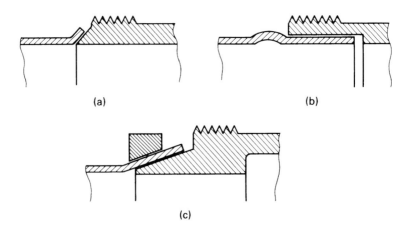

(a)   (b)

(c)

Fig. 20   Type B, manipulative compression joint

Type A

On these joints a compression ring usually made of metal and called an 'olive', is slipped over the pipe and then compressed on to it by tightening the cap on to the boss. The force of the compression distorts the olive and also the pipe when a metal olive is used, so forming a sound joint.

These fittings should generally be used on annealed tubes.

Type B

Type B fittings require the tube to be 'manipulated', that is, bent, usually by opening out the end, Fig. 20(a) and (c). The bent section of tube is gripped between the cap and the boss of the fitting.

These fittings should not be used on tubes above 28 mm size.

Making compression joints

The end of the pipe must be clean and not scratched or distorted. It must be squarely cut and all burrs should be removed. The cap should be tightened sufficiently to form a sound joint, but not over-tightened and the pipework should not be strained when the joint is made.

Some people prefer to smear the joint lightly with jointing compound before assembling it. This is not necessary but it does, perhaps, act as a lubricant between the surfaces as the joint is tightened.

Plastic Tube Joints

Plastic tubes may be jointed by:

- compression fittings, Fig. 19(c), Fig. 20(c)
- solvents, for PVC tubes
- welding or fusion, for PE pipes, using specially designed electric heaters or some other heating medium.

Pipe Bending

The direction of a pipe run can usually be changed by using a fitting, for example, a 'bend' or 'elbow' (see section on Pipe Fittings). In this way, corners may be turned and obstacles avoided. However, there are some times when fittings, which generally change the pipe direction by 90° (occasionally by 45%), do not provide the angle required and the pipe must be bent.

Pipe bending, or 'setting', as it is commonly called, has a number of advantages to offer over the use of fittings. These are:

- less resistance to the flow of gas or water
- neater
- less costly, generally, in both labour and materials
- fewer joints and potential sources of leakage
- irregular contours may be followed.

One point which must be borne in mind is that, if too many sets are used instead of fittings it may be very difficult to remove or disconnect the pipework on a subsequent occasion.

Various types of bends or sets are shown in Fig. 21.

Single and double sets are those most frequently required. For example, when setting out from a wall on to an appliance connection, when setting over a skirting board or when lining up a pipe run with an appliance.

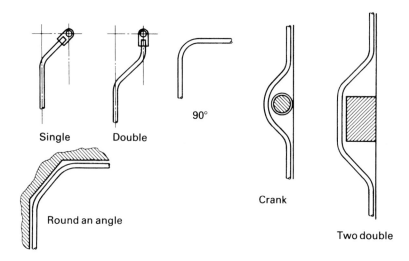

Single   Double

90°

Round an angle

Crank

Two double

Fig. 21   Pipe sets and bends

The 90° bend is not so frequently used, perhaps because it needs a bit more skill than fitting an elbow!

Crank sets may be used to cross over circular obstacles and rectangular ones require two double sets. A succession of single sets may be needed round a bay or a corner fireplace.

Manual Pipe Bending

The objective of pipe bending is to produce a bend with an even radius which follows the contours of the wall or obstruction accurately and as closely as is practicable. The bend should be neither squashed nor kinked.

Both light gauge copper and steel pipe up to 22 and 20 mm sizes, respectively, can be bent manually quite successfully, indeed most of the bends required on the district are produced in this way.

Medium and heavy weight steel pipe may be bent by means of a bending eye, Fig. 25(a), and without any loading to support the wall of the pipe.

Copper tube needs to be supported by bending springs but, being thinner and softer, may be bent round the knee, using suitable protection. Bending springs generally fit inside the pipe, but the very small sizes need external springs (Vol. 1. Chapter 13).

Whilst small, neat bends are ideal it is advisable not to attempt to bend to too small a radius. On steel pipe this is likely to lead to kinks and on copper pipe it may result in a bend which is difficult to adjust to the final dimensions and difficulty in removing the spring.

As a rough guide manual bends should be made to an inside radius of 5 times the outside diameter of the pipe. This is, for example, 110 mm for 15 mm steel which has an outside diameter of 21.6 mm and 75 mm for 15 mm copper tube. Since measurements are more easily taken from the beginning of a bend, it is necessary to work from the length of pipe to be bent, rather than the radius itself.

For a bending radius of $5 \times$ outside diameter (OD) the length of pipe to be bent is:

- $90°$ bend, length $= 8 \times OD$
- $45°$ bend, length $= 4 \times OD$
- $30°$ bend, length $= 3 \times OD$ (actually 2.6, so round off the answer downwards).

For example,

On 15 mm copper tube, which is a very commonly used size, the bending lengths are:

- $90°$ bend, length $= 120$ mm
- $45°$ bend, length $= 60$ mm
- $30°$ bend, length $= 40$ mm.

Marking out pipe for bending can be done in a variety of ways. One of the best methods is to make a drawing or a 'template' of the bend required. This is easily done in a workshop but seldom possible in customers' premises. So you need to be able to work without a drawing.

Actually, it is very simple. A folding 600 mm rule can be opened to the angle of set required. And any flat surface, like a wall or floor, can act as a straight-edge or base line from which measurements can be taken. (But do not put dirty pipe against the best wallpaper!) Alternatively, another length of straight pipe can be used when measuring the depth of a set.

Methods of marking out some simple sets to accurate dimensions are as follows:

90° Bends (Fig. 22)

Measure the distance into the corner, subtract the outside diameter of the pipe and make a mark.

Measure the bending length from this mark, $1/3$ forward, $2/3$ back and mark this on the pipe. For example, 15 mm copper tube will be 40 mm forward, 80 mm back, 120 mm total bending length.

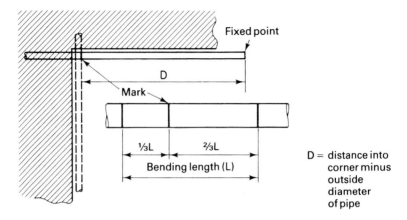

Fig. 22   Marking out 90° bend

Bend the tube exactly within these marks to an even radius and the bend will fit accurately in the corner.

Single Sets (Fig. 23)

For single sets measure from the fixed point to the start of the bend. Add on the bending length and bend the tube between the marks, to the angle required.

Double Sets (Fig. 24)

First, make a single set.

Hold the pipe against a flat surface or baseline and measure the depth of set required by a rule held at 90° to the baseline.

The mark on the pipe, from the edge of the rule, is the centre of the bend.

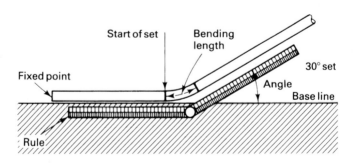

Fig. 23   Marking out single sets

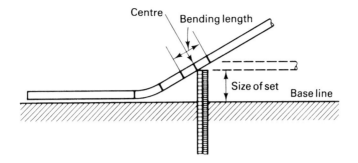

Fig. 24   Marking out double sets

Measure half the bending length on each side of the centre mark. For example, 15 mm copper tube is marked 20 mm on each side of the centre line, for a 30° set as shown.

Bend between the marks until both bends are parallel.

Bending Copper Tube

All manual bending calls for some measure of physical strength and this varies with individuals. Generally 6 and 8 mm tubes can be bent in the two hands, with the thumbs supporting the inside of the bend. 12 and 15 mm can be bent round the knee. To prevent injury, a leather knee-pad or a pad made of folded rag should be used. The ability to bend tubes of above 15 mm diameter will depend on personal strength.

Some people find it possible to bend 22 mm on the knee, others may find it necessary to anneal the tube first or bend it through a hole in a wood batten.

The procedure for bending is as follows:

● mark out the bend
● insert the spring so that it is located in the position of the bend required
● slightly over-bend the tube and then ease back to the desired dimensions, in order to free the spring
● pull the spring out, if necessary by inserting a tommy bar or stout screwdriver blade through the eyelet and turning it to contract the spring while pulling at the same time.

It is possible to make bends in tubes at a considerable distance from the ends. To do this, the spring must have a length of strong cord or wire attached through the eye so that it may be pulled out. Great care must be taken to avoid kinking the tube between the spring and the end or pinching the spring in the bend.

If this does occur it will necessitate cutting the tube to get the spring out and may damage the spring as well.

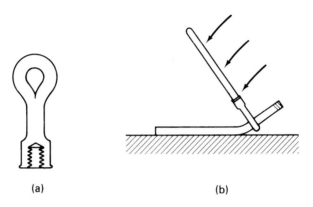

(a)   (b)

Fig. 25   (a) Bending eye; (b) bending method

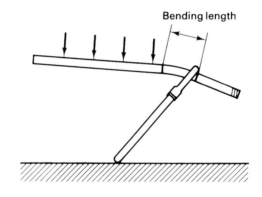

Fig. 26   Reverse bending method

Bending Steel Pipe

Up to 15 mm nominal bore pipe can be bent manually, using a bending eye, Fig. 25(a). 20 mm can also be bent manually, but it takes a great deal of strength and not everyone can do it successfully unless the eye is secured to a bench or truck and other pipes are used for leverage.

The bending eye is made of cast steel and is tapped to take a piece of pipe or a steel tubular handle. Figure 25(b) shows it being used to bend a piece of 15 mm pipe to form a 90° bend.

Bending must be spread evenly throughout the bending length. This is done by pushing down on the bending eye until the pipe touches the floor, Fig. 25(b). Then move the bending eye back about 25 mm and repeat until the pipe again touches the floor. Continue to the end of the bending length when the 90° bend will be completed.

This method has the advantage of safety and accuracy. There is less likelihood of the eye slipping and keeping the pipe on the floor avoids having a long length swinging about.

The bending eye may be used in the reverse position as shown in Fig. 26. This gives the operator more leverage, but it is more difficult to control. The method must be used when making the second bend of a double set.

Machine Bending

Using a machine not only takes the effort out of bending, it enables bends to be made to a smaller, neater radius. And all bends, for the same size of pipe, are exactly the same radius. The major advantage of a machine is that it enables bends to be made quickly and accurately to dimensions, with less waste. With manual bending it is often the practice to make a bend and then cut both ends of the pipe to the required lengths. With a machine it is usually possible to position the bend relative to one end of the pipe and then only cut the other to the dimension required.

The machines themselves were introduced in Vol. 1, Chapter 13. It now remains to describe methods of using them.

There are a number of ways in which pipe may be marked out for bending. A few simple, practical methods have been included.

Light Gauge Tube Bending

Details of the machine are shown in Fig. 27. The pipe to be bent is located in the fixed former as shown in Fig. 27(a). Measurements may be taken from:

- the outside edge of the former, which will correspond to the outside of the pipe, when bent
- the inside of the former, which will be in line with the inside edge of the pipe after bending.

Methods of making the various sets are as follows:

90° Bends

1. Measure the distance from the end of the pipe to the outside of the bend required, Fig. 28.

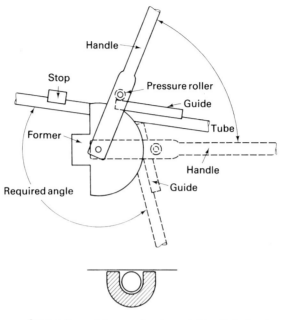

Section through former showing outside of tube level
with face of former

(a)

Fig. 27   Copper bending machine: (a) pipe in fixed former

2.   Mark a line across the pipe at this point. (The inside of the
     bend may be used if this is more convenient.)
3.   Place the pipe in the machine with the measured piece to the
     back and held in the stop.

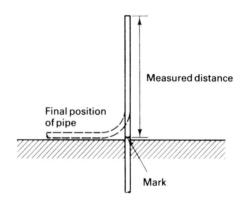

Fig. 28   Marking out 90° bend

4.   Pull the pipe into the former and adjust its position so that a square, held against the pipe and in line with the mark, is just touching the outer (or inner) edge of the former, Fig. 29.
5.   Place on the straight former or guide and make the bend by pulling on the handle until the pipe is bent to 90°.

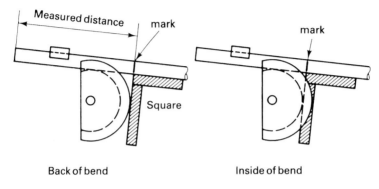

Back of bend   Inside of bend

Fig. 29   Locating pipe in the former

Single Sets

These need to be measured from the point at which the set will start, Fig. 30.

The mark is now made on the pipe at this point. It must then be placed in the former in line with the spot where the bend will begin.

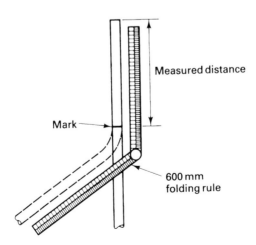

Fig. 30   Marking out single set

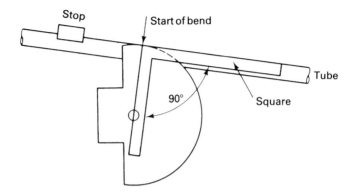

Fig. 31   Locating pipe in the machine

This spot can be found by placing a piece of pipe in the former and lining up a square with the pipe and the centre of the former, Fig. 31. Alternatively, a simple way of locating the pipe is as follows:

- place the guide on the pipe with its end in line with the marks on the tube
- insert the guide and tube into the machine until the guide just touches the roller, Fig. 32.

The angle of the bend may be judged by opening out a 600 mm folding steel rule to the angle required, Fig. 30.

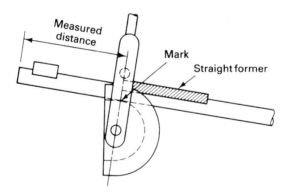

Fig. 32   Alternative method of setting up machine

Double Sets

To make a double set, first make a single set as above. It is usual to make the angle about 45° if the width of the set is 50 mm or over and 30° if it is less than 50 mm.

Then mark out the second set as shown in Fig. 33.

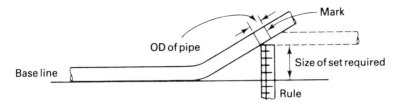

Fig. 33   Marking out double set

- hold the pipe against a flat surface or base line
- place the rule at 90° to the base line and measure the depth of set required
- mark the pipe where it touches the end of the rule
- add on one pipe diameter and make a mark on the pipe
- place the pipe in the machine, Fig. 34
- hold the rule vertically against the face of the former and adjust the pipe so that the second mark and the edge of the rule coincide
- set the tube until the second set is parallel with the first.

Crank Set

This is merely an extension of the double set as follows:

- make a single set
- mark out as for double set
- open folding rule to same angle as the single set
- make the centre bend to match this angle, Fig. 35
- place the outside of the centre bend on the base line with the straight pipe parallel to it, Fig. 36

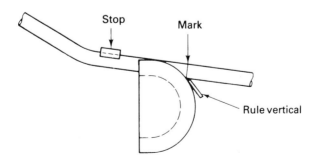

Fig. 34   Locating pipe in the machine

- mark off third bend as for a double set
- make the third bend to line up with the first.

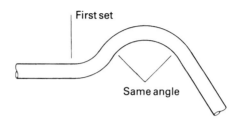

Fig. 35   Crank set

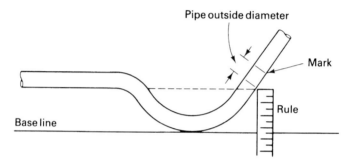

Fig. 36   Completing crank set

Angle Sets

To set around an angle:

- make a template or drawing of the angle, Fig. 37(a)
- lay the tube along the base line and make a pencil line on the tube to correspond with the first angle, Fig. 37(b)
- place the pipe in the machine with the pencil line against the outside edge of the fixed former, Fig. 37(c)
- set the tube to the angle required using a folding rule as a gauge
- mark out the second set using the hypotenuse as the base line, Fig. 37(d)
- set the tube to the required angle as before.

Bending by Hydraulic Machine

These machines are used for steel pipe. The smaller ones will bend pipes from 15 to 50 mm. The larger sizes can take up to 80 or 100 mm pipes.

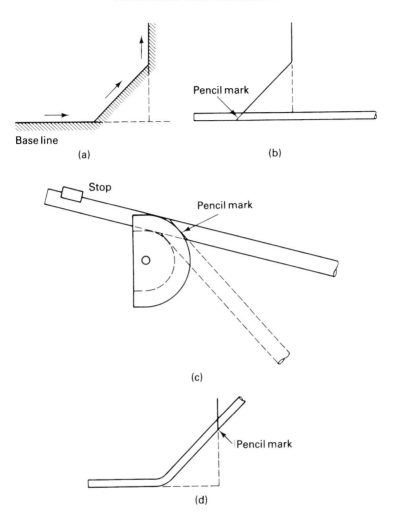

Fig. 37 *Marking out sets round an angle: (a) template; (b) marking out first set; (c) locating pipe in the machine; (d) marking out second set*

Because, on ram-type machines, the former moves to bend the pipes, it is necessary to base the system of marking out on the centre of the bend, that is, the centre of the former.

When making double or crank sets it is necessary to use a spirit level to ensure that the second or third bends are in the same plane as the first one.

Any pipes with welded seams should be placed in the machine with the seam uppermost.

90° Bends

When working from a fixed point, measure the distance from that point to the centre line of the final position of the pipe. Subtract the diameter of the pipe and mark that distance on the pipe, Fig. 38.

Place the pipe in the machine with the mark in line with the centre of the former.

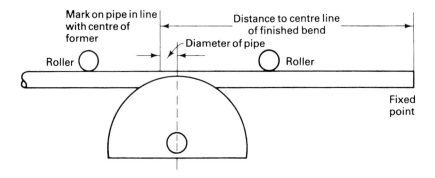

Fig. 38   Marking out 90° bend

Single Sets or Angle Sets

For sets of 45° or less use the following procedure:

- lay the pipe in the required position or against a template, Fig. 39
- lay a short off-cut to the angle and in the position required
- hold a rule with its edge against the off-cut and mark around the pipe where the corner of the rule touches it

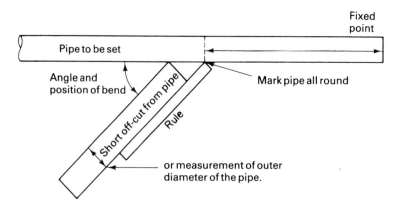

Fig. 39   Marking out single set

- place the pipe in the machine with the mark in the centre of the former.

Double Sets

Make a single set as above, then:

- hold the pipe against a flat surface or base line
- place a rule at 90° to the base line and measure the depth of set required, Fig. 40

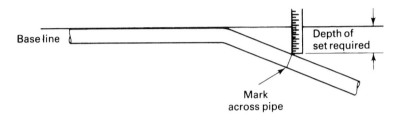

Fig. 40   Marking out double set

- mark the pipe where it touches the edge of the rule
- place the pipe in the machine with the mark in the centre of the former and level both ends
- bend the pipe until the second set is the same angle as the first and the pipe ends are parallel, Fig. 41.

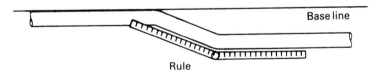

Fig. 41   Matching the angles

Crank Sets

To find the angle for the centre set, use two rules as in Fig. 42. Set a 600 mm folding rule to 587 mm for obstacles up to 50 mm diameter. The distance 'X' is approximately ¼ of the diameter of the obstacle.

If this fraction is increased, in order to make a tighter set, it will not be possible to replace the two outer formers in their correct position when the other two sets are made. If the formers are replaced in position for a smaller size of pipe there is a risk of damaging the machine.

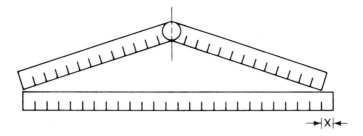

Fig. 42   Marking out crank set, angle of centre set

Having made the centre set in the position required, continue as follows:

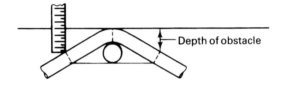

Fig. 43   Marking out crank set, position of side sets

- place the pipe against a flat surface or base line, Fig. 43
- make sure that the angles are the same on both sides of the set
- measure the depth of sets required and mark round the pipe at those points; depth of set equals depth of obstacle, plus clearance
- replace the pipe in the machine with one of the marks against the centre of the former and the ends level
- bend until the angle of the set corresponds to the angle set on a folding rule, Fig. 44
- make the third set to the same angle in a similar manner.

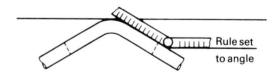

Fig. 44   Marking out crank set, angle of side sets

Pipe Fittings

(BS 143, and 1256: 1986, malleable iron and cast copper fittings for screwed pipe)

Fittings are specified by the following characteristics:

- size of pipe or thread – $1/2$, 15 mm, etc.
- type of thread – internal, female (f), external, male (m); taper or parallel, R, R_c, R_L, R_p
- type of material – malleable iron (mi), copper (cu)
- type of fitting – coupler, elbow, tee, etc.

Figure 45 shows three of the most common forms of fitting used.

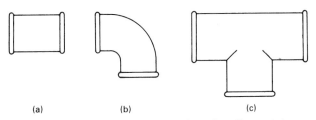

(a)   (b)   (c)

Fig. 45   Pipe fittings: (a) coupler; (b) elbow; (c) tee

These are:

- coupler, also known as a 'socket'. Used for jointing two straight lengths of pipe
- elbow, for turning a pipe run through 90°
- tee, for connecting a branch into a straight pipe.

Fittings which take one size of pipe only are specified by quoting that size once. For example, $1/2$ elbow or $3/4$ tee.

Reducing fittings with only two outlets are specified by quoting the larger size first. For example,

$$3/4 \times 1/2 \text{ elbow, } 1 \times 3/4 \text{ coupler.}$$

Tees are always specified by the two ends on the run first and the branch last. For example, $1/2 \times 1/2 \times 3/4$ tee. If there is a reduction on the run, the larger end is quoted first, for example, $3/4 \times 1/2 \times 1/2$ tee.

Figure 46 shows a number of less common fittings and the order in which the outlets should be quoted.

FIXINGS

Nails

Nails are used extensively in building construction and furniture. They are normally made from mild steel but some are available in other materials to suit special applications.

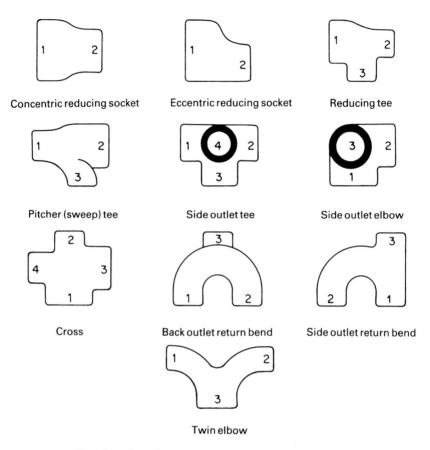

Fig. 46   Pipe fittings, method of describing outlets

Nails are specified by their length and form, for example, 50 mm oval brads. Quantities are traditionally ordered by weight, although some retailers may sell them by number.

Commonly used types of nails are as follows:

Oval Brad (Fig. 47)

A type of wire nail with an elliptical section and a small head which can be punched below the surface leaving only a small hole. It should not split the wood if driven in with major axis parallel to the grain. Used for general purposes it is available in lengths from 12 to 150 mm.

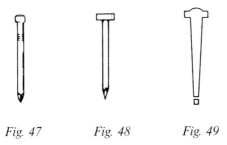

Fig. 47   Fig. 48   Fig. 49

Wire Nail (Fig. 48)

Also known as a 'French nail', this is used for rough work and has great holding power. It cannot be punched below the surface. Available sizes, 20 to 250 mm.

Cut Clasp Nail (Fig. 49)

Punched from sheet metal, it must be driven into wood with its long side parallel to the grain.

Floor Brad (Fig. 50)

A cut nail used for securing floorboards. It has a projection at the head and is tapered on one side only.

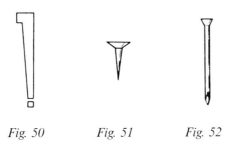

Fig. 50   Fig. 51   Fig. 52

Clout Nail (Fig. 51)

This has a large, thin, flat head which makes it useful for securing roofing felt or plaster boards.

Carpet Tacks

These are similar in shape to small clout nails but they are a cut nail and are always blued.

Masonry Pin (Fig. 52)

This is made from tempered alloy steel and is used for fixing timber to brickwork. The pins snap easily and the following precautions must be taken:

- wear goggles when fixing
- hit pins square to the end
- use light taps to drive pins home.

Wood Screws

Wood screws are made from steel or brass and may be plain or finished in a variety of ways including:

- blued
- bronzed
- chromium plated
- black japanned.

They range in size from No. 1, approximately 1.7 mm diameter, up to No. 32, approximately 12.7 mm diameter. Lengths may vary from 6 mm for No. 1s up to 230 mm for No. 32s.

When ordering wood screws specify:

- quantity
- length
- size number
- type of head
- finish or material.

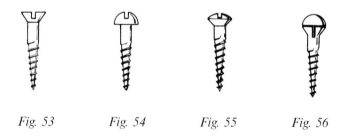

Fig. 53   Fig. 54   Fig. 55   Fig. 56

- For example, 100, 25 mm, No. 8, roundhead, brass.

Types of head are:

- countersunk, Fig. 53. For fixing wood or thick metal; the head finishes flush with the surface

- roundhead, Fig. 54. For securing thin metal parts, for example, pipe clips or appliance brackets
- raised head, Fig. 55. Normally used on plated screws for a better finish
- dome head, Fig. 56. Used to give a very neat finish, they are countersunk screws drilled and tapped to take the dome
- coach screws, Fig. 57. These have a square head to take a spanner. Available in large sizes and used for heavy work
- screw eyes, Fig. 58. Used for a number of purposes including providing a securing point for a rope lashing a ladder.

When a wooden panel has to be unscrewed fairly frequently it is an advantage to fit brass cups to take the countersunk screws which secure it, Fig. 59.

Fig. 57   Fig. 58   Fig. 59

Machine Screws

Machine screws are made to be screwed into threads tapped in holes in metal or plastic. Like bolts, they may sometimes carry nuts, but they have threads extending to the head, while bolts have a plain section before the thread starts. Metric sizes conform to BS 4183: 1967 (obsolescent), partially replaced by BS EN ISO 1580: 1994 and BS EN ISO 7045: 1994.

Types of Thread

There are many (actually 210) different types of thread still being used. So care must always be taken to ensure that the right thread of screw is being used for any particular tapped hole or nut. If it does not screw in easily by hand at first, it probably does not fit. Any attempt to tighten it with a spanner will probably ruin the thread on the screw and in the hole.

Types of thread in common use are as follows.

1.   British Standard Whitworth (BSW)

This has a fairly coarse thread with an angle of 55°. It is used for general work where it is not likely to be subjected to vibration and where a strong thread is required. Sizes are from $1/8$ in to 6 in.

2.   British Standard Fine (BSF)

Of Whitworth form but with a finer pitch, so less liable to come loose if subjected to shock or vibration, this has been used extensively on engines and machinery. Sizes $3/16$ in to $4^{1}/2$ in.

3.   British Association (BA)

This is based on a Swiss (Thury) thread and has an angle of $47^{1}/2°$. It is in metric units and some sizes may be interchangeable with metric screws. Used for screws below 6 mm diameter it is found on electrical components, instruments and small appliance parts. Sizes range from the smallest, No. 25, 0.254 mm diameter, pitch 0.07 mm, to the largest, No. 0, 6.0 mm diameter, pitch 1.0 mm.

4.   Unified Threads

Established in 1949 to provide interchangeability of components between America, Canada and the UK, this included three series of threads:

- UNC – coarse
- UNF – fine
- NEF – extra fine.

5.   ISO Threads

The International Standards Organisation threads to be used in engineering are as follows:

ISO metric coarse (BS 3643 Part 1, 1981, Part 2, 1981).

Sizes, up to 64 mm with varying pitch and a constant pitch series 1.6 mm to 300 mm.

ISO inch coarse (UNC) (BS 1580 Parts 1 and 2, 1962 (1985)).

Sizes, $1/4$ in to 4 in with varying pitch and a constant pitch series $5/16$ to 6 in.

ISO inch fine (UNF) (BS 1580 parts 1 and 2, 1962 (1985)).

Sizes, $1/4$ in to $1^{1}/2$ in with various constant pitches.

Metric Threads

Metric diameters are specified by the letter M followed by a number giving the diameter in millimetres. For example, M5 represents 5 mm diameter.

Types of Screw Head

Similar to the heads of wood screws, these are shown in Fig. 60. In addition, the cross-slotted or recessed heads available are shown in Fig. 61. These are often used in components which are assembled by power screwdrivers.

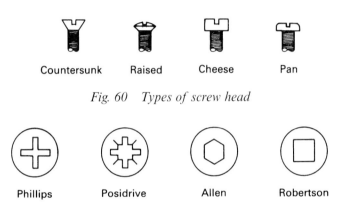

Countersunk   Raised   Cheese   Pan

Fig. 60   Types of screw head

Phillips   Posidrive   Allen   Robertson

Fig. 61   Recessed screw heads

Types of Screw

As well as the standard screw which has a parallel thread throughout its length, the following types have their particular uses.

Set Screw (Fig. 62)

Usually pointed or designed to fit into a specially shaped recess. Used to hold one component relative to another. For example, holding a pulley on a shaft.

Grub Screw (Fig. 63)

Used for the same purpose as the set screw, this has no projecting head and may be screwed below the surface of the component. The screw may have a slotted or a recessed head for an Allen key.

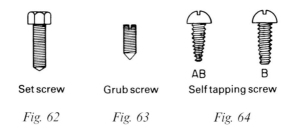

Set screw	Grub screw	AB B Self tapping screw
Fig. 62	*Fig. 63*	*Fig. 64*

Self Tapping Screws (Fig. 64)

There are several types of self tapping screws, the most common being AB and B, Fig. 64.

Type AB has a widely spread thread and gimlet point. It is used in light sheet metal, metal clad and resin impregnated plywood and soft plastics. It supersedes Type A which had a coarser thread and is now obsolete.

Type B has the same thread as AB but the point is blunter. It is used in light and heavy metal sheet, non-ferrous castings and other materials for which AB may also be used.

Tapping screws are available in numbered sizes, similar to wood screws and are detailed in BS 4174: 1972 (obsolescent), partially replaced by BS EN ISO 1749: 1994 and BS EN ISO 7049: 1994.

Ordering Machine Screws

When specifying metric machine screws quote:

- quantity (number)
- material
- type of head
- diameter and length
- plating.

For example, 10, steel, countersunk, M5 × 16 zinc plated.

Bolts

The length of thread on a bolt varies with the diameter. It may be from about 30% of the length of the bolt on small diameter bolts and could be 80% of the length of larger bolts.

Where the head of the bolt is large enough, it has ISO M and a number indicating the grade of steel, formed on the head. Nuts have a dot to indicate metric and another mark to show the grade. For example, the number 88 indicates a high-tensile steel. This is shown on a nut by a bar situated 120° clockwise from the dot.

The standard bolt has a hexagon head, Fig. 65. The dimensions of the head are related to the diameter of the bolt. The heads of BSW and BSF fasteners are the same for the same diameter of the bolt.

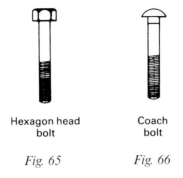

Hexagon head   Coach
bolt   bolt

Fig. 65   *Fig. 66*

Spanners designed for imperial sizes of fastener will not fit on metric fasteners; or vice versa.

There are a number of special bolts as follows:

Coach Bolt (Fig. 66)

This is used for securing timber. The head is domed and immediately below it is a square section which is pulled into the wood as the nut is tightened. This prevents the bolt from turning.

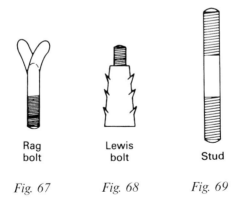

Rag   Lewis
bolt   bolt   Stud

Fig. 67   *Fig. 68*   *Fig. 69*

Rag Bolt and Lewis Bolt (Figs 67, 68)

Rag bolts and Lewis bolts are set in walls and floors. Both are secured with cement mortar or placed into the concrete in the position required.

Studs (Fig. 69)

One end is screwed into a tapped hole leaving the remainder projecting like a bolt. Studs are used where a flanged joint is required on the machined surface of an appliance or engine component.

Nuts

Nuts are usually square or hexagonal and they are usually made to the same dimensions as the head of the bolt on which they fit.

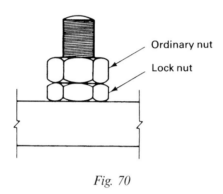

Ordinary nut

Lock nut

Fig. 70

Locknuts, Fig. 70, are half as thick as an ordinary nut. They are 'locked' by tightening against the main nut.

Slotted or castle nuts, Fig. 71, are used in conjunction with a drilled bolt and a split pin. When the nut is in the required position, the split pin is passed through both the hole in the bolt and the slot in the nut and then secured.

Wing nuts, Fig. 72, may be turned without the aid of a spanner and are often used to secure casings or panels which must be removed for servicing purposes.

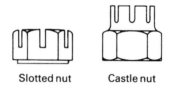

Slotted nut　　　Castle nut

Fig. 71

Captive nuts, Fig. 73, are manufactured from spring steel. They are usually available in three thread forms (BA, metric and self-tapping).

Wing nut

Fig. 72

They are particularly useful for panel fixing where the panels are
vitreous enamelled or of insufficient thickness for self-tapping screws.

Fig. 73   Two types of captive nut (self-tapping)

Washers

Washers are used, under the heads of bolts and nuts, to prevent
damage to the article and give the fastenings a flat surface on which
to bear, Fig. 74. They may be used to provide clearance distance
between two parts of an object when they would be called 'spacing'
washers.

Washers which incorporate a grummet of synthetic material, Fig.
74, are used on vitreous enamelled sheets to prevent excessive pressure being applied which might cause the finish to chip.

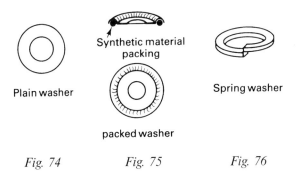

Synthetic material
packing

Plain washer   Spring washer

packed washer

Fig. 74   *Fig. 75*   *Fig. 76*

Spring washers, Fig. 76, are formed from rectangular section spring
steel. The sharp corners bite into the material and the nut, so locking

them together. Other washers also used to prevent nuts coming loose are the tooth lock washers shown in Fig. 77.

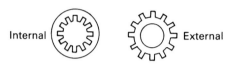

Tooth lock washers

Fig. 77

Rivets

Where a permanent assembly is required, rivets are used instead of nuts and bolts. Rivets for cold working are made of soft iron, mild steel, copper or aluminium. The common types of head are shown in Fig. 78.

Flat head rivets are used on thin plates and are commonly made of galvanized iron.

Roundhead, sometimes called snag head or snaphead, are used where countersinking would weaken the material and where a flush finish is not essential. A 'snap' is a shaped punch, used to form the head.

Countersunk rivets are used for general work when a flush surface is required. There are many other heads used for special jobs, including, cone, steeple, pan, mushroom, oval, globe and button.

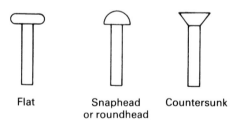

Flat   Snaphead   Countersunk
              or roundhead

Fig. 78   Rivets

The allowances to be added to the length in order to form a head are:

- roundhead, $1\frac{1}{4}$ times the diameter of the shank
- countersunk, equal to 1 diameter.

Pop rivets are used when only one side of a job is accessible, Fig. 78. They are inserted by means of a small, hand operated machine. Being hollow, they are not as strong as ordinary rivets.

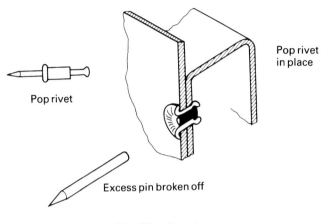

Fig. 79   Pop rivet

Pipe Fixings

There have been many types of pipe clips and hooks produced and a few of the more common types are described.

Hooks (Fig. 80)

Made of malleable iron, these are cheap and easy to fix in wood, stone or brick. They are not usually used on exposed pipe runs in domestic premises but are useful for holding pipe in chases or under floors.

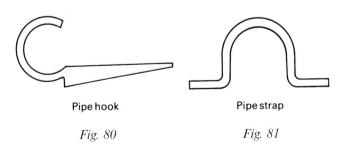

Fig. 80   *Fig. 81*

Straps (Fig. 81)

Pipe straps, or saddle clips, are usually of galvanised mild steel for steel pipe or copper for copper tube. They should be secured by two roundhead screws.

Spacing Clips (Fig. 82)

These are usually copper and are used where the pipe needs to be kept clear of the surface so as to prevent dirt collecting behind it.

Plastic Clips (Fig. 83)

There is a variety of plastic clips. Some are made in two parts and completely enclose the pipe. Others, like the one illustrated, allow the pipe to be forced into it and are made in one piece.

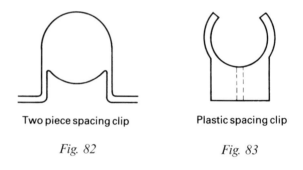

Two piece spacing clip   Plastic spacing clip

Fig. 82   *Fig. 83*

Split Pipe Rings (Fig. 84)

These are produced from malleable iron and are made either with extensions for cementing into walls or with back plates for securing to the surface. The plate and the clip may be screwed with $R_c\frac{1}{4}$ thread so that a piece of steel pipe may be used as an extension between the backplate and the clip.

Fixing Devices

Types of fixings and their application are as follows:

- solid walls   – compound fillers
  – wall plugs
  – masonry bolts
- hollow surfaces – toggles
  – anchors
  – battens.

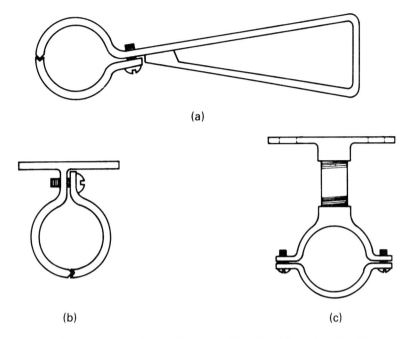

Fig. 84   Split pipe rings: (a) building-in; (b) school board; (c) with backplate and spacing nipple

Compound Fillers (Fig. 85)

These are putty-like compounds which are ideal for use where the hole is ragged, oversized or distorted. The filler is packed into the hole and a starting hole for the screw is made with the spike provided. When the filler hardens the screw may be driven in.

Wall Plugs (Fig. 86)

Made from fibre, plastic or metal, they are inserted into drilled holes and are expanded, by the screw, to grip the sides of the hole. These plugs are made in numbered sizes to fit the same number wood screw. The drilling, plug and screw must be matched for a satisfactory fixing.

Flanged plugs are used for bottomless holes.

Metal plugs, usually of white bronze, are made for use externally or where high temperatures could shrink fibre plugs or soften plastic.

In a tiled wall the plug must be inserted beyond the tile, Fig. 87. Otherwise the force of the expansion could crack the tile.

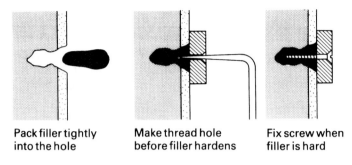

Pack filler tightly   Make thread hole   Fix screw when
into the hole   before filler hardens   filler is hard

Fig. 85   Compound filler

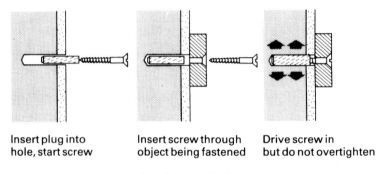

Insert plug into   Insert screw through   Drive screw in
hole, start screw   object being fastened   but do not overtighten

Fig. 86   Wall plugs

Masonry Bolts (Fig. 88)

These are used for fixing heavy objects. The hole is bored to the diameter of the plastic insert, which expands as the bolt is screwed home. Sizes range from 5 mm to 25 mm. There are various heads including hooks, studs and eyelets.

Toggles

Many ingenious devices are available for fixing to cavity walls. Two of the most common are gravity and spring loaded toggles, Figs 89, 90.

Gravity toggles have a swivel toggle which drops down when the bolt has been fed through the drilled hole.

Spring toggles have two spring-loaded arms which expand after being pushed through the hole.

Both toggles will be lost in the cavity if the screws are removed. So the screw must be put through the object to be fixed before the toggle is passed through the hole.

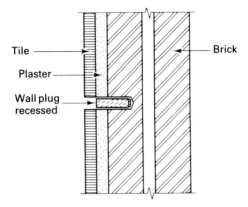

Fig. 87   Wall plug in tiled wall

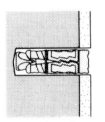

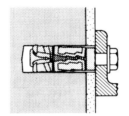

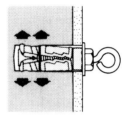

Bore hole to diameter
required

Insert bolt through fixture
and screw into anchor

Special heads include
hooks and eyelets

Fig. 88   Masonry bolts

Swivel toggle

Screw bolt through
fixture into toggle;
hold toggle on bolt

Push through hole
until toggle drops
into position, pull
back until toggle
touches wall

Pull back fixture
and tighten the bolt

Fig. 89   Gravity toggle

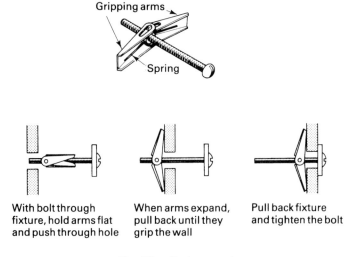

Fig. 90   Spring toggle

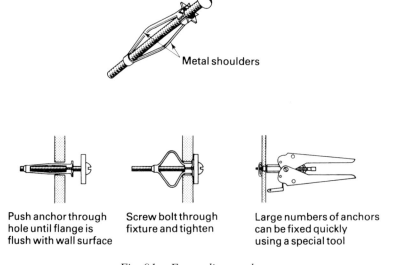

Fig. 91   Expanding anchor

Anchors

Anchors generally remain in place when the screw is removed. Figure 90 shows a collapsible anchor which has metal shoulders drawn against the inside surface as the screw is tightened.

The rubber anchor, Fig. 92, is expanded on tightening. It can be used on a solid wall.

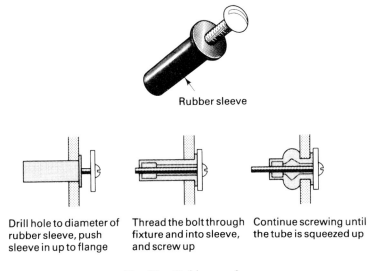

Rubber sleeve

Drill hole to diameter of
rubber sleeve, push
sleeve in up to flange

Thread the bolt through
fixture and into sleeve,
and screw up

Continue screwing until
the tube is squeezed up

Fig. 92   Rubber anchor

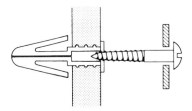

Fig. 93   Plastic anchor

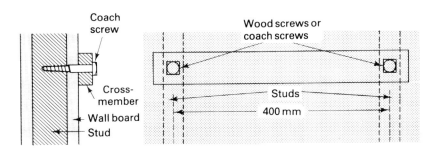

Fig. 94   Fixing to partition wall

Plastic anchors use wood screws, rather than machine screws. The
gripping arms are pulled on to the wall when the screw is tightened,
Fig. 93.

Battens

On lath and plaster walls or on plaster board, the wall cladding is not strong enough to support heavy weights, like appliances or meters. It is then necessary to fix a wood batten across the wall and on to the wall joists or 'studding'. Studs are usually 100×50 mm timbers fixed at 400 mm centres, Fig. 94. The position of the studs can be found by tapping the wall and listening to the difference between the hollow sound of the cavity and the dull, solid note on the stud. When you have roughly located the studs, check with a bradawl or a small drill before making a large hole.

CHAPTER 2

Building Construction

Chapter 2 is based on an original draft by H. V. Gowers and subsequently
updated by E. Thompson

Introduction

There are many different types of buildings. They range from flats to
factories, from castles to cottages. Broadly speaking, buildings may be
classified into three main groups:

- residential or domestic
- industrial or commercial
- public.

This chapter is principally concerned with domestic building work,
although some of the more general information may apply to the
other categories.

Building work in England and Wales is controlled by the Building
Act 1984. The Building Regulations 1985 made under it came into
effect in November 1985 and represented a significant departure
from previous ones. The 1985 Statutory Instrument consists of only
20 regulations which define terms, scope, and procedures. Technical
requirements were expressed in functional terms in two Schedules.
The Building Regulations 1991 revoked and replaced the 1985
Regulations and consolidated subsequent amendments to those regulations. The new regulations came into force on 1 June 1992 and have
subsequently been amended by the Building Regulations (Amendment) Regulations 1992, 1994, 1995, 1997, 1998 and 1999. The 1991
Statutory Instrument has 21 regulations and three schedules.
Approved Documents prepared by the Department of the Environment and published by HMSO give guidance on meeting the
requirements.

In Scotland the Building Standards Regulations were introduced in
1990 with subsequent amendments in 1992 and 1994 (Building Standard (Scotland) Amendment Regulations 1993 and 1994). They are
made under the Building (Scotland) Act 1959; technically their effect

is similar to the Building Regulations 1991. The Scottish Development Department publish guidance.

The National House Building Council operates a voluntary quality assurance scheme for house builders. NHBC publishes technical requirements, then inspects, certifies and guarantees the work of members on its register.

Building Procedures for Domestic Dwellings

Permission to erect a building must be sanctioned by the Planning Authority who will require details of the:

- size, complexity and use of the proposed building
- materials to be used in the construction
- location of the site and proximity to other buildings.

This ensures that the erection of a new building, or (in certain circumstances) the extension of an existing one, will not have an adverse effect on the environment or the living conditions of its immediate neighbours.

With few exceptions Building Regulations approval must be sought before building, altering, extending or changing the use of a building or controlled services within a building. Under 1991 regulations this does not always require the deposit of full plans.

A person wishing to undertake any of the above operations may either seek local authority approval or engage an Approved Inspector. There are then two options. If the work does not apply to a designated building under the Fire Precautions Act 1971 or a work-

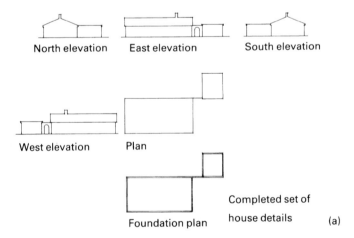

Fig. 1(a)   House plans: outline elevations and plan

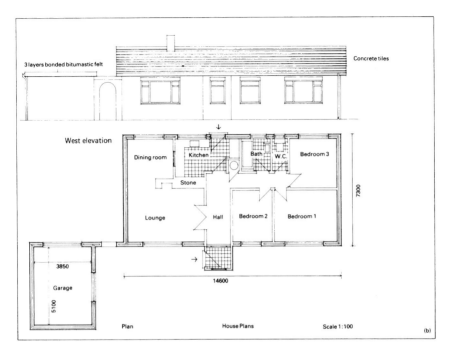

Fig. 1(b)   West elevation, detailed plan

place subject to the Fire Precautions (Workplace) Regulations 1997, the local authority may be given a building notice; detailed plans are not then required. The second option is to deposit full plans as under former regulations. When full plans are deposited with the local authority and approved, the builder can be confident that work done in accordance with them will conform with building regulations.

In all circumstances the local authority will require some basic information which will include the:

- intended use of the building and its size and position
- site boundaries, street widths and details of other buildings on the same site
- provisions to be made for drainage.

If full plans are deposited they must be in duplicate and include:

- accurately drawn scaled details of the building showing elevations, sections and plans, to a scale of not less than 1:100 (10 mm = 1 m), Fig. 1
- a block plan showing the disposition of the proposed building in relation to other property in the area, to a scale of not less than 1:1250 (10 mm = 12.5 m), Fig. 2

- a key plan of the locality, to a scale of not less than 1:2500 (10 mm = 25 m), Fig. 3.

Figure 1(a) shows the four elevations of the building in outline, together with an outline plan and foundation plan. It serves to indicate the aspect which the building will present when viewed from each direction.

Figures 1(b) and 1(c) are the usual house plans which show the layout of the rooms and the general construction of the building. Dimensions shown are in millimetres. Plans of this type could be used to determine the location of any gas installation and appliances. They could also be used for the calculation of heat requirements and the design of a suitable heating system.

Figures 2 and 3 are based on the local survey sheets and show that the proposed building is in a rural area in mid-Wales with the nearest dwelling about 50 m away.

These notices and plans enable the local authority to assess the validity of the structure, the soundness of the construction proposed and its effect on the local environment.

The builder will require more detailed information in the form of a specification. This is a written account of the work to be done and the

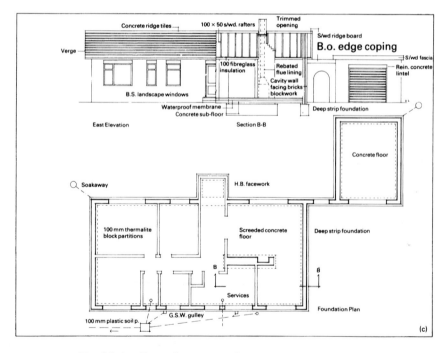

Fig. 1(c)   East elevation and section, foundation plan

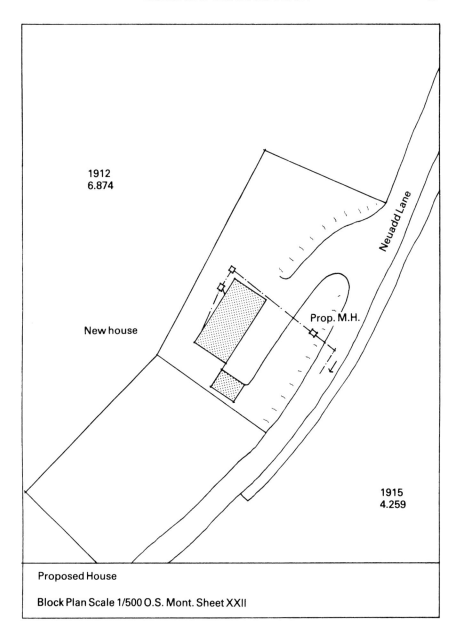

Fig.2   Block plan

materials to be used. The specification is set out under the various trade headings, that is, concreter, bricklayer, carpenter and joiner,

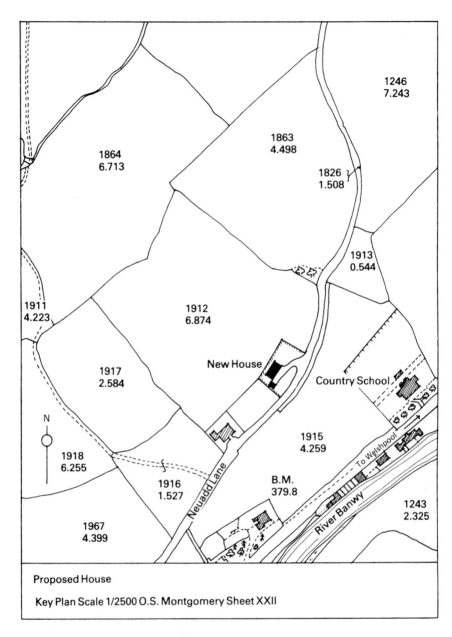

Fig. 3   Key plan

plumber, painter and decorator, etc. It informs the builder exactly what each tradesman has to do and the materials he must use to do it.

In addition, the builder may require larger scale drawings to a scale of 1:50, or in some cases for complex work, full size details.

For a large scale building operation, a 'Bill of Quantities' will also have been prepared by the quantity surveyor. This is an accurately compiled list of all materials and operations, specified under the various trade headings. It is prepared with reference to a document called 'The Standard Method of Measurement'. This ensures that all the information for the varied techniques used in construction is given in terms which will be understood by all interested parties, in all parts of the country.

The Bill of Quantities, which may run into several books, is made available to all builders who wish to tender for the contract. The builders' estimators then put their own prices against the various items in the bill in order to arrive at their final estimated tender price for the contract.

In practice, this procedure is both complex and expensive. It requires the services of quantity surveyors, estimators, workers-up and cost accountants. In recent years progress has been made towards reducing the amount of time and labour involved by the use of computers and electronic calculators.

Work may commence on site before plans are passed provided it conforms with regulations. If it conforms with plans which are subsequently rejected, the local authority can require it to be demolished.

Sequence of Building Work

The sequence of erection of a domestic dwelling may vary according to the locality, the design and type of building and the site conditions. The sequence described here is based on the erection of the traditionally brick built bungalow already illustrated in Figs 1 – 3.

1.   Preparation of Site

The proposed site and the levels are established. The taking of levels is necessary in order to determine the eventual level of the building in relation to the road, sewer or other properties in the area, Fig. 4.

If the ground is fairly level, the operation is simplified, but if the site is sloping or undulating, the taking of levels becomes significant.

The levels of the selected points are related to a fixed, common datum level. This could be the crown of the road, as in Fig. 4, or the top of the manhole cover. Failing this a temporary, but firmly concreted, post could be fixed at a convenient distance from the proposed excavation.

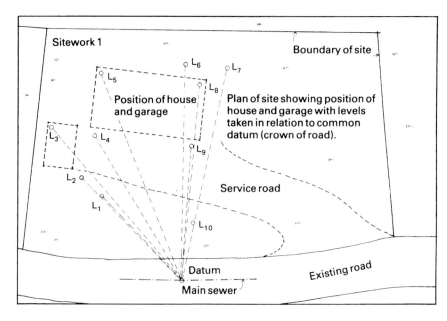

Fig. 4   Levels (sitework 1)

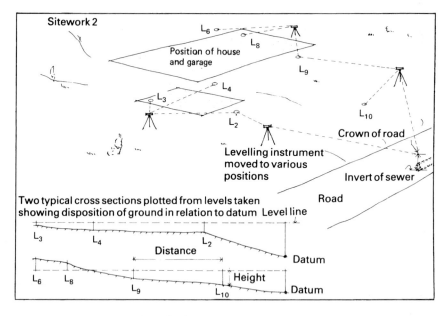

Fig. 5   Method of taking levels (sitework 2)

The measurement of the heights of the points relative to the datum is carried out using an optical instrument and a graduated staff. The instrument is usually a 'dumpy level' or a quick-set level. It has a spirit level mounted on a small telescope which is fitted with a grid of fine black cross-lines, similar to those in a rifle sight. The instrument fits onto a tripod and may be traversed through 360°.

The staff is a wood or metal telescopic rule, with bold metric markings printed on it which can be easily read through the telescope of the dumpy level.

Figure 5 shows how the level and staff are used. The instrument is set up and levelled so that the telescope will traverse in a horizontal plane. The staff is placed on one levelling point and the instrument traversed to read the height. The staff is then moved to the next point and the telescope again traversed to read that height. The difference in the two readings is the height of one point above the other. The results of the survey may be recorded as shown.

The site is marked out using wooden pegs driven into the ground, the top soil is removed and the trenches dug out for the foundations.

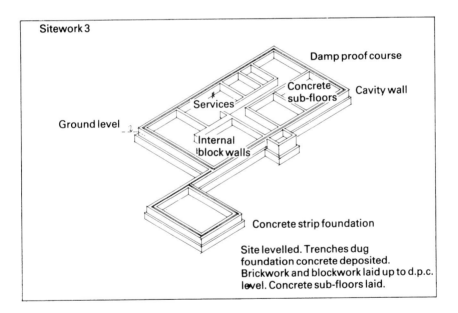

Fig. 6   Foundations (sitework 3)

2.   Foundations

The concrete for the foundations is then placed and levelled. Where the site is sloping, the foundations will be stepped, in multiples of a brick height. The spaces between the foundation trenches are levelled and a hardcore fill is spread evenly over the area to a minimum depth of 100 mm. This is covered by at least 100 mm of oversite concrete to form the sub-floor (see section on Floors). A waterproof membrane may be placed either on a blinding on the hardcore or between the concrete sub-floor and the finishing screed. Figure 6 shows the work carried out up to this point.

Positions for gas, water, electricity and other cable services may be provided before the site concrete is poured. Ducts or channels may be left in the sub-floor to accommodate the pipes or cables (see Chapter 4). Ducts carrying gas services must be ventilated.

3.   Walls

Brickwork and blockwork is now commenced. The internal and external door frames are built in as the work progresses.

When the walls have reached the top of the door and window frames, the lintels are placed in position. These may be of metal or reinforced concrete. Sometimes concrete lintels are cast in situ.

At first floor level, the bridging timber joists are spaced and levelled into position. Openings would be trimmed to accommodate hearths, pipe ducts and staircases, if required.

The construction is continued to eaves level, Fig. 7. In a two-storey building, the window and door frames would be built in as before. If

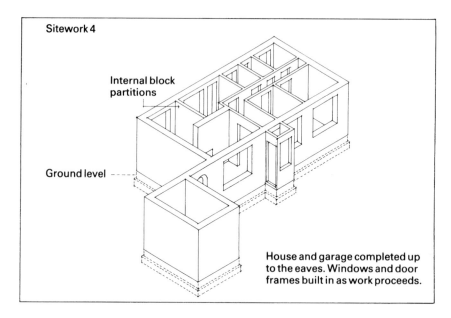

Fig. 7   Construction to eaves level (sitework 4)

the property has a pitched roof, the gable ends are built up as the roof carpentry is being fixed, any openings for chimneys being trimmed.

4. Roof

As soon as possible the roof is made watertight or, at least, covered with felt and battens. The windows are glazed and the panes clearly labelled to prevent damage to the glass. Internal work can then proceed. The roofer fixes and lays the slates or tiles and the plumber prepares and fixes the flashings and generally makes the roof watertight.

5. Internal Work

Immediately the building is waterproof, the internal work begins. This includes:

- internal flooring
- partitions
- staircases
- gas, water, electricity, other cables, waste, soil and security installations
- plastering
- internal joinery – cupboards
                    – skirtings
                    – architraves
- internal decoration.

6. External Work

The external plumbing is completed. Soil pipes are connected to the main sewer. Paths are completed and the house is decorated externally.

7. Completion

Finally, the premises are swept out and cleaned and inspected prior to occupation.

CONSTRUCTION OF DOMESTIC DWELLINGS

Recognition of Construction

Buildings, like fashions, are constantly changing and the problem of giving guidance to installers of building engineering services appears at first sight to be insurmountable. Fortunately, despite the fact that buildings look different, their basic requirements are the same. So, although the installer may have to deal with a wide range of building styles, a knowledge of certain, recognizable characteristics should enable him to avoid any pitfalls.

All building work, however complex, has to comply with certain basic rules. Fundamental constructional principles are common to all types. For example, every building needs support from the ground. The type of support will vary depending on whether the building is either taller and heavier or shorter and lighter and whether the ground is stable or unstable.

Recognition of the type of construction can often be established by careful observation and an ability to 'read the signs'. This ability may be developed by acquiring a broad knowledge of the basic methods and materials of building construction and their particular characteristics.

Domestic or residential dwellings may be classified into three main groups:

- traditional buildings, constructed of brick or stone, or a combination of both
- timber-framed buildings, constructed of braced timber frames covered entirely with timber cladding, or part brick and part timber
- contemporary system building, constructed of pre-fabricated units using a variety of materials including glass fibre, reinforced concrete, steel and plastics.

Some modern materials are deliberately designed to simulate traditional materials. This simulation may be difficult to detect and in contemporary buildings it is advisable to check carefully before cutting into the fabric.

PARTS OF A DOMESTIC BUILDING

Foundations

Adequate support for a building is achieved by spreading the weight over a large area of ground. The extent of the area will vary according to the amount of dead weight requiring support and the ability of the ground to sustain it without giving way.

Where the ground is very unstable and has a soft substrata, support will only be found at a considerable depth. This requires the foundation to be built of 'piles'. These piles are pillars formed of mass concrete or steel reinforced concrete set in holes in the ground. They gain support from a lower hard strata or from frictional resistance from the surrounding soil. Piles are often used to support tall, heavy buildings.

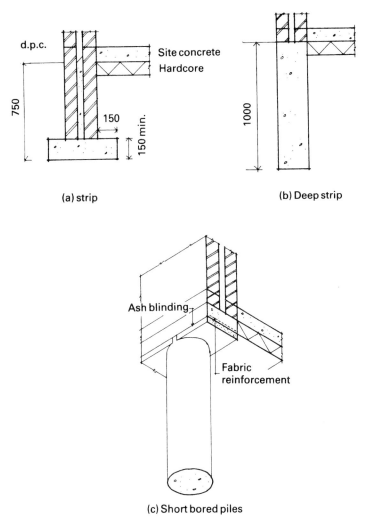

Fig. 8   Foundations: (a) strip; (b) deep strip; (c) concrete piles

Generally, domestic dwellings use the simpler foundations shown in Fig. 8. The most common is the concrete strip foundation shown in (a). This is usually at least 150 mm thick and it extends about 150 mm each side of the wall which it supports. The strip is laid in a trench at least 450 mm deep; on shrinkable clays this minimum depth should be

900 mm. When the wall has been built to damp course level the cavity is filled with concrete to ground level as shown.

These strips of mass concrete form a continuous network under the main walls of the building.

Deep strip foundations, shown in (b) are rarely more than 1 metre in depth or wider than the wall supported.

In areas where the substrata is suspect or where subsidence could occur, the foundation used is usually a continuous reinforced concrete raft. This covers the entire area of the building. Alternatively, the strip foundation could be reinforced and linked to short, bored concrete piles, as in (c).

Very old traditional houses usually have brick footings widening the area of support up to twice the width of the wall. These rest on mass concrete foundations up to three times the width of the wall.

Floors

Ground Floors

The ground floor of the building may be either hollow or solid. Hollow floors may be:

- suspended timber
- precast concrete beams
- suspended reinforced concrete.

Suspended timber floors either span the load-bearing walls, Fig. 9(a), or rest on dwarf brick walls, Fig. 9(b). In some modern houses this traditional form of construction is still used, but the distance between the underneath surface of the floor and the concrete is much less than in older properties, for reasons of economy. The advantage of the wooden floor is its warmth and resilience.

The underfloor area must be kept ventilated and the presence of air bricks or gratings in the external walls is usually a sign of a wood floor. The air bricks are just below the damp-proof course, near to ground level.

'Sleeper walls' made of honeycomb brickwork are often built under the floor to support the joists near to the centre. This effectively halves the span of the joists and allows shallower timber (100 × 50 mm) to be used. Reinforced concrete suspended floors are used where the fill is more than 600 mm deep, Fig. 9(c).

Solid floors are more economic to construct where the fill is less than 600 mm. Figure 10(a) shows a typical solid floor. This consists of a floor finish on a 50 mm sand and cement screed which has been levelled over a waterproof membrane laid over the site concrete.

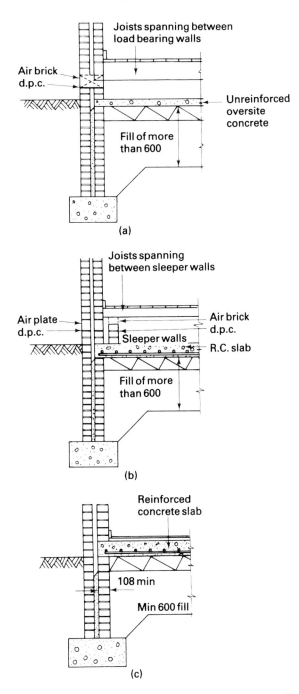

Fig. 9 *Suspended floors: (a) joists spanning load-bearing walls; (b) joists spanning sleeper walls; (c) reinforced concrete slab*

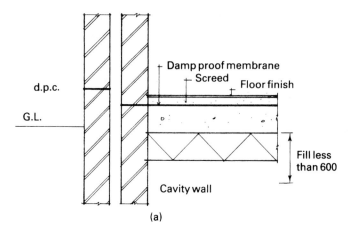

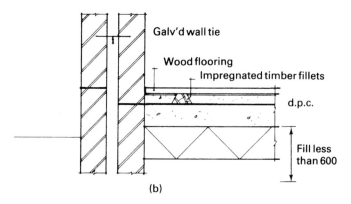

Fig. 10   Solid floor: (a) concrete with 50 mm screed; (b) timber on concrete

With the development of 'early grinding' and 'power floating' techniques for finishing the oversite concrete, screeds of 12 to 20 mm may be laid monolithically, that is, as part of the oversite. In this case the waterproof membrane is laid directly on to the infill hardcore or on to a blinding laid on the hardcore. Installation pipes cannot be located in monolithic screeds and alternative positions need to be found. These have included siting pipes in or under the site concrete but this was later discontinued in favour of PVC sleeves or ducts. Current recommendations must always be followed.

The floor finish could be wood block strip or mosaic laid in bitumen on the screed. The increasing use of fitted carpets has influenced building construction. Timber fillets are often incorporated into the edges of the screed to provide a fixing for a carpet, if subsequently required. In older properties, the floor finish could be quarry tiles, or magnesite compound.

To reduce heat losses through the floor, insulating material may be laid between the screed and the concrete sub-floor.

Another type of floor construction, shown in Fig. 10(b), is a combination of two previous methods. Pressure-treated timber bearers are fixed to clips embedded in the concrete, and floorboards are nailed to the bearers. There is no ventilation and the absence of air bricks in the outer wall is a sign of this type of construction.

Joists are usually fixed at a standard spacing of 400 mm between centres.

Upper Floors

The upper floors of the building are usually supported on deep timber joists (200 × 50 mm) spanning from wall to wall. The joists rest on wood wall plates or galvanised hangers and are covered with tongued and grooved floorboards, plywood or blockboard.

Ceilings in older properties usually have riven sawn laths nailed across the joists and covered with lime and hair plaster. In modern buildings, plasterboard forms the 'lathing' and is covered with a thin skin of ordinary plaster.

There are many variations in floor construction. Joists may be lightweight steel beams covered with wood panels and used with a ceiling of fibreboard panels clipped into place. Alternatively, wood box panels with plywood facing may be used to form a series of stressed skin beams. If there is any doubt about the construction, ask the builder or refer to the plans. If this is not possible it may mean cutting it to find out!

Walls

The walls of a building may be constructed of timber, concrete, steel, glass fibre, stone or brickwork. Of these materials, brick is the most commonly used.

An indication of the thickness and the construction of the main brick wall may be gained from the 'bond' or face pattern of the brickwork. The common bonds are shown in Fig. 11.

English bond (a) is used for solid walls of 220 and 335 mm thickness. It is stronger than Flemish bond.

Flemish bond (b) is also used for solid walls of 220 and 335 mm thickness. It is considered more decorative than English bond.

English garden wall bond (c) is used for solid walls of 220 mm thickness. It is almost as strong as Flemish but uses fewer of the more expensive facing bricks.

Stretcher bond (d) is used for 105 mm brickwork and for 260 mm cavity walls. It is the simplest and most widely used bond.

Solid brickwork will still be found in older properties, but energy conservation requirements ensure that new dwellings have external walls built in cavity construction with two or more leaves or, if solid,

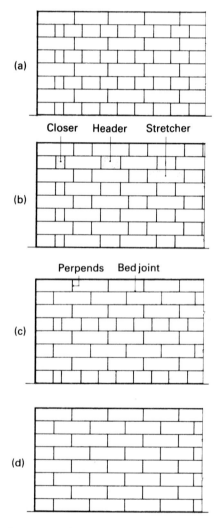

*Fig. 11   Brick bonds: (a) English bond; (b) Double Flemish bond;
(c) English garden wall bond; (d) Stretcher bond*

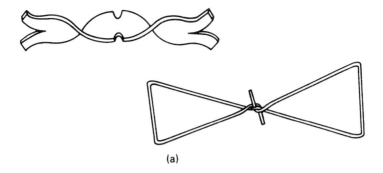

(a)

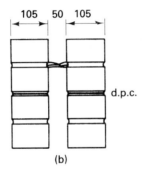

(b)

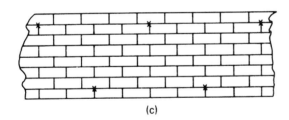

(c)

Fig. 12   Cavity walls: (a) wall ties; (b) section through wall;
(c) ties staggered

using special thermal blockwork; an added layer of insulation will usually be included either as a fill or partial fill in the cavity or added to the inside face of the inner leaf. Inner leaves are likely to be built of 100 mm thick lightweight concrete bricks rather than brickwork.

An alternative form of construction consists of a loadbearing timber frame. This may have a single leaf of brickwork as an outer, weather-resisting skin or some form of timber or plastic cladding. Thermal insulation – often fibreglass – is incorporated within the depth of the frame and plasterboard provides the inner lining. Two vital features of timber frame construction are, first, a vapour barrier positioned on the warm side of the insulation: it is most important that this membrane is not punctured. The second feature is a water repelling breather paper fixed to the weather side of the timber frame. Fig. 13 depicts details of typical timber frame walls. Fig. 13(a) shows a ground floor section with an outer brick leaf and Fig. 13(b) shows a wall section often used at first floor level.

Some important features are:

Breather paper (membrane)	– The semi-permeable sheet tacked on to the outside of timber sheathing. It protects the timber from rainwater but allows air and water vapour to pass through it within the structure.
Cavity	– The gap between inner and outer leaves of external walls which impedes penetration of rainwater and assists drainage.
Inner leaf	– The innermost load bearing structural section of an external wall – the timber frame with plasterboard, vapour barrier, insulation, sheathing and breather paper.
Insulation	– For example, glass fibre or rock fibre located in the timber frame.
Outer leaf	– The outermost section of a wall cladding. Usually brickwork but sometimes tiles, weatherproofed boarding, etc.
Separating wall	– The internal wall (party wall) separating two adjacent dwellings. This should not be confused with internal partition walls which occur between adjacent rooms within a single dwelling
Sole plate	– The timber member placed over the damp-proof course and fastened directly to the brick or concrete inner leaf. It is the first timber component to be fixed and thus forms the base on to which the whole timber frame is erected. It is pressure treated with a chemical preservative.
Studs	– Vertical structural members of the timber frame. They usually occur at centres spaced 600 mm

apart (sometimes 400 mm). About 80% of studs used in Britain have finished sizes 89 mm × 38 mm.

Sheathing — Sheet material (usually plywood, sometimes fibreboard) nailed to the timber framework. It constitutes a fundamental structural feature.

Nogging — A short length of timber inserted into a timber frame.

Vapour barrier (vapour control layer or vapour check) — The membrane that prevents moisture laden air from passing into the framework structure where it may cool and condense to give damp patches. It is usually a polythene sheet located immediately behind the plasterboard (see Fig. 13). In some circumstances, the vapour barrier may be an integral part of the lining in the form of a sheet bonded to the back of the plasterboard (e.g. blue polythene or silver coloured metallised plastic).

Masonry cavity walls and brick clad timber frame constructions use galvanized mild steel wall ties to link the outer leaf to the inner leaf or the timber frame; these are illustrated in Fig. 12.

To give masonry walls increased resistance to wind-driven rain in exposed locations, the outer surface may be cement rendered and given a textured or thrown finish. Indeed, the appearance of a rendered finish may be preferred to fair-face brickwork.

For a time, factory made, non-traditional forms of construction were in vogue; they employed a load-bearing framework of storey-height precast concrete panels; the frames were clad in precast concrete, sheet metal or brick panels. Internal partitions were either of concrete or preformed, hollow plaster panels. Plasterboard on timber stud partitions was frequently used in dwellings.

Damp-Proof Course

To prevent dampness entering the building from the ground or 'rising damp', a water barrier, made from a thin layer of impervious material is used. This is known as a 'damp-proof course' (d.p.c.). In addition to the waterproof membrane already laid in the foundations a continuous d.p.c. is inserted in all the walls. The d.p.c. can be seen in Figs 8, 9, 10, 12 and 13.

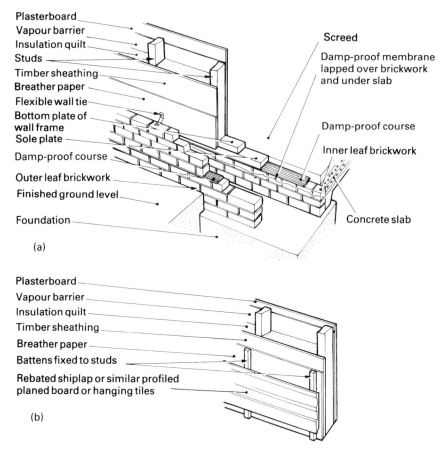

Fig. 13   Timber frame walls: (a) brick outer leaf and (b) tile or timber
          outer leaf

The materials used include:

- lead
- asphalt
- copper
- bituminous felt
- PVC
- slate.

Very old properties may not have a d.p.c., while in some dwellings, slate or two courses of blue engineering bricks were used.

The d.p.c. must be laid at a height of at least 100 mm above ground level. In the case of sloping sites, it must be stepped down to the next level of bricks, the vertical section preventing water penetrating between the two levels.

Property which is built in waterlogged ground has a continuous 'tank' of impervious material, usually asphalt, reinforced to prevent rupture of the d.p.c. by water pressure.

All openings in walls, for example, doors and windows, also require a d.p.c. to prevent moisture affecting the joinery or bridging the cavity. Any projections formed in the roof, such as chimney stacks, flue pipes, lantern lights or skylights, also need to be weatherproofed by a d.p.c. or a flashing. In old properties without a d.p.c. rising dampness can be prevented by injecting a chemical d.p.c., usually a silicone into the masonry wall of a building, as described in BS 6576: 1985 Code of Practice for installation of chemical damp-proof courses.

Lintels

All openings in the main walls, for example, doors and windows, have a lintel to support the weight of the fabric or brickwork above the opening (Fig. 14).

Lintels may be made of:

● brick arch
● reinforced concrete
● rust-proofed metal
● stone.

Sometimes, in the case of stone or brick, the lintel also provides a feature. In most cases, however, the lintel is hidden by plaster or a false brick face.

Lintels are required for some large openings for balanced flue ducts.

Roofs

The essential function of a roof is to protect the interior of the building. It must be wind-resistant and watertight. The roof also serves as an architectural feature and can give elevation to the building. Sometimes the space in the roof is utilised to house the central heating unit or to provide additional rooms. Natural light may be introduced through a dormer window which allows the height of the ceiling to be maintained into the window area. The common terms associated with roofs are illustrated in Fig. 15.

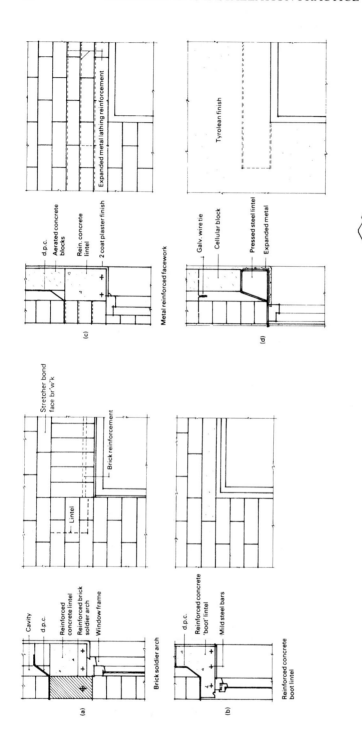

*Fig. 14   Lintels: (a) brick soldier arch; (b) reinforced concrete 'boot' lintel;
(c) reinforced concrete with metal reinforced facework;
(d) galvanised pressed steel lintel*

Roofs may be as shown in Fig. 16:

- flat
- mono-pitched
- double-pitched
- mansard or curb.

Flat roofs may be constructed of reinforced concrete or timber joists covered with tongued and grooved boarding. They may be protected by:

- zinc
- copper
- lead
- three layers of bituminous felt
- asphalt.

To ensure drainage of surface water the surface of the roof is laid to a slight fall of between 2° and 10°. Felted and asphalt roofs are covered with stone chippings.

Thermal insulation is incorporated within the thickness of flat roofs and is laid between the rafters of pitched roofs. To reduce the risk of condensation within the roof space, the Building Regulations (F2) require this space to be ventilated.

Sloping roofs may be mono-pitched but double-pitched roofs are more usual. Structural members may be cut to size and assembled on site with rafters, ceiling joists, purlins and, perhaps collars or, very commonly today, factory made trussed rafters are used. With the latter, timber sections are smaller, pitches flatter and trusses spaced at 450 mm or 600 mm centres. It may be more difficult to accommodate water storage and header tanks within a trussed rafter roof space. Either type of pitched roof may be covered with slates, concrete tiles or clay tiles over the felt and battens.

The ridge of the roof is protected by 'ridge tiles' and other junctions at valleys, hips and verges are finished off using other purpose-made components. Any projections through the roof are made watertight by flashings of lead, copper or a proprietory plastic material, such as Nuralite. Guttering and down pipes collect and carry away the rainwater.

Curb roofs have two different degrees of pitch, meeting at the curb, or side of the ridge. They are also called 'mansard' roofs, after a French architect who popularised this particular style.

Fireplaces and Hearths

Building Regulations require fireplaces and hearths and the chimneys serving them to be built of such materials and in such a manner that

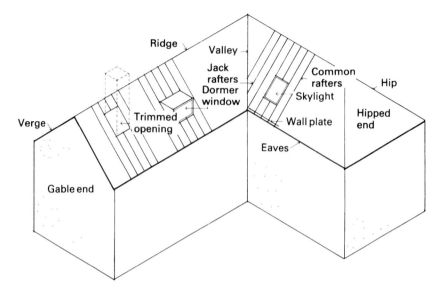

Fig. 15   Common roofing terms

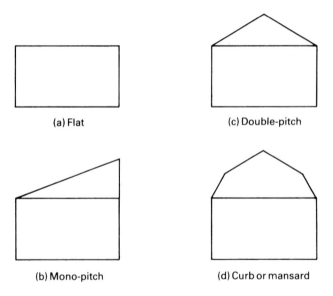

Fig. 16   Types of roofs

fires or appliances will burn safely. Flue gases must be discharged properly to the outside of the building and precautions must be taken to ensure that the heat of the appliance or the flue gases cannot set the building on fire.

Fireplaces and chimneys for open fires and stoves are usually constructed of brick or concrete blockwork. Hearths for open fires consist of a concrete constructional hearth with a superimposed hearth on top; when associated with timber floors regulations stipulate where or how closely timber may be placed relative to hearths, fireplace openings and chimneys. Similar precautions are applied to combustible materials and flue pipes.

Since 1965 regulations have required flues to be lined but older properties may be found to have no lining. When fitting some gas-burning appliances to discharge into such a flue, a lining may have to be inserted.

Industrialised Building and Unit Construction System Building, High Rise Flats

There are many systems available and the choice of system may be dictated by availability, experience and cost. Invariably the construction consists of a basic framework of reinforced concrete (or steel) stanchions linked to structural floors. The walls are usually non-load-bearing. They serve to protect the occupants from the elements and to divide up the spaces between the floor.

Alternatively, the building may be 'monolithic' in structure. That is, with the walls and floors constructed as one unit.

Whatever system is encountered, it will differ considerably from traditional methods of construction. System building makes use of component parts, including:

- plastic-faced panels
- aluminium extrusions
- stressed skin wood panels
- lightweight concrete panels.

All these units are factory produced to exacting limits. The 'curtain walls' as they are called require careful installation to make them weatherproof and windproof and to allow for expansion and contraction. Often, because of the height at which they are installed, special provision has to be made to combat wind pressure and structural swaying.

Figure 17 shows details of the framed construction and the infill panels which form the curtain walling.

The floors may often be made of a patent design of hollow terra cotta pots or tiles or of hollow reinforced concrete panels. These are laid between the structural members of concrete or steel, grouted in with cement and sand.

A wood mosaic or strip floor may look traditional, but its sub-floor will not be. False ceiling panels which clip into position serve to hide services and present a flat surface for decoration.

Great care must be taken when attempting to bore holes into the fabric and special fixing devices may be required to secure pipework or appliances. Hollow concrete panels which are accessible from one face only necessitate the use of toggles or anchors.

It is usual for the main services to be carried in a continuous vertical shaft as shown in Chapter 3, Fig. 3. Access panels are provided at each floor for inspection and routine maintenance.

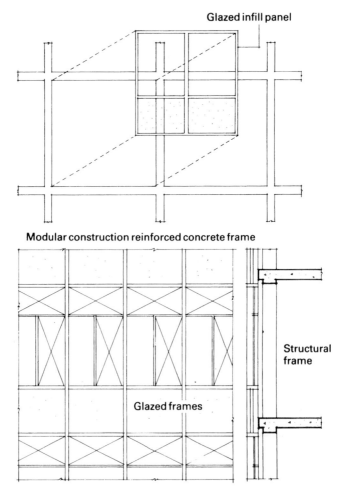

Fig. 17   Curtain walling

External work may be carried out from an adjustable cradle suspended from a continuous rail system around the roof of the building. Cradles have strong fixing points for clipping on safety harnesses. The manufacturers of industrialised building systems aim to produce an economic modular construction which can be built under ideal conditions in the factory and then varied on site, to suit the client's needs.

So far, these modules have only been proved to be economic in large scale operations and where the needs are standardised. Tower office blocks and high rise flats are good examples.

The move towards general standardisation was given greater impetus by the introduction of metrication to the building industry. Opportunity was taken to standardise building components and working tolerances on site.

An increased use of standard unit components could reduce the considerable amount of time now spent on cutting and fitting. This is costly in both labour and materials. Tradition dies hard, however, and it may be some time before modular co-ordination is fully accepted and implemented.

INSTALLING BUILDING ENGINEERING SERVICES

Making Good

During the installation of additional services in a building it is inevitable that the fabric will be damaged or disturbed. So there will be a need to carry out minor repairs or replacements. It is important to keep these to a minimum, for, apart from the cost, it is very difficult to match successfully the finish on weathered surfaces. Fortunately, a range of portable drilling equipment, including tipped drills, core drills, and percussion drills, has been developed. This has reduced the use of hammers and chisels which produced large, unsightly holes requiring extensive patching. Where possible, any replacement materials should be matched with the original material. They can be 'aged' by tinting the mortar and rubbing the surface with rag.

Wherever possible avoid points of entry into the building which will be very obvious. A little expense in resiting them can be offset by the saving in making good with special materials and by eliminating obtrusive unsightly patches.

Cutting Joists

Cutting into any part of the fabric which has structural significance should be avoided.

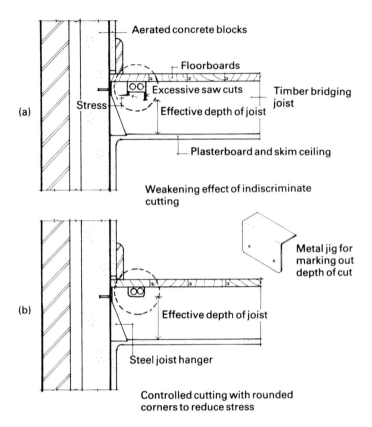

*Fig. 18   Timber flooring: (a) cutting joists, effect of indiscriminate cutting;
(b) correct method of cutting notch*

Although floor or roof members may have to be cut to allow pipes
or cables to pass through they must not be weakened to an extent
which would cause collapse. Often a 'dead' load may not give
evidence of strain. Subsequent live loads – a party, for example – can,
however, provide sufficient stress to cause failure (see Vol. 1, Chapter
4).

Timber joists are particularly vulnerable to indiscriminate cutting.
Despite the factor of safety used in calculating their depth, serious
structural failure can occur if this is significantly reduced.

If a pipe run must cross a line of joists then the depth to which the
material is removed must be kept to an absolute minimum. It must
never exceed 12½% of the depth (Chapter 3, section on Pipes in
Wood Floors).

A common method of notching joists, used by installers, is to make
two saw cuts in the top of the joist and remove the material between

with a chisel. If the cutting is not done carefully and accurately it can result in excessively deep cuts which greatly reduce the effective strength of the joist, Fig. 18(a). The corners of the notch are also weak.

A better method is to bore two holes at the minimum depth required and then remove the waste with saw and chisel. The rounded corners of the notch so produced are structurally more sound, Fig. 18(b). The illustration shows two pipes in a notch. Where a notch is required for a single pipe it may be made by drilling a hole slightly larger than the pipe and removing the top waste with a saw.

Where a number of joists have to be notched for the same size of pipe, it is worth making a small jig so that they can be quickly bored to the same depth.

Where it is possible, a hole bored through the centre of a joist is the least harmful. Any holes made along the neutral axis have only a slight effect on the strength of the joist. This method is frequently used for electric cables which cross joists. It can also be used for soft copper microbore tubing.

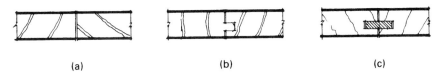

(a)   (b)   (c)

Fig. 19   Types of flooring joint: (a) butt joint; (b) tongue and groove; (c) loose tongue

Removing Floorboards

Taking up floorboards without damaging them is seldom easy. First check the boards are tongued and grooved by inserting a knife blade in the joint. Common joints are shown in Fig. 19.

If the boards are plain, with a simple butt joint, punch the nails through and lever the board up with a floor-board lifter. Figure 20 shows a board held up by chisels or screwdrivers after being partly raised in this way. The board can then be cut across, over the centre of a joist, and removed as required. The cut may be made with a floor saw (Vol. 1 Chapter 13). With a modern, powered circular saw it is possible to cut boards over the centre of a joist without levering the boards up. Great care must be taken to ensure that the other services – electricity cables or water pipes – are not damaged when cutting through the board. Powered saws with a facility for adjusting the depth of cut are available and are recommended for this type of work. On upper floors, in older property, it is advisable to lift the boards without attempting to punch the nails through. The plaster keys between the laths could well be broken and the ceiling might fall if vibrated by hammering.

With wood tongued and grooved boards it is possible to cut the tongues with a power jig saw or a pad saw. The boards may be levered up and cut as before. In some cases it will be necessary to cut the board at the edge of a joist as in Fig. 21. When replacing the board it is necessary to fit a batten fillet to the joist to provide a support.

With steel tongued and grooved boards and with some boards which are laid with secret nails it is necessary to destroy the piece of board to be raised. Methods of nailing boards are shown in Fig. 22. All types of floorboard can be nailed in the normal way with the nails visible. Any of the tongued types may, if required, be secured by secret or hidden nails.

There are, of course, other types of wooden floors which may be found in very old properties, usually in the major cities. These include double floors and framed floors usually used for spans over 5 m in width. Joists were often strengthened by fitting X shaped struts between them. This 'herring' strutting prevented the joists from vibrating or buckling.

In very old property, as in the very newest, it may be necessary to make a trial cut into the fabric in order to establish the form of construction.

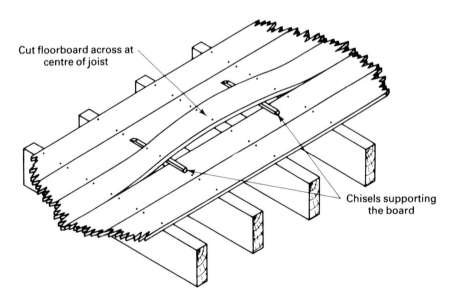

Cut floorboard across at centre of joist

Chisels supporting the board

Fig. 20   Lifting plain boards

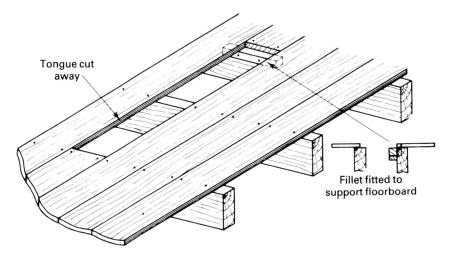

Fig. 21   Lifting tongued and grooved boards

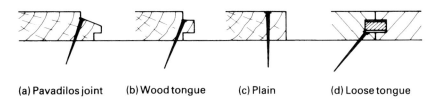

(a) Pavadilos joint   (b) Wood tongue   (c) Plain   (d) Loose tongue

Fig. 22   Methods of nailing boards

Chipboard and Plywood Floors

Many modern domestic premises have floors made from chipboard or occasionally plywood. This type of floor is fitted in large sheets – approximately 2.5 metres × 2.0 metres. Where the joist centres are 450 mm apart board around 19 mm thick are used but if the centres between joists are increased to 600 mm then the material thickness is increased to approximately 22 mm.

The safety precautions to be taken when removing floorboards should be heeded when working on chipboard or plywood floors. When fitting pipework etc. under this type of floor it is not necessary to take up a full sheet and it is preferable to remove small sections of the floor.

Where cuts have to be made that are not supported by a joist it will be necessary to fit a batten to the joist or some additional cross-bracing between the joists to carry the replaced floor sections. Fig. 23(a), (b) and (c) illustrate the fitting of cross-braces. After replace-

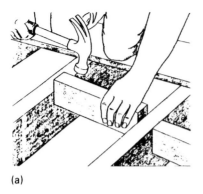

(a)

The cross-bracing should be a hammer fit between the joists.

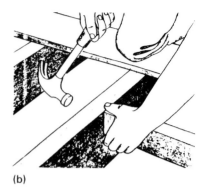

(b)

The cross-bracing should be held in position with wire nails driven through the joists into the end of the bracing.

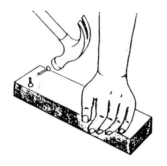

(c)

An alternative method of fixing a cross-bracing is to drive nails, at an angle, through the bracing into the joist.

Fig. 23   Fitting of cross-braces

ment the sections should be drilled and countersunk then secured with woodscrews to battens, joists or cross-braces.

Building Regulations Requirements

Building Regulations make requirements for heat producing appliances in Part J of Schedule 1. There are three requirements:

J1 *Air Supply*: Adequate air for efficient and safe combustion and for the efficient working of any flue-pipe or chimney must be provided.

J2 *Discharge of Products of Combustion*: Adequate provision must be made for discharging products of combustion to the outside air.

J3 *Protection of the Building*: Appliances and flue-pipes must be so installed and fireplaces and chimneys so constructed as to reduce to a reasonable level the risk of the building being set on fire as a result of their use.

It will have been noted that these requirements are written in broad functional terms: their interpretation into practical details is a matter for the Building Control Officer (BCO) in each local authority. However, the Secretary of State for the Environment has issued guidance in the form of an Approved Document; the title in this case is *Heat Producing Appliances* and it is numbered J1/2/3 (amended 1992 version).

The Approved Document (AD) is in four sections. Section 1 covers general provisions, Section 2 solid fuel burning appliances with a rated output up to 45kW. Section 3 deals with gas burning appliances with a rated output up to 60kW and Section 4 oil burning appliances with a rated output up to 45kW. Section 3 includes cooking appliances, balanced flued appliances, decorative fuel effect appliances and other natural draught open flued appliances.

The advice about air supply is given in terms of room size, areas of openable window and areas of permanent ventilation opening related to type and size of appliance.

Requirements for flue-pipes and flues are to be found in this section. In some situations only balanced-flued appliances are permitted. It should be noted that the Gas Safety (Installation and Use) Regulations 1998 which stated that only a room-sealed appliance may be fitted in a private garage, are quoted in this 1992 Approved Document. The Gas Safety (Installation and Use) Regulations 1998 no longer includes this regulation. Advice is given on the siting and protection of terminals.

Also covered in this section are brick chimneys, blockwork chimneys, flexible flue liners, debris collection spaces and hearths.

Section 3 states that, as an alternative approach, the requirements may also be met by following the relevant recommendations in the following British Standards – BS 5440: Installation of flues and ventilation for gas appliances of rated input not exceeding 60kW (1st, 2nd and 3rd family gases).

Part 1: 1990 Specification for installation of flues.

Part 2: 1989 Specification for installation of ventilation for gas appliances.

BS 5546: 1990 Specification for installation of gas hot water supplies for domestic purposes (2nd family gases).

BS 5864: 1989 Specification for installation of gas-fired ducted-air heaters of rated input not exceeding 60kW (2nd family gases)

BS 5871: Installation of gas fires, convector heaters, fire/gas boilers and decorative fuel effect gas appliances.

- Part 1: 1991.
- Part 2: 1991.

● Part 3: 1991.

BS 6172: 1990 Specification for installation of domestic gas cooking appliances (1st, 2nd and 3rd family gases).
BS 6173: 1990 Specification for installation of domestic gas catering appliances (1st, 2nd and 3rd family gases).
BS 6798: 1987 Specification for installation of gas-fired hot water boilers of rated input not exceeding 60kW.
Other Approved Documents which involve the gas industry are B, F, G and L.
B has 5 sections and covers Fire safety. Information is given on the locating of meters, appliances and gas service pipes in protected stairways and gas service pipes in protected shafts and includes ventilation requirements.
F is in 2 sections and covers Ventilation. F1 is Means of ventilation and F2 Condensation in roofs.
G has 3 sections and covers Hygiene. G1 is Sanitary conveniences and washing facilities. G2 is Bathrooms and G3 Hot water storage.
L is in 1 part and covers Conservation of fuel and power.
The J and B approved documents are dated 1992, the F and L 1995.

British Standards

The Approved Document makes frequent reference to British Standards and Codes of Practice. The most useful ones are the following:

BS 5258   Safety of domestic gas appliances: The following parts of this BS are relevant to the regulations:
- Part 1   Central heating boilers and circulators
- Part 2   Cooking appliances (obsolescent)
- Part 4   Fanned circulation ducted air heaters
- Part 5   Gas fires
- Part 6   Refrigerators and food freezers
- Part 7   Storage water heaters
- Part 8   Gas fire/back boiler combined appliances
- Part 9   Combined appliances fanned-circulation ducted-air heaters/circulators
- Part 10   Flueless space heaters (3rd family gases)
- Part 11   Flueless catalytic combustion heaters (3rd family gases)
- Part 12   Decorative log, etc. effect appliances (2nd and 3rd family gases)
- Part 13   Convector heaters
- Part 14   Barbecues (3rd family gases)
- Part 15   Combination boilers
- Part 16   Inset fuel effect gas fires (2nd and 3rd family gases)
- Part 17   Direct gas-fired tumble driers.

BS 5386   Specification for gas-burning appliances
- Part 1   Instantaneous water heaters – domestic
- Part 2   Mini water heaters (2nd and 3rd family gases)

Part 3   Cooking appliances – domestic
Part 4   Built-in cooking appliances – domestic
Part 5   Instantaneous water heaters with automatic output
          variation (2nd and 3rd family gases)
Part 6   Domestic cooking appliances with forced convection
          ovens.

BS 5440   Code of Practice for flues and air supply
          Part 1   Flues
          Part 2   Air supply

BS 5546   Code of Practice for installation of gas hot water supplies for domestic
          purposes. (1st, 2nd and 3rd family gases).

CHAPTER 3

Internal Installation

Chapter 3 is based on an original draft by S. Burrows

Introduction

The British Standard 'Glossary of Terms used in the Gas Industry' does not recognise the term 'internal installation'. It refers instead to 'installation pipes'.

However, the internal installation is universally understood to be an installation within premises, including the meter, governor and any other control devices, between the meter control and the points to which the appliances are connected, Fig. 1.

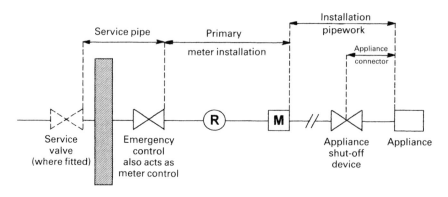

Fig. 1   Internal installation

This is, at the moment, at variance with the continental definition which has their internal installation starting at the meter outlet and including the appliances. Although the British definition includes the meter, this chapter deals with the pipework. The installation of domestic meters is covered in Chapter 4.

The Gas Safety (Installation and Use) Regulations 1998 define installation pipework as any pipework for conveying gas for a particular consumer and any associated valve or other gas fitting but it does not mean:

(a)   a service pipe or
(b)   a pipe comprised in a gas appliance, or
(c)   any valve attached to a storage container or cylinder; or
(d)   service pipework.

The installation is also called the 'carcass' or house carcass. This term is deprecated by the Glossary but it has persisted and is specifically applied to the pipework installed in new houses whilst they are under construction. 'Carcassing' is generally understood to mean working on a building site, installing pipework in the properties before the floors are laid or the walls are plastered.

The installation of pipes in new buildings is ideally planned at the drawing board stage, when the meter and appliance positions and the pipe runs may be indicated on the plans. Taking-off materials from the plans and costing can also be done at this time. The installation is phased in with the work of the other building crafts so that all the building engineering services can be installed immediately the shell of the building and the floor joists are completed.

The installation should be carried out in accordance with BS 6891: 1998. It must also comply with the Gas Safety (Installation and Use) Regulations 1998 and current Building Regulations. The requirements of these documents are embodied in the practices recommended in this Chapter.

Both the person carrying out the installation and the employer are equally responsible for ensuring that the Gas Safety Regulations are satisfied. And any person offending against the regulations can be prosecuted.

Pipe Runs

Materials

Although the use of materials varies throughout the country, those most commonly used for domestic internal installations are:

- steel pipes, medium weight, BS 1387
- light gauge copper tubes, BS EN 1057: 1996

Where steel pipe is used for the main installation, any extensions or appliance connections are usually made in copper tube.

The factors affecting the choice of a particular material are:

- possible need for mechanical strength to resist damage
- need for protection from corrosion
- appearance of visible pipe runs
- overall cost, including fittings and labour

Pipe Sizing (Natural Gas)

The method of determining the size of pipes for a domestic installation was described in Vol. 1, Chapter 6. The main points again are:

- pipes should be sized from the maximum gas rate of the appliances, bearing in mind the maximum demand likely at any one time
- allowance should be made for any future additional load or extension
- the pressure loss between the outlet of the meter and the appliance should not normally exceed 1 mbar
- the main gas supply should generally be not less than 20 mm steel pipe or 22 mm copper tube
- branches from the main supply should generally be not less than 15 mm steel or copper.

The effect of fittings in a pipe run is given in Table 8. The pressure loss through the fitting is expressed as an additional length of pipe to be added to the length of the run.

A table giving the pressure losses in lengths of steel pipe was included in Vol. 1, Chapter 6. A similar table for light gauge copper tube is provided in Table 8(a).

TABLE 8   Effect of Pipe Fittings on Gas Flow

| Nominal pipe size* | Additional length to be added, metres | | |
	Tees	Elbows	90° Bends
Up to 25/28 mm	0.5	0.5	0.3
32/35 mm to 40/42 mm	1.0	1.0	0.3
50/54 mm	1.5	1.5	0.5

The smaller of each pair of dimensions refers to steel pipe, the larger to light gauge copper tube.

Planning Internal Installations

Location and Route

There are a number of general principles which must be borne in mind when planning internal installations.

TABLE 8(a) Discharge in a **Straight Horizontal Copper Tube*** with 1.0 mbar differential pressure between the ends, for gas of relative density 0.6 (air = 1)

Size of tube (mm)	Length of tube (m)									
	3	6	9	12	15	20	25	30	40	50
	Discharge (m³/h)									
6	0.12	0.06								
8	0.52	0.26	0.17	0.13	0.10	0.07	0.06	0.05		
12	1.5	1.0	0.85	0.82	0.69	0.52	0.41	0.34	0.26	0.20
15	2.9	1.9	1.5	1.3	1.1	0.95	0.92	0.88	0.66	0.52
22	8.7	5.8	4.6	3.9	3.4	2.9	2.5	2.3	1.9	1.7
28	18	12	9.4	8.0	7.0	5.9	5.2	4.7	3.9	3.5
35	32	22	17	15	13	11	9.5	8.5	7.2	6.3
42	54	37	29	25	22	18	16	15	12	11
54	110	75	60	51	45	38	33	30	26	23
76.1	280	190	150	130	120	98	86	78	66	58
108	750	510	410	350	310	260	230	210	180	160

*To specification of Table X, BS EN 1057: 1996

These include:

- the route should avoid any positions where the pipe could be liable to damage, either during the building operations or when the property is finally occupied
- the fire resistance of the building must not be impaired
- the route should, as far as possible, avoid the need to cut into load-bearing walls or joists
- pipes should be concealed, where possible, but provision should be made for access
- pipes should not be run in floors in which under-floor heating is installed
- there should be a space of at least 25 mm between a gas pipe and any other service with preferably 50 mm from an electrical supply
- when the spacing requirements are impracticable the pipe should be either PVC wrapped or a panel of insulating material should be interposed.
- pipes should be at least 150 mm away from electricity meters and associated excess current control or fuse boxes.
- pipes may not be run in the cavities of cavity walls
- pipes passing through a cavity must take the shortest route and pass through a gastight sleeve
- pipes are normally only run in floors or roofs, dropping down to appliances on the floor below, when no other route is available on the lower floor
- where pipework is installed under foundations footings etc. adequate steps should be taken to protect the pipework in the event of movement of the structure or ground. (New regulation 19(5), Gas Safety (Installation and Use) Regulations 1998).
- with dry gas it is unnecessary to provide condensate receivers but dust traps may be required at the foot of large risers
- where pipes are required to be buried in the structure, prepared ducts or chases should be included in the plans so that they may be formed during erection
- pipes should never run diagonally across walls or floors, they should normally follow the line of the walls and be kept close to them.

Usually installations in new properties will have been planned after consultation with the builder or architect. Always check with the Clerk of Works or General Foreman before starting work on a site. Any alterations to the agreed routes for the pipework or the actual installation should first be discussed with the site official.

Appliance Points

The type of appliance and its ideal position dictates the location of the 'point' to which the appliance will be connected. Where the specific appliance is known, the maker's instructions can provide the information. As a general rule the point should be positioned so that the final connection is kept as short and as unobtrusive as possible.

When installing appliance gas points make sure that:

- the point is square to the surface through which it protrudes
- the cap or plug is fitted so that it may be removed without undue force
- when the cap or plug is being removed no other fitting can become unscrewed, for example, a socket or nipple
- the termination point is at a suitable height and position for the intended appliance.

Some approximate terminal positions are as follows:

Cookers	— 600 mm	from floor level
Refrigerators	— 100 mm	above floor level
Central Heating Appliances	— 450 mm	above floor level
M.P. Instantaneous Water Heater	— 900 mm	above floor level
M.P. Storage Water Heater	— 100 mm	above floor level

When the position of points has been agreed with the builder, only minor alterations are permissable without special permission.

INSTALLING PIPEWORK

Pipes in Walls or Solid Floors

Pipes are often run in recesses in brickwork, blockwork or reinforced concrete. If the recess or channel is purpose-made or formed when the concrete is poured, it is called a 'duct'. Large, vertical ducts are used in blocks of flats to carry all the services up through the building.

Channels which are cut in brick or block walls, after construction, are called 'chases'. The chase or duct must be sufficiently deep to allow for the pipework and its protective covering to be accommo-dated and then covered by the usual rendering and plaster finish, Fig. 2. A chase must be cut vertically or horizontally, never diagonally.

Some building blocks are made from breeze or clinker and are corrosive. Pipes chased into these blocks must be suitably sheathed, wrapped or coated.

Ducts must be waterproof and well ventilated, Fig. 3. They should not bridge fire barriers. Pipes must be clear of other services and must be accessible. Ducts for large risers in multi-storey buildings are dealt with in Vol. 3, Chapter 1.

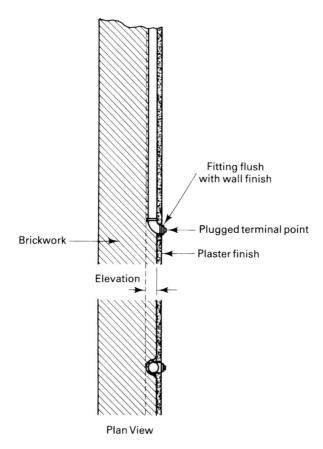

Fitting flush
with wall finish

Plugged terminal point

Brickwork

Plaster finish

Elevation

Plan View

Fig.2   Pipes chased in wall

In reinforced concrete floors pipes may be laid in prepared channels and covered with cement mortar. Frequently pipes are laid in the 50 mm screed which covers the floor. There should be a minimum of 20 mm cover over the fittings, and pipes up to 25 mm diameter can be accommodated. Pipes should not be laid in the structural concrete or close to reinforcement. Structural floors should not be chased without first consulting the builder.

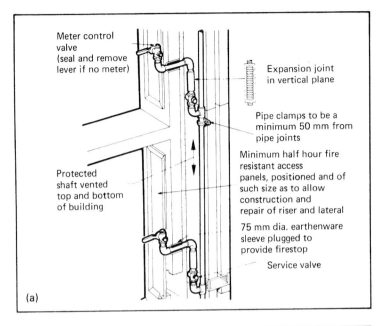

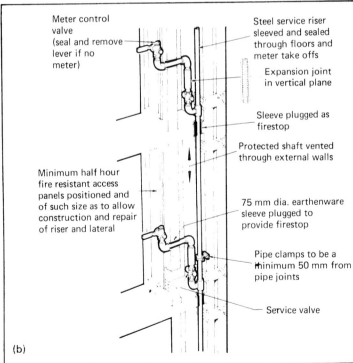

Fig. 3   Service risers in ducts: (a) continuous shaft; (b) sectional shaft

In hollow floors, made of pot tiles or hollow sections, pipes are usually laid in the cement screed or in a position indicated by the architect.

Pipes of less than 15 mm diameter should not normally be buried. The exception is:

- the pipe between a control cock and the back of a hearth to provide a concealed connection to a hearth fitting gas fire.

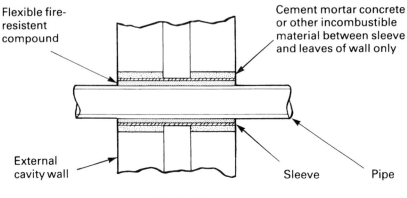

Fig. 4　Pipe sleeve

When pipes pass through solid floors or load-bearing or cavity walls, they must be in a sleeve, Fig. 4. Sleeves normally extend through the structure and finish flush with the surface at either side. An exception to this is when the sleeve is in a floor which may be washed down or made with a corrosive screed finish. In these cases the sleeve should project at least 25 mm above the surface.

Sleeves must be set into the structure with cement mortar and the gap between the pipe and the sleeve should be sealed at one end with a flexible fire resistant compound, the other end being left open to prevent any escaping gas accumulating in the space between pipe and sleeve. No pipe joint should be located in a sleeve.

Pipes in Wood Floors

Pipe runs under suspended wood floors may be supported as shown in Fig. 5. The spacing of the supports should be as given in Table 9.

Where the floor has a ceiling below, pipes crossing the joists must be laid in notches, Fig. 6(a). These must be cut to ensure that the weakening of the floor is minimal, Chapter 2, Section on Cutting Joists.

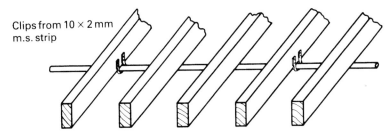

Clips from 10 × 2 mm m.s. strip

Suggested method of securing pipes under ground floor joists where working room is restricted

(a)

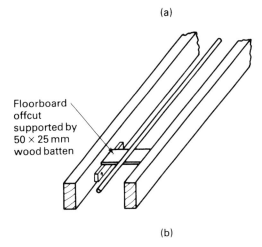

Floorboard offcut supported by 50 × 25 mm wood batten

(b)

Fig. 5 Supporting pipe: (a) pipe crossing below joists; (b) pipe running with joists

Notches should be:

- located not more than one-quarter of the span from an end support
- not deeper than $12\frac{1}{2}\%$, or approximately $\frac{1}{8}$, of the depth of the joist
- not cut into joists of less than 100 mm nominal depth
- preferably 'U' shaped, that is, with radiused corners
- positioned, where possible, under the centre of floorboards to avoid damage to the pipe from nails or water dripping on to the pipe through the joints between the floorboards.

The largest outside diameter of pipe which may be laid in notches in joists 100 mm high is:

$$12\tfrac{1}{2}\% \text{ of } 100 \text{ mm} = \frac{12.5}{100} \times 100 \text{ mm}$$

$$= 12.5\text{mm} \ (12\text{mm pipe})$$

So 12 mm copper tube is the largest size which may be used.

Where access to the pipe may be required, the floorboard should have the tongue and lower half of the groove removed, Fig. 6(b). It should be marked to indicate that the pipe is laid underneath and it should be secured with countersunk woodscrews. The screw length should be about $2\tfrac{1}{2}$ times the thickness of the board. For a board nominally 20 mm, a 50 mm screw is required. Usually the screws would be No. 8 or No. 10, the larger size being used for soft woods.

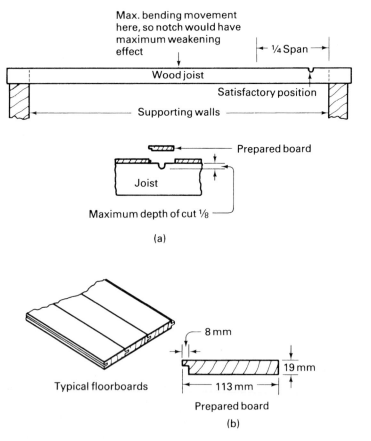

Fig. 6   Pipes crossing on joists: (a) notch in joist; (b) covering floorboard

Pipes on the Surface

Although, as a general rule, pipes should be concealed, there are some situations where running pipes on the surface cannot be avoided. In these cases care must be taken when selecting the material and the route.

Copper tube has neat and compact fittings and its outside diameter is smaller, for the same carrying capacity, than steel pipe. So it has the advantage of a better appearance. When painted to tone in with its surroundings it is generally more acceptable.

The route for exposed pipework should make it as unobtrusive as possible. On, or just above, skirting level may be the best location for a horizontal pipe. It may then be obscured by furniture or fitments. Vertical pipes may be sited in angles of walls or against architraves.

Whether pipes are clipped close to the wall or held clear depends on the particular situation. Generally pipes are spaced clear of walls so that dirt cannot lodge behind them.

Pipework must be securely fastened and adequately supported. Table 9 gives the maximum spacing recommended between pipe supports.

TABLE 9 Maximum Intervals Between Pipe Supports

Pipe material	Nominal size (mm)	Interval for vertical run (m)	Interval for horizontal run (m)
Steel	Up to 15	2.5	2.0
Steel	20	3.0	2.5
Steel	25	3.0	2.5
Light gauge copper	Up to 15	2.0	1.5
Light gauge copper	22	2.5	2.0
Light gauge copper	28	2.5	2.0

BS 6891: 1988

Protection Against Corrosion

Where pipes are installed in damp or corrosive situations, they must be protected with a suitable, durable covering (Vol. 1, Chapter 12). This may be applied in the factory by sheathing, wrapping or painting, or on site, by wrapping or painting. Care must be taken when bending, jointing or fitting the coated pipes into position, to avoid damaging the protective covering. After assembly, all joints and bends should be carefully wrapped and pipes must be clean and dry before the wrapping is applied.

Assembled pipework must be tested for soundness before the wrapping is carried out.

The amount of overlap on wrapping tape should never be less than 12 mm and should always be at least half the width of the tape. This gives a protective layer of two thicknesses of tape and is a good safeguard against irregular wrapping.

Where pipes are to be buried, they must either be made of corrosion resistant material, or be suitably protected. Channels should be clear of debris and the pipe supported so that it may be completely surrounded by mortar.

Magnesite or magnesium oxychloride floor screeds are particularly corrosive. Pipework to be buried beneath this material must be copper, with a factory-bonded plastic sheath. It must be set in a 3:1 sand/cement mortar with a cover of a least 25 mm. Where pipes emerge they should be protected by a sleeve, Fig. 7.

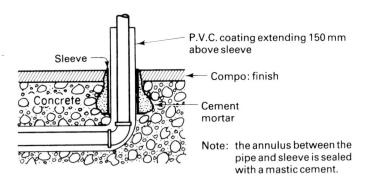

Fig. 7   Pipe sleeve in composition floor

Where steel pipes are run on walls or ceilings, they should be spaced clear of the surface. This ensures that the pipe may be properly painted and prevents contact with any damp material.

Working in Customer's Premises

There are approximately 20 million gas users in the UK of which British Gas plc has approximately 17 million. Some are pleasant and easy to work for, others are awkward, fussy and difficult. But they have all paid their money for your work, your skill, your specialist knowledge. If you disappoint them there are other fuels and other firms. They can easily end up as someone else's customer.

When a customer is having another appliance installed there are two areas in which problems can arise. One concerns the appliance, the other the property.

Anyone who has just bought something new wants to be told that they have made a wise choice. They want to feel that they have bought the best, or the best value. They are likely to ask you, as the expert, for your opinion, when what they really want is confirmation.

Of course, you may not care for that particular appliance. But if you are foolish enough to share that opinion with your customer you are in for trouble!

In the first place, they certainly will not thank you for implying that they were stupid to buy that model. And secondly, they will always be slightly dissatisfied with it and always be looking for flaws.

Unfortunately it is a fact that when the first man – the installer – has put a customer off an appliance, no one can ever really persuade them that he was wrong.

The property can also prove to be a source of troubles for the unwary. When a gas supply has to be run in existing, occupied premises there are ornaments and furniture to move, floor covering to roll back and floorboards to lift. All of these operations must be done carefully, to avoid damage occurring, but above all watch out for the accident that is waiting for you to come along.

You can easily be blamed for spoiling something which has already been damaged or is ready to fall apart if you touch it. There may be cracked glass or china ornaments, scratched furniture, stained carpets or cracked wash basins. Keep a look out for these things and draw the customer's attention to them before you start work.

The main points to remember when working in customer's premises are:

- be careful what you stand on, use steps if possible
- if steps are not available make sure that the customer does not mind your using an alternative, such as a chair or a stool and that it will easily take your weight
- never stand on lavatory seats and do not put any weight on basins or shelves
- move furniture carefully clear of the work area
- if you need to stack small items on the bed always turn back the eiderdown or duvet first and protect the covers
- roll up carpets or other floor coverings with the pattern outside so that the edges will not subsequently curl up (and avoid cracking any old lino)

- before starting work on upstairs floors make sure that the ceiling is in good condition and there is nothing suspended from it which could be damaged by vibration
- before raising floorboards check to see whether any have been lifted before, keep clear of electric cables and water pipes: the position of switches, sockets and central heating radiators often indicate where they are likely to be run
- if nails have not been punched down through the floorboards, pull them out as soon as the board is raised, they are a hazard while they are sticking out
- do not leave open floorboards unattended, put them back if you have to leave the job for any length of time
- when working on an existing installation, make sure that any mixture of gas and air in the pipes will not be ignited by a blowlamp; if necessary, purge the pipes of gas
- any meter should be temporarily isolated from the supply and the open ends capped or plugged
- while the pipework is in progress, keep the ends of the pipe covered by plugs or caps to keep out dirt and debris
- test the pipes for soundness before finally replacing the floorboards or painting or wrapping pipes
- avoid making dirty marks on paintwork, walls or floors, do not touch them when your hands are greasy and kneel on a mat or a dust sheet if your shoes mark the floor covering
- at all times be polite to your customers; whether they live in a palace or a slum they are entitled to the same consideration
- finally, no job is complete until the place is tidied up and everything is back where it belongs.

Testing for Soundness

Low Pressure Gas Pipework up to 28 mm (DN 25) in Domestic Premises (2nd Family and LPG/Air Mixture gases)

BS 6891: 1998 gives details of the actions to be taken when testing a domestic installation for soundness. The procedures to be followed are:

- where appliances are connected to the installation pipework being tested, check that all operating taps, pilots, etc. are turned off
- connect a pressure gauge to the system, either to a suitable pressure test point where the installation is connected to a gas supply or to one branch of a test tee piece (Vol. 1, Chapter 13).

The other branch of the test tee piece must be valved to permit air to be pumped into the installation

- slowly pressurise the installation (with gas or air as appropriate) to 20 mbar (make sure that the pressure does not exceed 23 mbar as this may cause the meter governor to 'lock-up')
- allow 1 minute for the temperature to stabilise. Any pressure loss at this stage must be rectified before a further test is carried out. If a pressure rise is observed at this time, either the temperature of the gas or air inside the system is rising and a longer period of stabilisation is needed or the means of isolating from the pressure source (meter control or emergency control etc.) is leaking/ 'letting-by', this must be rectified before the soundness test is continued. If the meter control or emergency control is passing gas ('letting by') this must be reported to the gas suppliers emergency service immediately
- note any pressure loss in the next 2 minutes and check that there is no smell of gas
- for the soundness test to be deemed satisfactory there must be no smell of gas and where the appliances are isolated from the installation being tested, there must be no pressure drop during the 2 minute test period. Where the appliances cannot be isolated and the operating taps, pilots etc. have been turned off, there must be no smell of gas and any pressure drop during the 2 minute test period must not exceed the permitted tolerance given in Table 10.

TABLE 10   Maximum Permissable Pressure Drop During a 2 Minute Test Period for Existing Installations with Meters up to U 16, 16 m³/h (565 ft³/h)

Meter Designation		Permissable Pressure Drop	
Case reference Approx.	Capacity/ Revolution		
	(ft^3)	(mbar)	(in w.g.)
U6 (D05 or D07)	Up to 0.071	4.0	1.6
E6 Ultrasonic meter	n/a	8.0	3.2

Note: The pressure drops given relate to average lengths of pipework in domestic gas installations. In a domestic installation where the installation pipe is connected to pipe sizes larger than 28 mm copper or DN 25 steel, reference should be made to Institution of Gas Engineers publication IGE/UP/1 or IGE/UP/1A.

Some modern cookers with a drop-down lid over the hotplate are fitted with a device which cuts off the gas supply to the hotplate when the lid is closed. When testing the soundness the lid must be in the open position to ensure that the hotplate taps are included in the test.

New installations must be tested for soundness on completion. Before any alterations are made, or any extension is connected to an existing gas installation both the existing pipework and the meter must be tested for soundness. Any extension should preferably be tested before being connected to the existing pipework. After the extension has been connected the soundness test must be repeated on the whole installation.

All soundness tests must be carried out before applying protection (wrapping, painting etc.) to the pipework.

The pipework, between the means of isolating from the pressure source (meter or emergency control etc.) and the meter regulator, must be pressurised and tested with leak detection solution before the whole system can be deemed sound. This action is necessary because a governor subjected to a pressure greater than its adjusted working pressure can 'lock-up' (the valve is held against the valve seating, stopping the flow of gas). When this occurs and the pressure gauge is fitted anywhere on the outlet side of the governor, any loss of pressure (leakage) on the inlet side of the control will not register on the pressure gauge. The pressure test does not include any pipework downstream of the appliance shut-off devices or the appliance operating taps. These must be tested with leak detection fluid and where necessary with the appliance burners on and lit.

Purging

Low Pressure Gas Pipework up to 28 mm (DN 25) in Domestic Premises (2nd Family and LPG/Air Mixture Gases)

When testing for soundness has been carried out and the installation has proved to be sound, then the air must be cleared from the meter and the pipework.

As gas enters the installation, some of it mixes with the air and the mixture may be explosive. Precautions must be taken to prevent the mixture being ignited. The main points to remember are:

- make sure that there is good ventilation so that the gas can disperse, open an outside door or window
- do not operate any electric switches
- do not allow any smoking or naked lights
- purge all installation pipes commencing at the point(s) furthest from the meter
- the system will be purged when the meter has passed a volume of gas not less than 5 times the capacity per revolution of the meter, for meters up to U16 size

- light up any appliances already installed and check that the flames are normal when purging has been completed.

The capacity per revolution is marked on the meter, either on the index or on the badge. For a U6 meter it is 2 dm^3 (0.071 ft^3)/revolution. So to purge a domestic installation with a U6 meter it is necessary to pass 5×2 dm^3 = 10 dm^3 or 0.01 m^3. In cubic feet this is $5 \times 0.071 = 0.355$ ft^3. On completion of the test and purge all the re-sealed test points must be tested with leak detection fluid.

Installations which are connected to meters larger than U16 size should be purged in accordance with the British Gas publication IGE/UP/1 Purging Procedures for Non-Domestic Gas Installations. This is described in Vol. 3, Chapter 1. Testing for soundness and purging of LPG installations is covered at the end of this chapter.

Both testing for soundness and purging are requirements of the Gas Safety Regulations.

Electrical Bonding

Permanent Bonding

Where it is necessary to fit a permanent electrical bond to the gas installation pipes, the work must be carried out by a competent electrical contractor in accordance with BS 7671 (IEE Wiring Regulations). The connection should be made to a clamp on the gas pipework at the outlet of the meter (Vol. 1, Chapter 9). The clamp should be positioned as near to the meter as possible and, in any case, not more than 600 mm from it. The type and size of the conductor is specified in the Institution of Electrical Engineers 'Regulations for the Equipment of Buildings'.

Temporary Bonding

Because it is possible for the gas installation pipes to act as an earth return path for stray currents when faults occur on appliances, it is necessary to maintain continuity at all times. Indeed, the Gas Safety Regulations require that a temporary continuity bond shall be used whenever a gas pipe, pipe fitting or meter is disconnected, replaced or removed.

The temporary bond was illustrated in Vol. 1, Chapter 13.

It consists of not less than 1.2 m of single core tough-rubber-sheathed (TRS) flexible cable with a cross-sectional area of at least 10 mm^2, 250 V grade, to BS 6500. The cable should be fitted with an insulated clip or a clamp at each end.

The procedure for using the temporary continuity bond is as follows:

- fit the continuity bond before disconnecting any pipe or fitting
- ensure that the clips make good electrical contact; if necessary, clean the pipe and remove paint, rust, dirt or pipe wrapping
- fit the first clip on the upstream side (the service pipe side) of the work, Fig. 8(a)
- fit the second clip on the downstream side (the appliance side) of the work, Fig. 8(b)
- carry out the disconnection and replacement of the meter or pipework taking care not to disturb the temporary bond, Fig. 8(c)
- when the work has been completed, remove the clip on the downstream side first, Fig. 8(d).

Keep a watch for sparking when fitting the second clip or disconnecting the gas supply. This will only occur if there is an electrical fault. If it happens turn off the electricity supply and ensure that any accumulation of gas has dispersed before removing the bond.

Check that the fault is not on any gas appliance and advise the customer to call in an electrical contractor.

Fitting Tees into Existing Supplies

Installing an additional appliance in occupied premises usually entails cutting into the existing pipework and inserting a tee. Where this is done immediately on the outlet of the meter it is a simple matter to disconnect and reconnect the supply and to accommodate the tee either by shortening the standpipe or by adjusting a flexible meter outlet. When inserting tees into a pipe run the procedures are as follows:

- in all cases, first test the existing installation for soundness
- always fit a temporary continuity bond before cutting or disconnecting the pipe or meter
- disconnect and cap off meter outlet.

Light Gauge Copper Tube

When sufficient movement can be gained on both pipes a normal tee can be used. If the copper tube is not held rigidly at both ends and is sufficiently long for one end to be pulled out of line with the other or 'sprung', a slip tee can be used.

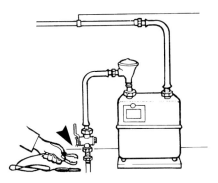

(a) Fit the first clip upstream of the work.

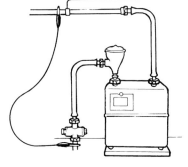

(b) A temporary continuity bond must be fitted before any part is disconnected.

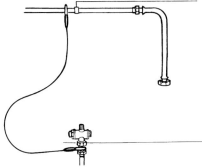

(c) Carry out the work without disturbing the temporary continuity bond.

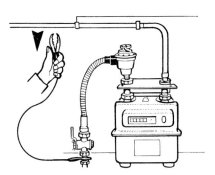

(d) When the work is complete remove the downstream clip first.

Fig. 8   Temporary continuity bonding

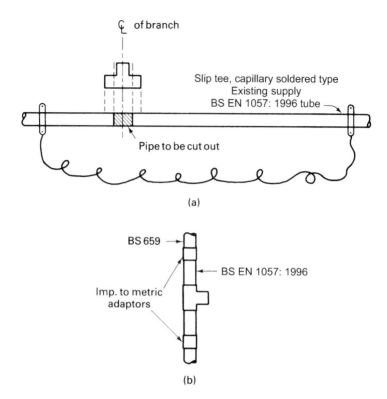

Fig. 9   Tee fitted in copper supply: (a) not rigid, BS EN 1057: 1996 tube;
          (b) BS 659 tube

This tee is a capillary soldered fitting without internal shoulders so that it can slide along the pipe. The tee is inserted as shown in Fig. 9(a):

- mark the position of the centre line of the branch
- place the tee on the centre line and mark the length of pipe to be cut out
- fit the temporary continuity bond
- disconnect and cap-off meter if necessary
- cut the pipe and prepare the ends and the tee for soldering
- spring the pipe to one side and slip the tee onto one end
- slide the tee back over the other pipe end and line it up with the centre line
- insert the branch and make the joints.

If a metric fitting has to be inserted into an Imperial supply adaptors or a socket forming tool may be required as in Fig. 9(b).

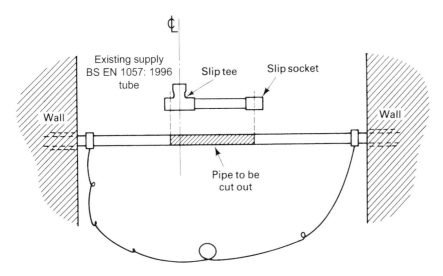

Fig. 10   Tee fitted in rigid copper supply

Where both ends of the supply are rigidly held and the pipe cannot be sprung it is necessary to use a slip tee and a slip socket as in Fig. 10. The procedure is similar to the previous one except that both fittings may now be slipped on to the short length of tube which is cut from the piece removed from the main supply. The tee and the socket can then be slid back over the prepared pipe ends.

Take care that the fittings are satisfactorily positioned on the pipe so that good joints can be made.

Steel Pipe

If one end of the pipe can be sprung, then the tee may be inserted using an ordinary longscrew with coupler and backnut. The method is shown in Fig. 11.

- mark the centre of the branch
- make the joint between the tee and the longscrew
- place the tee over the centre line and mark the length of pipe to be cut out, Fig. 11(a)
- fit the temporary continuity bond
- disconnect and cap-off meter if necessary
- cut the pipe, remove and thread the two ends
- reassemble the ends into the fittings, spring the pipe to one side and screw on the tee and longscrew, Fig. 11(b)
- run the coupler on to the pipe end and seal with the backnut and grummet.

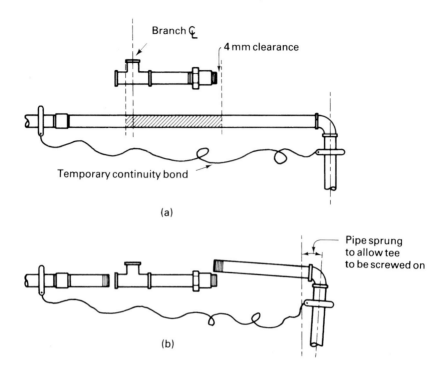

*Fig. 11   Tee fitted in steel pipe, not rigidly held: (a) marking out;
(b) fitting tee*

Where there is an elbow (or tee) at one end of the pipe, as in this case, there is actually no need to spring the pipe. Instead, remove the elbow and remake the joint. Then tighten the elbow again, stopping before it is fully tightened so that the short length of pipe can be inserted and will not be obstructed by the tee and longscrew which have been fitted to the other pipe end. Finally tighten the elbow, bringing the pipe end into line with the longscrew and then complete the joint as before.

If the pipe is rigidly held in position at both ends, a double longscrew must be used, Fig. 12. A space nipple or a hexagon nipple should be screwed into the tee and the longscrew connected temporarily to it. The tee can then be lined up with the centre line of the branch and the pipe marked out for cutting. If a double longscrew is not available, the job can be done by using two ordinary longscrews, one screwed into each side of the tee.

When the pipe has been cut out, the ends may be threaded by using ratchet stocks and dies and holding the pipe ends against turning with a chain wrench.

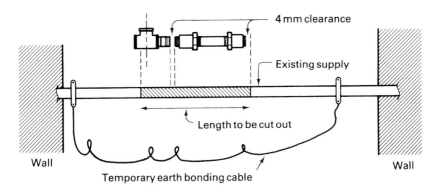

4 mm clearance

Existing supply

Length to be cut out

Wall

Wall

Temporary earth bonding cable

Fig. 12   Tee fitted in rigid steel pipe supply

Concealed Connections

Connections to cookers and refrigerators are usually made behind these appliances and so they are automatically concealed. In the case of space heaters and all types of gas fires, there is a greater need for concealment to improve the appearance of the installation and special treatment is necessary.

There are several ways in which a concealed supply can be run to the back or underneath of a gas fire.

These include:

- fit the control cock in a floor trap and chase the supply through the hearth to the back of the builder's opening
- conceal the supply within a hollow plinth, leaving only the control and a short length of pipe visible at the side of the hearth
- fit a sleeve through the side of the chimney breast to carry the supply, Fig. 13:A
- chase the supply into the brickwork on the face of the chimney breast and cover it with the skirting board, Fig. 13:B
- Fig. 14 shows a concealed supply, 8 mm plastic covered soft copper pipe, run through the base of the chimney breast. It is terminated with a $Rc^{1}/_{4} \times R^{1}/_{4}$ ball valve which should be screwed directly into the union connection of the appliance. Fires with a high input rating should be fitted with larger bore plastic covered copper pipe.

Regardless of the method used there must always be a means of shutting off the supply of gas at the appliance inlet. (The Gas Safety (Installation and Use) Regulations 1998, Regulation 26(6)).

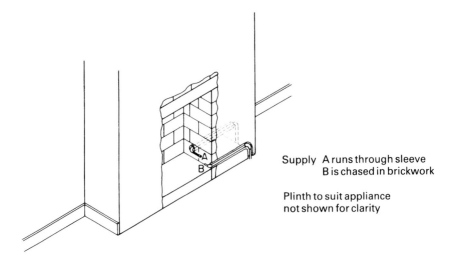

Supply A runs through sleeve
        B is chased in brickwork

Plinth to suit appliance
not shown for clarity

*Fig. 13   Concealed connection in fireplace opening: A: sleeve through
chimney breast; B: chase behind skirting board*

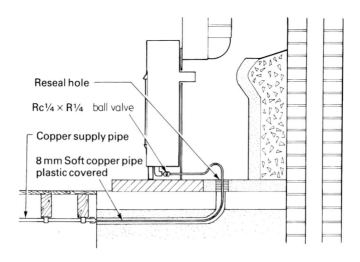

Reseal hole

Rc¼ × R¼ ball valve

Copper supply pipe

8 mm Soft copper pipe
plastic covered

Fig. 14   Concealed connection through base of chimney breast

FAULTS IN INTERNAL INSTALLATIONS

The faults which can occur in internal installations can be divided
into four main categories:

- mechanical failure
- gas escapes

- inadequate gas pressure
- fluctuating or excessive gas pressure.

Mechanical Failure

In this section are included:

- pipe clips or meter brackets becoming loose
- pipes becoming bent, flattened or otherwise damaged
- pipework corroding
- control cocks seizing up.

The failure of clips or brackets to stay secure can be due to faulty workmanship in their installation. It may also be caused by the customer expecting them to support additional weights. Pipes should never be run in positions which invite their use as handrails, coat-hangers or supports of any kind.

Where pipes have been damaged, consideration should be given to re-routing the pipe run.

This also applies to corroded pipework. It can, of course, be remedied simply by cleaning and re-wrapping or painting. Control cocks should be eased and greased periodically. The greases used may contain graphite, molybdenum disulphide or silicone.

The meter/emergency control presents a special problem as it is usually not possible to turn off the gas and it may have to be serviced under live gas conditions. A number of tools for this purpose were described in Vol. 1, Chapter 13.

In some cases the work is carried out in association with British Gas Transco who will temporarily shut off the gas supply.

As an alternative, the meter control may be eased by disconnecting the outlet supply, slackening the plug slightly, and introducing penetrating oil on to the plug through the outlet.

Gas Escapes

New Installations

When a test for soundness on a new installation or extension shows that there is a leak of gas it is usually a simple matter to locate the escape by means of a leak detection fluid. Applied around the joints, this indicates the escape as the leaking gas bubbles through the liquid. The size of the bubbles gives an indication of the rate of leakage.

Existing Installations

Where pipes are not exposed, on existing installations, locating the escape can prove more difficult. It may be necessary to use a leakage detector as described in Vol. 1, Chapter 14.

When dealing with an existing installation, test all the exposed pipework first, particularly in the vicinity of any smell of gas. Make sure that the escape is not on any of the appliances.

If possible, split up the installation and test one half. If that is sound, then the leak is in the other half.

On a large installation it may be necessary to do this several times to locate the leaking section.

Where installations are buried in solid floors there is no alternative but to cut off the faulty pipework at either end, cap or plug the ends and run a completely new installation.

If a gas leakage detector is available it may be used to determine the concentration of gas in premises where gas is escaping and to explore the confined spaces in the structure.

Samples of the atmosphere should be taken above head height and as near to the ceiling as possible. It must be remembered, where an installation uses LPG, that LPG is heavier than air and must be sampled below knee level.

In enclosed spaces of less than 1 m × 1 m floor area, only one sample from the centre of the space is necessary. (In vertical ducts it may be necessary to take a number of samples in order to get a representative indication of the situation.)

In spaces larger than 1 m × 1 m it is necessary to take a number of readings to get a realistic picture of the concentration. Do not take readings within 0.5 m of the walls. Make a record of all the readings, the positions from which they were taken and the time.

Cavity walls can be checked by taking samples from the cavity ventilator bricks from around window or door frames.

Underfloor spaces should be sampled through the joints between floorboards and skirtings. Tests can also be made where pipes pass through the boards, or at the ventilating air bricks. If floor joints are tight, test holes can be drilled through the boards but must be plugged after testing.

Tests should be repeated at intervals to discover the rate of change of concentration and so the severity of the escape.

Dealing with Emergencies

British Gas has developed detailed procedures to be followed by all personnel in the event of any emergency.

However, the Gas Safety Regulations place a legal obligation on the customer to turn off the gas, in the case of a suspected gas escape and to notify British Gas plc or the supplier in the event of the escape continuing.

When an escape or a 'smell of gas' is reported it must be investigated immediately. The order of priority is:

- ensure the safety of people
- ensure the safety of property
- locate and repair the escape, leaving the installation safe.

If you are called to premises where an escape has been detected, take the following precautions:

- do not smoke and warn others about smoking
- do not use any electric bells or switches
- ventilate the premises, open doors and windows in the rooms affected
- ventilate any enclosed spaces where there is a smell of gas
- extinguish any naked lights and warn the occupiers against their use
- if possible, turn off the meter/emergency control
- it can be dangerous to enter enclosed spaces or cellars where there is a high concentration of gas and it may be necessary to seek assistance or use breathing apparatus
- it may be advisable to evacuate the premises if any persons are being adversely affected by the gas or if, in your opinion, there is a danger to life or property
- as soon as possible, test the internal installation for soundness and locate and repair any leak indicated
- if gas is entering the premises from outside, report this immediately and try to establish the extent of premises affected.

Where there is reason to suspect that gas has escaped within or into a building and access cannot be gained immediately, then forcible entry must be made by the employee of the gas supply company. The presence of gas may be suspected because of gas detection instrument readings or other relevant information.

If possible an independent witness, e.g. a police officer, should be present before forcing entry. However, this should not delay the search or interfere with priorities.

Care must be taken to cause only the minimum of damage and the premises must be left no less secure than before making entry. (Gas Safety (Rights of Entry) Regulations 1996).

A 'smell of gas' may be due to a number of other causes.

A smell similar to that of gas may be caused by:

- poor combustion or inadequate flueing of gas appliances
- sewer gas
- some paints and creosote
- fuel vapours
- fresh coal.

However, if a smell is reported, always test the installation for soundness. Sometimes a smell will persist in a kitchen even when the test is sound. This may be due to leaks on the outlet side of the appliance taps controlling supplies to ovens, timers, thermostats, grills and pilots. Appliance integral connections should be checked with leak detection fluid.

Inadequate Gas Pressure

In Vol. 1, Chapter 6, it was shown that the pressure loss in an installation varied with the:

- quantity of gas flowing
- length of the pipe
- diameter of the pipe.

In fact, the pressure loss is:

- directly proportional to the square of the quantity; when the quantity is doubled, the pressure loss increases $2^2 = 4$ times
- directly proportional to the length of pipe; double the length, double the pressure loss results
- inversely proportional to the fifth power of the diameter; if the diameter is halved, the pressure loss increases $2^5 = 32$ times.

In an internal installation, a complaint of inadequate pressure, which means an excessive loss of pressure, could be due to a number of causes including:

- faulty installation design, pipes too long or of inadequate diameter for the quantity of gas required
- additional appliances added to a previously adequate installation
- partial stoppage in the pipes due to a build up of rust or dirt or debris left in during installation
- faulty or overloaded meter
- faulty meter regulator or choked filter
- inadequate or partially blocked service pipe
- inadequate district pressures.

To locate the cause use a 'U' gauge. Check the standing pressure and then, with the appliances turned on, test the working pressures at the following points:

- on the inlet of the regulator to find the pressure loss through the service pipe
- at the outlet of the meter to check the pressure thrown by the regulator and the pressure loss through the meter

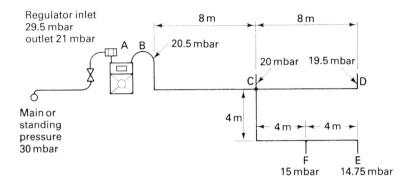

Working pressure at points B and D
indicate clear pipes between these points
Working pressure at F indicates excessive
pressure loss between C and F due to partial stoppage

Fig. 15   Locating partial stoppages

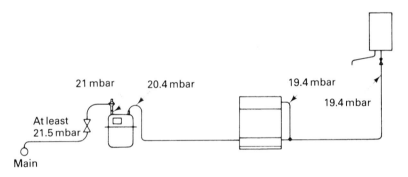

Fig. 16   Normal working pressures

- at intervals through the installation to check that the pressure loss is proportional to the length; any sudden increase in pressure drop indicates a stoppage, Fig. 15.

The pressure losses which might be expected are shown in Fig. 16 as follows:

- in the service pipe, 0.5 mbar
- through the meter, 0.6 mbar
- in the internal installation, 1 mbar between the outlet of the meter and any appliance point.

The faults may be remedied by renewing any inadequate installations or exchanging defective meters, filters or regulators. Partial stoppages can be cleared by a service-cleaning unit and district pressure problems must be reported and referred to the Distribution Department.

Fluctuating or Excessive Pressures

Fluctuations in pressure can occur as the gas load varies. They are more likely to be due to a faulty gas meter or to faults in district regulators causing a fluctuating district pressure. Checking standing and working pressures on the service and the meter will indicate the cause.

The meter regulator should prevent any excessive pressure reaching the internal installation. However, damage to the valve and diaphragm assembly or simply dirt on the valve or seating can result in the regulator failing to lock up. On rare occasions the district pressures may be accidentally increased so that they are higher than the inlet pressure for which the regulator is designed. When this happens the regulator may let by and the pressure in the installation may become excessive.

In cases of emergency, turn off the meter/emergency control and report the situation immediately. When the fault is in the meter regulator, it must be repaired or exchanged.

General

When any alterations are made to a gas installation, the altered part must conform to the Gas Safety Regulations. It may not be necessary to update the whole installation but other associated requirements may need to be met, for example, the flueing and ventilation arrangements when an appliance is exchanged.

LPG (Family 3 Gases)

A gas installation supplying family 3 gases can be divided into two or three sections, each section being subjected to a different pressure:

Section 1, between the tank or cylinder and the 1st stage regulator is subjected to pressures between 2 and 9 bar.

Section 2, where fitted, is between the 1st and 2nd stage regulators and is subjected to pressures of approximately 0.75 bar. This section is not included in some installations, especially cylinder types, where the 1st stage regulator reduces the pressure to that required at the appliance (37 mbar for propane and 28 mbar for butane).

Section 3 is between the 1st or 2nd stage regulator and the appliances and is subjected to pressures of 37 mbar for propane and 28 mbar for butane.

The design considerations and testing of the high pressure stage will be dealt with in Chapter 4.

The size of pipe in the internal installation (low-pressure stage) is determined by the maximum gas rates of the appliance(s) to be connected. In general, for satisfactory operation when the branch length does not exceed 3m, the sizes of connecting pipes to appliances should be as specified in Table 11.

For pipe runs longer than 3m, reference should be made to the table in BS 5482: Part 1: 1994. A pressure drop of 2.5 mbar is permissible between the outlet of the pressure regulator and any draw-off point, when the installation is subjected to maximum load.

TABLE 11 Pipe Sizes (Family 3 gases)

Appliance	Size of pipe
	(mm)
Cooker (large)	12
Cooker (small)	10
Hotplate (2 burner)	6
Lighting	6
Sink storage w/heater	8
Sink instantaneous w/heater	12
Multipoint w/heater	15
Refrigerator	6
Central heating boiler	12
Space heater (small)	6

Note: In the case of steel pipe, the size relates to bore. The nominal size of copper is related to the outside diameter (see Table Y of BS EN 1057: 1996: Part 1).

The equivalent lengths of pipe for fittings are as follows:

Elbow or tee	0.6 m
Connector or 90° bend	0.3 m
Globe valve $1/2$ (15 mm)	1.0 m
Globe valve $3/4$ (20 mm)	1.4 m
Gate valve $1/2$ (15 mm)	0.12 m
Gate valve $3/4$ (20 mm)	0.18 m.

To estimate the size of the main pipe run in an internal installation, add the square of the sizes of all the branch connections and take the square root of this total; an example is given below:

Appliance	Size of branch pipe (a)	a^2
	(mm)	
Cooker	10	100
Multipoint w/heater	15	225
Refrigerator	6	36
Space heater (small)	6	36
		Total 397

Diameter of main pipe run = $\sqrt{397}$ = 19.9 approximately

** size of main would be 20 mm (steel) or 22 mm (copper)

Testing for Soundness of Low Pressure Stage (Internal Installation)

Two tests are explained in detail.

(i)   Of the installation before appliances are connected and
(ii)   The complete installation.

(i) Testing before appliances are connected

- connect testing tee (Vol. 1, Chapter 13)
- cap or plug all other open points in the installation
- attach 'U' gauge to one cock on the tee and a pump or other suitable instrument to the other
- turn on both cocks on the tee
- inject air into the installation until the 'U' gauge registers 45 mbar
- turn off the cock which supplied the air
- leave for 5 minutes for temperature to stabilise
- note pressure on 'U' gauge
- turn off 'U' gauge cock
- leave for further 5 minutes
- turn on 'U' gauge cock, pressure should not have fallen.

If the pressure has fallen, restore pressure and check each joint with leak detection solution. Repair where necessary and repeat test until sound.

Fig. 17   Standard nozzle pressure point adaptor

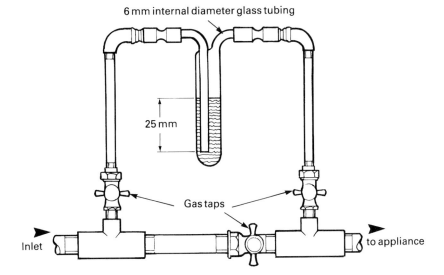

Fig. 18   Bubble leak indicator

(ii) Testing of complete installation

After proving the installation sound, connect the appliances and the gas supply to the installation and proceed as follows:

- turn off all appliance taps
- turn on gas supply
- turn on each appliance in turn and purge the installation of air, keeping a light near enough to the stream of gas/air issuing from the burner to ignite it when the mixture is right. Close each tap in turn.
- ensure that the pressure regulator is operating at the design pressure of the appliances by opening the gauge cock of the testing tee and observing the 'U' gauge
- turn off the gas supply, open an appliance operating tap until a pressure drop of approximately 0.5 mbar is observed, then immediately close the appliance operating tap
- allow 5 minutes for temperature stabilisation
- note the 'U' gauge reading
- turn off 'U' gauge cock
- wait 5 minutes turn on 'U' gauge cock. If the reading is lower than the reading previously taken there is a leakage in the installation which must be located and repaired.

When testing an existing installation, connected to a gas supply, Test (ii) 'Testing of a complete installation' must be applied. If there is no pressure point fitted in the installation the 'U' gauge may be attached to any of the injectors on an appliance by removing the jet and fitting the pressure test point adaptor and washer shown in Fig. 17.

Besides the use of the 'U' gauge in Test (i) 'Testing the installation before appliances are connected', BS 5482 : Part 1 : 1994 also describes the use of 'the bubble leak indicator', Fig. 18.

In this test, the glass bubble leak indicator is fitted across a gas valve at the beginning of the installation. A 'U' gauge is fitted at a suitable test point. Air is supplied at the inlet and maintained at an appropriate constant pressure to the installation, through the gas valve on the indicator. When the pressure has been slowly raised to the test value, open the taps to and from the glass tubes of the indicator and shut the gas valve. The rate of bubbles emerging in the indicator indicates the leakage rate.

The installation is considered to be sound if the leakage rate does not exceed 85 cm^3/h over a period of one minute.

A rate of 6 bubbles a minute through the indicator should be equivalent to 85 cm^3/h but this should be checked before the indicator is used.'

The Liquified Petroleum Gas Industry Technical Association (LPGITA) have also introduced a Code of Practice No. 22 – LPG Piping System Design and Installation, which describes methods of testing installations for soundness.

CHAPTER 4

Meters and Services

Chapter 4 is based on an original draft by W. McManus

Introduction

The gas service engineer is not expected to lay or repair service pipes or mains but in any case his work brings him into contact with the business end of the service pipe and he must be able to recognise faults which can effect the safety of the public or the operation of the customer's appliance.

The service pipe, usually shortened to 'service', is the pipe from a main to a meter/emergency control. Where it is connected to two or more meter/emergency controls it is called a joint, dual or multiple service. If it is branched to feed more than one property it is known as a 'teed service'. A vertical pipe is called a 'service riser'. It may supply one or more meter/emergency controls and have lateral connections at appropriate floor levels in high-rise buildings.

Services may supply natural gas at a range of gas pressures. In domestic services the pressure is usually about 30 to 50 mbar. Pressures in mains carrying natural gas may vary from about 30 mbar up to 100 bar. They are classified as:

- low pressure   – up to, but not above, 75 mbar
- medium pressure   – above 75 mbar, but not above 2 bar
- intermediate pressure   – above 2 bar, but not above 7 bar
- high pressure   – above 7 bar.

Although the majority of domestic services in the UK are operating on low pressure, medium pressure is being used more frequently. Low pressure services to single family dwellings should not be generally less than 20 mm diameter.

Medium pressure supplies to new housing estates may be fed to a centralised regulator installation and then by low pressure mains and services to the individual premises. Alternatively, natural gas may be supplied by medium pressure services to individual regulators (Vol. 1, Chapter 7).

Intermediate pressure services are rarely used for domestic supplies. They may be used for industrial or commercial applications.

High pressure is normally confined to national pipelines or regional grid networks. Very occasionally a high pressure service may be taken directly from a transmission line to a regulator installation on a large industrial complex.

SERVICE LAYING

Materials

New services carrying natural gas are now usually polyethylene (PE). Steel, often with a factory applied plastic sheathing was previously, and in some circumstances still is, occasionally used. They may be connected to mains which could be made from PE, cast iron, ductile iron or steel.

Other materials have been used for services. Lead pipe was not uncommon in some mining areas, to allow for subsidence. Soft, heavy gauge, copper pipe is an effective material but its cost has become prohibitive. Wrought iron or mild steel pipes have been used, protected by a variety of coatings and wrappings.

The types of joint used are as follows:

PE Pipe

Electrofusion jointing: This type of joint is used extensively by British Gas plc on fittings for PE pipework. Incorporated into the inside of each fitting is an electrical coil which is connected to two terminals on the outside of the fitting. Also on the outside of the fitting is one or more small holes which indicate when satisfactory fusion (jointing) has taken place and a label which indicates the fusion and cooling times. In addition to the fitting, a clamp to hold the fitting in position, is required. The power for jointing is supplied by a 110 V mobile district supply (a vehicle) or a portable generator. The 110 V supply is fed to an electrofusion control box, which sets the fusion time and has a 39.5 V output for connection to the fitting. Fig. 1 shows a PE tapping tee. The tee is fused onto the main then the service is connected to the tee by the reducer/coupler. It is then possible to test the service for soundness before screwing down the steel cutter to drill a hole in the main. The cutter, because of an internal thread, retains the plug of PE from the main end and is then screwed up again to allow gas into the service pipe. Fig. 2 shows a section through a PE main and tapping tee.

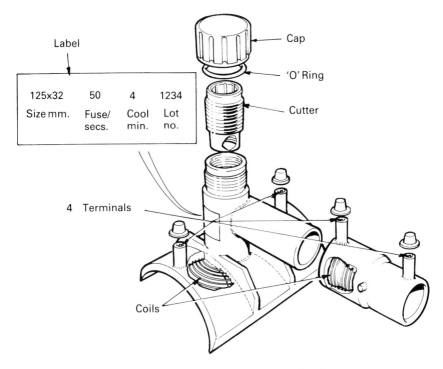

Label

| 125x32 | 50 | 4 | 1234 |
| Size mm. | Fuse/ secs. | Cool min. | Lot no. |

Cap

'O' Ring

Cutter

4 Terminals

Coils

Fig. 1   Electrofusion tapping tee and reducer

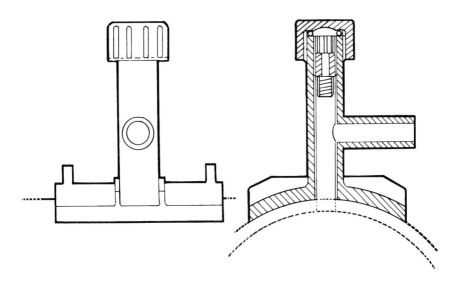

Fig. 2   Section through a PE main and tapping tee

Butt fusion welding: This method is used to join pipe ends together (Chapter 1: Plastic tube joints). The pipes are cut and the ends prepared before being placed in a clamp. A special tool is inserted to heat the ends, which are forced together when the tool is removed. Butt fusion joints are used for larger sized pipes above 90 mm diameter.

Compression joints: These may be used on the service tee connection to an iron or steel main. Fig. 3 shows a service tee screwed into an iron main. The connection for the PE service is by compression joint (Chapter 1 Compression joints) with the added support of a PE anti-shear sleeve.

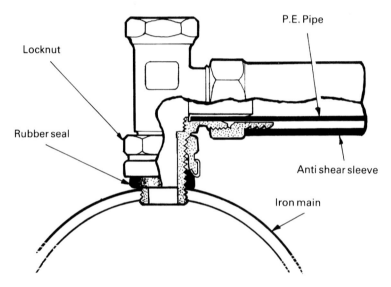

Fig. 3   Service tee with anti-shear sleeve (Kontite)

A compression joint is also used on the house entry tee (transition tee). Fig. 4 shows a house entry tee which has been factory fitted with a BS 1387 DN 20 heavy grade steel pipe, protected by a 32 mm PE pipe. It is made up by the manufacturers as a one piece item for use above ground through the wall service entries and terminates with a BS 21 R¾ end for direct connection to a meter/emergency control or elbow, depending on the meter position. The PE service pipe is connected to the inlet of the tee by a compression joint. The nut of the compression joint has a plain extension tail to accept a glass reinforced plastic (G.R.P.) sleeve which protects the whole of the exposed PE service pipe on the external wall of the property.

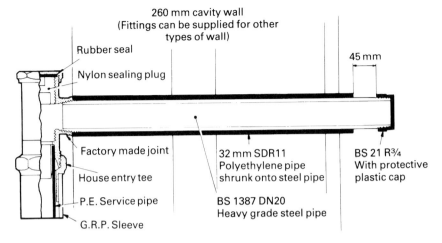

Fig. 4   House entry tee with through-wall service pipe (Kontite)

Another fitting used to terminate a PE service pipe is a meter box adaptor which is used with the British Gas plc standard meter boxes. A section of a meter box adaptor is shown in Fig. 5.

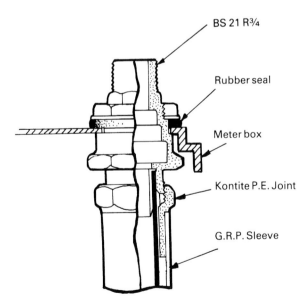

Fig. 5   Meter box adaptor (Kontite)

Steel Pipe

Screwed joints. Screw threads to BS 21 (Chapter 1: Steel pipe) are used on both low and medium pressure services to join pipes up to 50 mm diameter. Above this size the pipes are usually flanged or welded.

Compression joints. These are generally only used outside the building and underground. Fig. 6 shows a service tee used to connect a steel service to a steel main. The outlet connection from the tee to the service pipe is by a compression joint.

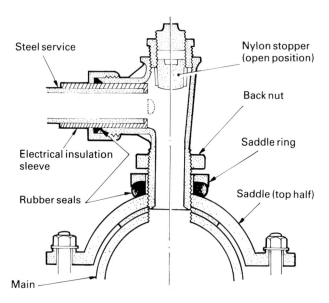

Steel service

Nylon stopper (open position)

Back nut

Saddle ring

Electrical insulation sleeve

Rubber seals

Saddle (top half)

Main

Fig. 6   Service tee connection for a steel service from a steel main

Flanged joints. Used on larger services, these joints are usually employed for above ground connections to regulators, valves or meters (Vol. 3, Chapter 1).

Welded joints. Welding is carried out on larger services and service entries to buildings higher than four storeys. Partial-penetration (p.p.) welding, which is a simple form of arc welding, may be done by distribution personnel who have been trained and are competent to carry out this type of welding.

Installing Services

Service pipes must be installed in accordance with the Gas Safety Regulations 1998, Part A(2) and should comply with the British Gas Code of Practice for Distribution.

Wherever practicable, the service pipe should be laid:

- in a straight line
- at right angles to the main
- directly to the meter position.

The route chosen should avoid:

- unstable structures
- unventilated voids or cavities
- ground liable to subsidence
- hazardous locations where the service may either be damaged or cause damage, for example, adjacent to electrical equipment or pipes or containers of flammable substances.

The layout of a typical service is shown in Fig. 7.

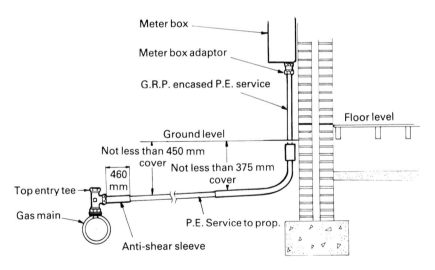

Fig. 7   Layout of typical service pipe

The trench for the pipe should be excavated to the appropriate depth, the normal minimum depth of cover in private property is 375 mm with, wherever possible, an even gradient towards the main. Where it is necessary to lay at less than 375 mm cover, special protection must be provided.

The normal minimum depth of cover in public roadways should be 450 mm. There must be at least 75 mm of soil between the top of the pipe and the bottom of any road foundation. The pipe bed should be even and solid throughout its length. Services should enter the building either above ground and floor level by rising externally, or above the site concrete within the building but avoiding passing either through unventilated voids or below load-bearing foundations, Fig. 8. Where services pass through solid floors, cavity or load-bearing walls they must be sleeved.

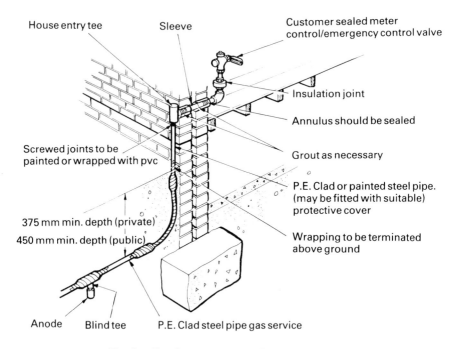

Fig. 8   Steel service entry above floor level

Most low pressure domestic services terminate with a meter/ emergency control within the building on an external wall, the exceptions to this are:

- where an external meter box is fitted
- where an external meter is used.

Where a PE service is run up to an external meter box, Fig. 9, it should be sleeved with glass reinforced plastic to ground level then protected by a preformed PVC bend to bring it to a horizontal position.

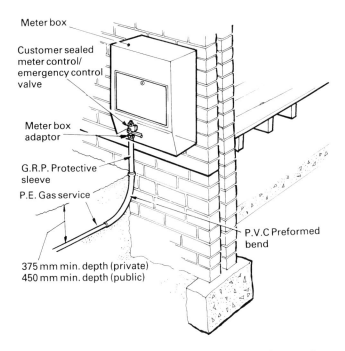

Meter box

Customer sealed
meter control/
emergency control
valve

Meter box
adaptor

G.R.P. Protective
sleeve

P.E. Gas service

P.V.C Preformed
bend

375 mm min. depth (private)
450 mm min. depth (public)

Fig. 9   Typical PE service to surface mounted meter box

Control Valves

Service Valves

In addition to the meter/emergency control, valves are generally fitted on services when:

- the diameter of the service is 50 mm or more
- the pressure is above 75 mbar
- the service supplies two or more primary meters in a building
- the building or its occupation constitutes a hazard, e.g. fried fish shop, dry cleaners, etc.

Valves are usually located outside, but near to the property boundary and should be fitted with a surface box and cover. Typical valve positions for teed or joint services are shown in Fig. 10.

Where teed services have been installed a joint service badge, Fig. 11, should be attached to each service immediately below the meter/emergency control. This acts as a warning to personnel that more than one property will be affected if the supply is interrupted. Access must be gained to all premises to turn off appliances before any work on clearing or repairing a service can be started.

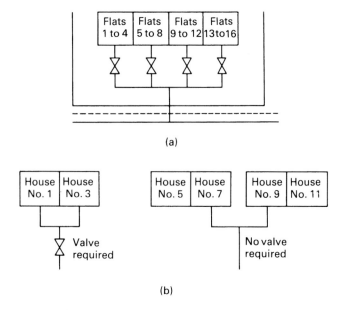

Fig. 10   Typical valve positions: (a) flats; (b) joint houses

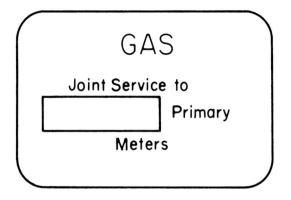

Fig. 11   Joint service warning label

Meter/emergency Control

An approved type of control cock or valve must be fitted as close as practicable to the meter inlet connection. The domestic meter control was described in Vol. 1, Chapter 11 and a typical cock is shown in Fig. 12. This consists of a brass body and plug. The plug has a square head fitted with a malleable iron or steel key secured by a split pin. A screw is located in the body below the top of the plug. When tightened, the screw can lock the plug in the 'off' position.

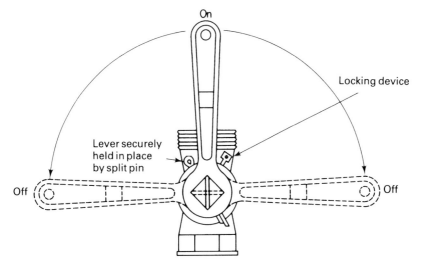

Fig. 12   Meter/emergency control

It is required practice (Gas Safety Regulations) to fit the meter/ emergency control when the service is installed. After testing and purging the service the control is locked and sealed in the following manner:

- turn the control off
- tighten the locking screw
- remove the split pin and the key
- fit a wire seal through the holes in the plug head and the locking screw
- fit a cap on to the control outlet thread.

When a meter is fitted, the installer reverses the operations as follows:

- remove the cap and wire seal
- unlock the plug by loosening the screw
- fit the key and secure it with a split pin
- turn on the gas supply.

A meter/emergency control should always be fitted with the female thread on to the service pipe. A union or a flexible connection is then screwed to the male end. The control must be easily accessible and the key fitted so that when it is vertical, the gas is on and when moved downwards as far as possible, the gas is turned off (Gas Safety (Installation and Use) 1998 Regulation 9(2)).

The nut and niting washer must be clear of the wall so that the plug may be eased and, if necessary, the control removed.

Before fitting a control it should be checked to ensure that there is no grease or dirt in the gasway. The plug should be loose enough to turn easily but not so loose that it can be turned accidentally or by the weight of the key.

High Rise Buildings

Service risers in multi-storey buildings may be:

- heavy weight steel to BS 1387, with screwed threads to BS 21 (for buildings up to four storeys)
- Carbon steel to BS 3601, welded to BS 2640 or 2971 (for buildings above four storeys).

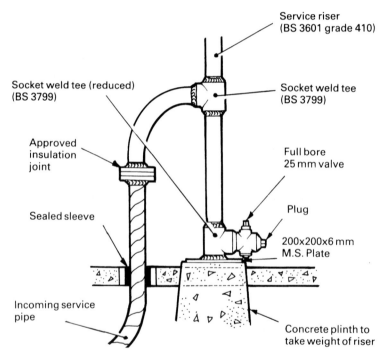

Fig. 13   Service riser support for buildings higher than four storeys

Individual risers are normally not above 108 mm diameter. They should be run through a ventilated fire-resistant shaft with sealed panels at each floor to give access for maintenance to the service and service valves, see Chapter 3, Fig. 3.

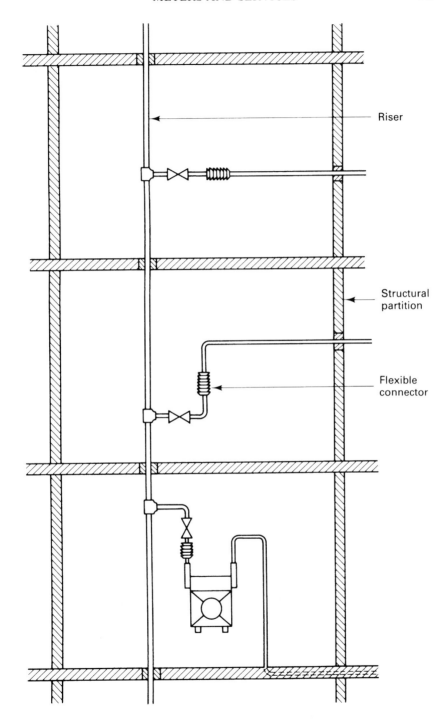

Fig. 14   Lateral connections

The riser must be adequately supported at its base and should incorporate a dust trap, Fig. 13. It must rise vertically without change of direction and be supported at suitable intervals in such a way as to allow thermal expansion.

Service laterals, that is, the horizontal pipes connected to the riser at each floor, should incorporate a valve and a means of disconnection immediately adjacent to the riser. They should have flexible connections to compensate for differential thermal expansion, Fig. 14.

Differential expansion can occur when:

- riser and lateral are of different materials with different rates of expansion; for example, steel riser, copper lateral
- temperature in the riser shaft is different to that outside, so riser and lateral are at different temperatures and expand at different rates.

To avoid differential expansion, the riser should not be fitted close to hot water or steam pipes or in any position where it might be subjected to fluctuating temperatures.

The supply to the building should be fitted with a service valve outside the premises and in an accessible position. The gas pressure should be governed to within the low pressure range before the service enters the building.

Earthing and Insulating Services

Cross-bonding of gas, water and electrical supplies and protective multiple earthing (PME) were dealt with in Vol. 1, Chapter 9.

Fire Risk

Under certain conditions an electric current can produce sparks when a service pipe is being connected or disconnected. To prevent this occurring, temporary continuity bonds are used, in the same way as on internal installations.

Corrosion

This can occur at any time when a flaw develops in the protective coating of a steel service. A stray current passing to earth through this point will cause electrolytic corrosion and finally failure of the service (Vol. 1, Chapter 12).

To prevent this occurring an anode is fitted to the service pipe, in the following manner. A blind tee (a tee with the outlet branch connection blanked off) is inserted into the pipe below ground, the tee

Fig. 15   Plastic PME insulators

Fig. 16   Metal PME insulators

is wrapped and the anode is screwed into the branch and left exposed. The location of the anode can be seen in Fig. 8 and the 'blind' tee and anode in Fig. 17.

It is essential that all new steel services are electrically insulated from internal installation pipework. This is done by fitting an insulation joint in the service above ground and upstream of the meter/emergency control, Fig. 8. An existing service constructed mainly of PE but terminating in steel, part of which is buried may also be electrically insulated by fitting an insulating joint at the outlet of the meter/emergency control.

Insulators may be either plastic or metal.

The type normally used is the plastic insulator, Fig. 15. This is made from a fire-retardant thermoplastic and it is fitted directly on to the outlet of the meter control. To prevent the insulator being subject to stress it must always be installed next to a flexible or semi-rigid connector.

Metal insulators, Fig. 16, are used in rigid meter fixes or where the meter installation is required to be fire resistant.

In the metal insulator a mica gasket is used to provide insulation and an 'O' ring forms the gas seal. The two threaded metal flanged components are held together by rivets which are insulated from one flange by a mica washer and plastic collars. The washer is protected by a metal backing washer and the fitting is given a protective coating to prevent electrical tracking over the surface and moisture penetration.

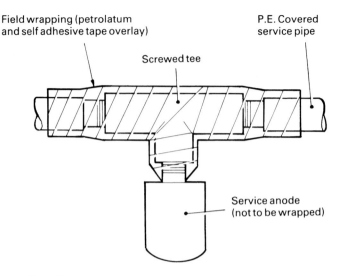

Fig. 17   Installation of an anode to a service pipe

Testing Services for Soundness

Services should be tested for soundness as specified in the British Gas Code of Practice for Distribution, Section 5, Module 5.5 – Pressure Testing.

The tests should be carried out as follows:

Low pressure – low rise dwellings

- (a) This service is tested with air, with the apparatus shown in Fig. 18, attached to either the meter/emergency control or the mains service tee.
- (b) The meter/emergency control is left open and the opposite end of the service to the test apparatus, blanked off.
- (c) The pressure is raised to 100 mbar (no temperature stabilization required).
- (d) The test duration is 5 minutes, no fall in pressure is allowed.
- (e) On completion of a successful test, close the meter/emergency control and remove test apparatus or the blank and check meter/emergency control for internal soundness.
- (f) The test pressure should be released from the end of the service opposite to where the test apparatus was connected, to prove the whole service was tested.

Low pressure – high rise dwellings (more than four storeys)

(a) The service risers should be tested before connecting to the main and before fitting the meter/emergency control. Each service is capped and a test pressure of 350 mbar applied for 10 minutes to the riser and service outside the building, no pressure loss is allowed.
(b) A 100 mbar test is then applied for 10 minutes with the meter/emergency controls fitted and open but capped off, no pressure loss is allowed.
(c) The service is then connected to the main and a pressure test of 100 mbar applied from the service tee. The test is for 5 minutes with no pressure loss allowed.

Medium pressure

The test is from the main to the inlet valve of the service regulator installation. The test pressure is 3 bar for 5 minutes, no pressure loss

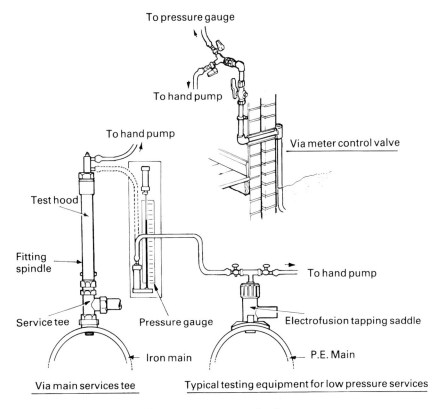

Fig. 18   Typical testing equipment for low pressure services

is allowed, the gauge used is a Bourden tube type. On completion of the test all exposed screwed, welded or flanged joints, not included in the test, should be checked with leakage detection fluid before and after the installation is purged and commissioned.

Commissioning (Purging) Services

Services up to and including 63 mm diameter

Services may be purged directly with gas after being tested as follows:

- place a terminating flame trap safely outside the building and connect a purge hose to the meter/emergency control;
- make sure the flame trap is manned;
- ensure all other live services to the premises have been isolated;
- pressurise the service with gas, open meter/emergency control and purge;
- the purge is complete after two successive Gascoseeker readings at the test points indicate at least 90% gas in air;
- close and cap the meter/emergency control and screw down the securing screw unless a meter is being fitted immediately. If there is a service isolation valve, leave it in the open position;
- check the mains connection and the plug/cap for leaks. Make sure that the leakage detection solution is washed away thoroughly and the pipe dried and wrapped.

Additionally, for high rise buildings:

- purge the highest lateral first and the remainder as shown in Fig. 19;
- purge services to atmosphere from the meter/emergency control, via a purge hose fitted with a flame trap safely positioned outside the building;
- leave the service and emergency control valves in the 'open' position.

Service Insertion

An old steel service pipe may be used as a carrier for a PE service pipe. At least 1 m of the old service pipe must be removed leaving enough pipe beyond the outside wall to fit an anode. The method of installation is shown in Fig. 20.

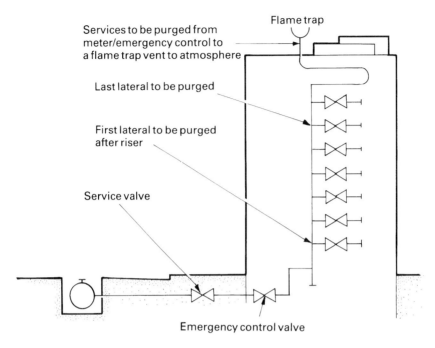

Fig. 19   Purging services in a high rise building

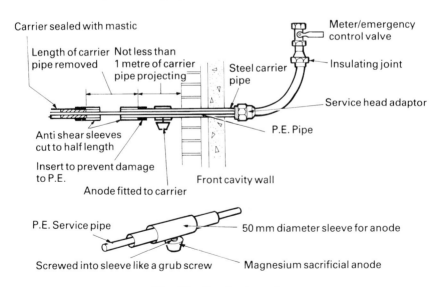

Fig. 20   Service insertion

METERS

Introduction

The construction and operation of domestic gas meters was covered in Vol. 1, Chapter 8. It remains for this chapter to deal with handling, installing and exchanging meters.

Although a number of the older types of meter may still be found on the district, the meter in general use is the unit construction meter. The domestic meter range is almost totally catered for by the U6 meter, but this will change in future due to the recent development of the Ultrasonic (E6) meter. (Vol. 1 Chapter 8). This is an electronic meter with the same rated capacity as the U6.

TABLE 12 Sizes and Ratings of 'U' Series Meters

Meter Designation	Maximum Flow Rating (Rated Capacity)	
	m^3/h	ft^3/h
U2.5	2.5	88
U4	4	141
U6	6	212
U10	10	353
U16	16	565
U25	25	883
U40	40	1413
U65	65	2296
U100	100	3530
U160	160	5650
U250	250	8830
U400	400	14130

The complete proposed range of unit construction meters is given in Table 12. The following points should be noted:

- U2.5, U4 and U10 are not generally used for measuring natural gas in the UK but are used on the continent as domestic meters.
- U2.5 and U4 are used for measuring LPG in this country and on the continent.
- U16 may be used for large domestic or small commercial installations.
- The two largest sizes, U250 and U400, are rarely used, turbine or rotary displacement meters normally being preferred.

One-Pipe Meters

A one-pipe meter with coaxial connections was developed a number of years ago by G. Kromschröder A.G. of Osnabrück. These meters were generally used in Europe but were never popular in the UK. They had several advantages over the familiar two-pipe meter. Their major disadvantages, however, outweighed the advantages. If the coaxial sealing washer was removed or became perished or damaged then the meter would pass unregistered gas to the outlet.

The one-pipe meter could be more easily installed than the then current tin case meter with lead connections and blown joints. It was particularly suitable for fitting in 'batteries', supplying a number of flats.

Handling, Storing and Transporting Meters

Some of these points were made in Vol. 1 but they are probably worth repeating.

Meters appear to be very robust steel boxes with only the glass of the index likely to be damaged.

Unfortunately, nothing is further from the truth. The inside mechanism, on which the accuracy of the meter depends, is by no means as tough as the case. The valves and their seatings or grids are perhaps the most vulnerable parts.

As you know, meter valves are semi-rotary slide valves, moving in an arc over their seatings.

They are commonly phenolic, self lubricating and are held on to the grids only by their own weight. They can be lifted off the seatings a short distance if the meter is tipped up or shaken.

If the valves are lifted up, dirt or dust can get between the valves and their seatings. As the valve moves, the dirt can cause scratches and damage. If the dirt sticks to the valves or seatings it will hold them apart. If any of these things happen, the meter will pass some unregistered gas.

Because of this meters should always be kept in an upright position when being stored or transported. They should not be dropped or shaken about. It has been recommended that a meter which has been dropped or jarred should be retested before being fitted.

When a meter is not actually connected, the inlet and outlet must be sealed at all times. This will prevent dirt getting into the meter and it will prevent any mixture of air and gas in a used meter from being ignited. After removal a meter should be capped before transportation. Diaphragm meters above 11.3 m^3/h (400 ft^3/h) capacity should first be purged, either *in situ* or after moving to a safe place.

Finally, take care that the case is not scratched or dented. If the paint is damaged the steel may rust and the case become perforated. Although the Ultrasonic E6 meter has no valves or moving parts it should still be treated with the same care as other meters.

Meter Locations

A primary meter must be fitted in a location that complies with the Gas Safety (Installation and Use) Regulations 1998. The meter position must be as close as practicable to the point where the service pipe terminates. The siting of the meter must provide easy access for:

- installation and servicing of the associated controls
- installation, exchange or removal of the meter
- reading the index
- operation of the coin mechanism and removal of the cash box on prepayment meters.

A prepayment coin meter must be installed inside a dwelling although an alternative position can be considered for a prepayment meter accepting tokens (see Vol. 1, Chapter 8). The recommended height of the coin/token or Gascard slot is between 750 and 1400 mm. The site for any gas meter must be well ventilated and it should be spaced at least 25 mm away from surrounding walls. It should not be in contact with any cement or magnesite compound floor. Special precautions will need to be taken where a meter is fitted above a floor that must be frequently washed down as in a butcher's or fishmonger's shop.

Under no circumstances should a meter be sited:

- close to a source of heat or where it will be subjected to extremes of temperature
- where it might be exposed to accidental damage
- where it causes an obstruction
- in a damp or corrosive atmosphere, such as directly under a sink (where such a location cannot be avoided special precautions may have to be taken by applying a protective coating to the meter and associated pipework/controls.)
- in a food store or larder
- where there is a risk of damage to it from electrical appliances.

In buildings with three storeys or more (two or more floors above the ground floor) no meter should be installed in or under a stairway or in any passage which provides the only means of escape in the event of fire.

In buildings with one or two storeys (less than two floors above the ground floor) the meter should preferably not be installed on the escape route but, if this is unavoidable, the meter and its controls should:

- be of fire-resistant steel construction, or

- be housed in a fire-resistant compartment having automatic self-closing doors, or
- include a thermal designed to cut-off when the temperature exceeds 95°C.

Thermal Cut-Off

This was described in Vol. 1, Chapter 10 and another type of cut-off is shown in Fig. 21.

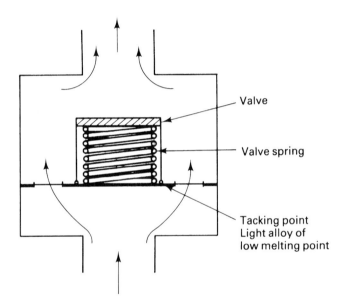

Fig. 21   Thermal cut-off valve

Cut-offs are available in 20 and 25 mm (R¾ and R1) sizes. They should generally be fitted downstream of the meter control to avoid working with 'live' gas.

The valve closes when the solder melts at an ambient temperature of 95°C (the Gas Safety (Installation and Use) Regulations 1998). This temperature is high enough to prevent the valve operating if exposed to direct radiation from the sun, through a window.

Meter Compartments or Boxes

Various types of meter housings have been used in the past and the main requirements of an internal meter compartment are:

- adequate ventilation
- construction of non-combustible material, to BS 476
- dimensions sufficient to allow installation and exchange of the meter and installation, servicing and exchange of the meter control, filter governor or thermal cut-off as appropriate
- provision of an adequate fire resisting and insulating barrier between the meter and any adjacent electrical apparatus.

It has, for some time, been considered desirable to provide a means by which the meter index could be read without the meter reader having to enter the building. Glass bricks set into external walls and letter-box type apertures have been tried. The most common method used is the external meter compartment or 'meter box'.

Typical meter boxes used by British Gas plc come in three types:

- built-in   – Fig. 22
- surface mounted   – Fig. 23
- semi-concealed   – Fig. 24.

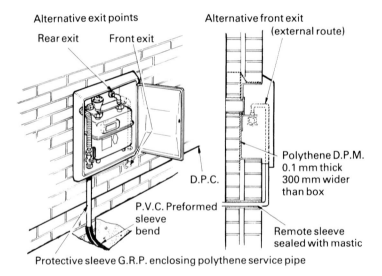

Fig. 22   Built-in meter box

They are designed and constructed so that access can only be gained by using a special key which is made available to the customer. If there is a leakage of gas inside the box the design and installation prevents entry of gas into the building and the cavity of its external wall.

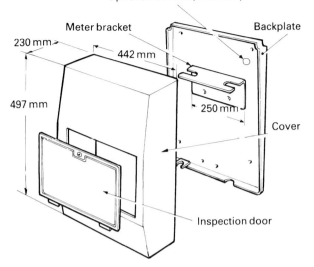

Fig. 23   Surface mounted meter box

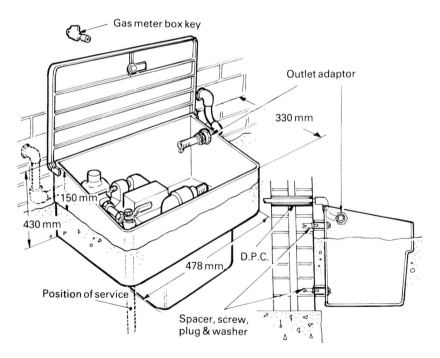

Fig. 24   Semi-concealed meter box

The service pipe is run on the face of the outside wall to terminate in a boss to which is fitted the meter/emergency control. The meter is suspended from a meter bar by its union connections. Semi-rigid stainless steel tubing is used between the insulator (if fitted) or the meter/emergency control and the filter/governor. The outlet connection terminates at an elbow facing into an installation pipe duct entering the building.

A major advantage of the external meter box is that no uncontrolled gas is taken into the premises and, in an emergency such as a fire or a supply failure, authorised persons have easy access to the meter/emergency control. With the introduction of the E6 meter new, smaller meter boxes are being developed.

Meter Supports

Meter supports must be:

- able to take the full weight of the meter and its associated controls
- arranged so that the meter is prevented from touching walls or floor
- fitted securely and level by fixings appropriate to the materials to which they are secured.

Provided that these requirements are met, almost any type of support can be used. Figure 25 shows some typical arrangements. In all cases a wood or rubber stud is used to keep the meter clear of the wall.

One type that was commonly used was the two independent brackets, Fig. 25(c).

These were made from 25 mm × 7 mm steel strip and had rubber sleeves into which the meter studs rested. The brackets which were available in various sizes, were fitted so that they were hidden behind the meter and so provided a neat and effective support.

Meter brackets/bars of the type shown in Fig. 26 are now more often used.

The object of this fitting is to allow the service and the internal installation pipes to be run, before the meter is fitted and to leave the meter connections ready for its subsequent installation.

Current unit construction meters have internal reinforcement which allows them to be suspended by their unions so allowing meter brackets/bars to be used. The type of meter bracket/bar most commonly used consists of a piece of 3 mm plate bent at right angles and with two U-shaped cut-outs in the horizontal section. These cut-outs allow the specially-constructed union liners to recess into the meter

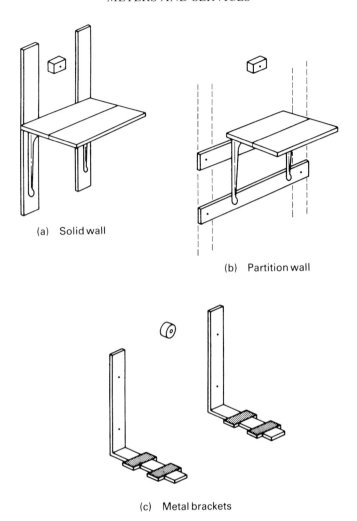

(a)   Solid wall

(b)   Partition wall

(c)   Metal brackets

Fig. 25   Meter fixing brackets: (a) solid wall; (b) partition wall; (c) metal brackets

bracket/bar so supporting the meter. The inlet and outlet liners are secured to the meter bracket/bar by bolts.

Meter Connections

Tin-case meters, which have soldered side pipes and union bosses, must be connected by non-rigid materials such as flexible stainless steel.

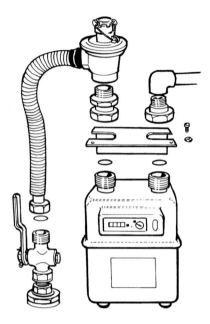

Fig. 26   Standard domestic meter fix using a meter bracket/bar

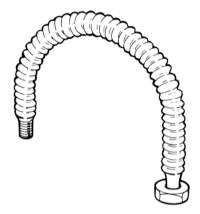

Fig. 27   Semi-rigid stainless steel connection

Lead pipe with 'blown' joints was used for many years but its use is now prohibited under the Gas Safety (Installation and Use) Regulations 1998. However, as these regulations are not retrospective installations containing lead connections may still be encountered. Semi-rigid stainless steel connections, as shown in Fig. 27, are now being used where a lead connection needs to be replaced. They are also used in locations where fire-resistant connections are required.

Steel cased meters may be fitted with rigid connections using mild steel or copper tube.

Fixing and Exchanging Meters

Fixing Meters

Before connecting the meter, check that the service is clear and the internal installation is sound and free from any stoppages.

Ensure that the correct type of meter support is used and the meter is fitted upright and level. Avoid straining the meter bosses when fitting the connections.

The regulator and meter must be fitted as close to the meter/ emergency control as possible. Some earlier meter governors fitted in certain regions of British Gas also included a filter unit containing a paper filter to prevent dust and other foreign bodies from entering the installation. Treatment of natural gas has generally made the installation of filters unnecessary; however, if a filter has to be used due to special circumstances, it must be fitted upstream of the meter regulator.

Fixing the E6 meter

1. Check that there is a reading on the Liquid Crystal Display (LCD). If not, return the meter and get a replacement.
2. Position the meter so that the LCD can be read easily.
3. Install the meter as normal using the standard meter bracket. Leave enough space below the bracket in case the meter needs to be replaced with an electronic token meter (ETM) at a later date.
4. Fill in the meter details on the job document. Remove the battery label from the outside of the meter case and stick to the job document. Fill in the emergency label.
5. Ignore any warning flags at this stage. These will clear when the display advances through zero (00000.000).
6. Test for soundness and purge as if the meter were a U6 model.
7. Leave the leaflet about the E6 meter with the customer.

Note: Warning flags appear on the meter screen and give information about meter faults or tampering.

A temporary continuity bond must be used when installing or exchanging any meter (Chapter 3: Electrical bonding). The procedure is as follows:

- connect one side of the bond to the gas service pipe which may need to be cleaned to ensure good metal-to-metal electrical contact
- holding the insulated part, place the other clamp on to a clean section on the internal installation; under no circumstances touch the metal of the clamp until good contact has been made
- tighten the clamp on the internal installation
- connect the meter
- remove the continuity bond, downstream end first.

Testing and Purging Meters

The procedures for testing and purging internal installations, including meters, were given in the previous chapter. Every newly installed meter and any new or altered installation pipework must be tested for soundness before being purged and put into use. Meters are usually tested with the installation pipework. If tested separately, check the joints with leak detection fluid when the meter is finally connected.

When purging a bypassed meter, ensure that the bypass is also purged. When a meter is being purged check throughput of gas to ensure that no explosive mixture remains and the meter is registering satisfactorily. After purging a prepayment meter, check the operation and setting of the coin mechanism.

If appliances are installed, light the burners and check that the flames are normal. Check the pressure from the meter regulator and ensure that it is sealed to prevent the adjustment being altered.

Exchanging Meters

Meters should be replaced by others of the same type and size, except where it may be necessary to fit a steel-case meter in place of a tin-case to meet safety regulations requirements, or a larger or smaller capacity meter is required to supply an altered installation.

The following procedure should be adopted when exchanging a domestic meter:

- check with the customer and turn off all appliances
- turn off the meter/emergency control and test for soundness
- attach temporary continuity bond, first to the service, then to the outlet
- disconnect, remove and cap off the existing meter
- fit new union washers, if necessary, and check that supports are secure
- connect meter
- test for soundness

- purge the meter
- relight pilots or appliances
- check the test dial to ensure that the replacement meter is registering
- remove the continuity bond, first from the outlet and then from the service
- read and record the index of the old and the replacement meters.

Exchanging E6 meters

The existing installation must be updated to the requirements for new installations by using the meter bracket to give rigidity.

- remove the existing meter as normal
- carry out steps 1 to 7 for a new E6 installation (page 161).

Primary and Secondary Meters

Primary Meter

A primary meter is a meter which is connected to the service pipe, the index of which forms the basis of charge for all gas used on the premises.

Secondary Meters

Secondary meters are subsidiary meters which measure gas used by separate appliances or in separate parts of the premises. All the gas passing through the secondary meters must already have passed through a primary meter.

Secondary meters, or 'check meters' as they used to be known, are installed where it is necessary to obtain prepayment or a record of gas used in a particular room or part of the property. For example, in hotel bedrooms, bed-sitters, colleges and apartments.

In the past a secondary meter was usually a small, compact, prepayment meter supplying gas to a fire, gas ring or hotplate. Now, however, a secondary meter may be either credit or prepayment and can supply a number of appliances. Special meters are no longer manufactured and a secondary meter will be one of the range used as primary meters.

On installations with secondary meters, all the gas used on the premises is paid for by the owner on the basis of the primary meter reading. The owner then charges his tenants or guests on the basis of their secondary meter readings. The charge made by the owner must not exceed the maximum allowed by Ofgas*. This maximum is reviewed periodically.

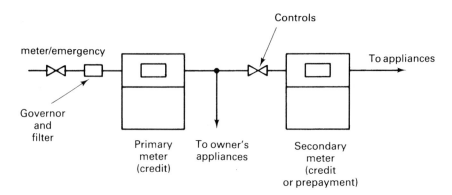

Fig. 28   Installation with secondary meter

Installing Secondary Meters

Most of the recommendations for the installation of primary meters also apply to secondary meters. The basic installation consists of a service pipe supplying a primary meter via the usual controls. The primary meter outlet is the inlet to the secondary meter, the outlet of which supplies the appliances, Fig. 28. The owner's appliances (if any) would be supplied from a point before the secondary meter.

In some cases the secondary meters may be installed in a different building to the primary meter. If the supply passes underground it should comply with the recommendations for gas service pipes.

Where a filter is fitted on the inlet to the primary meter, there is generally no need to fit a filter on the secondary meters.

When secondary meters are installed, the primary meter must never be a prepayment meter. This is because, if prepaid gas had run out at the primary meter and was then reinstated, there is a risk that appliances supplied by the secondary meter would be turned on and would then emit unburnt gas.

*Ofgas (The Office of Gas Supply) is an independent regulatory body set up under the Gas Act 1986. It has responsibility for monitoring and enforcing the authorisation granted to British Gas plc by the Secretary of State for Energy (now the Department of Trade and Industry) and for issuing authorisations to other suppliers of gas through pipes in Great Britain.

Warning Notices

Warning notices are strategically placed to warn either the customer or anyone working on the installation of the potential hazards. The wording should give clear and concise information but not unduly alarm the customer.

The following permanent notices should be prominently displayed on or near to the primary meter or meters.

1. Escapes (Fig. 29)

On every meter installation, there should be a notice to indicate that the customer should:

- shut off the gas supply in the event of an escape on the premises
- where gas continues to escape after the supply has been shut off, as soon as practicable, notify the gas supplier and
- not re-open the supply until all necessary steps have been taken to prevent gas from escaping.

2. Joint Services (Fig. 30)

Where a service pipe is installed which supplies more than one primary meter in the same, or in different premises, there should be a joint service notice.

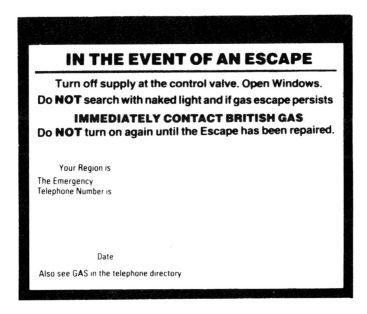

Fig. 29   Escapes notice

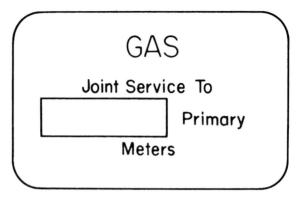

Fig. 30   Joint service notice

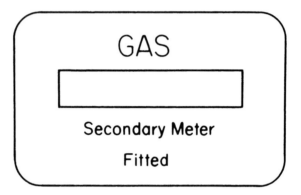

Fig. 31   Secondary meter notice

Warning !
This gas inlet valve must be fully opened before starting any gas compressor or gas engine and must not be shut or partially closed while any such plant is in operation otherwise meter and plant will be damaged.

Fig. 32   Notice at meter

3. Secondary Meters (Fig. 31)

Where secondary meters are installed, a notice should be fitted at the primary meter. In addition a line diagram, of permanent form, must

---

Warning!
Before starting this compressor always see that the inlet valve on the gas meter is open or the meter will be damaged.

Fig. 33   Notice at compressor

be displayed on or near the primary meter and on or near all emergency controls connected to it. This diagram must show the configuration of all pipes, meters and emergency controls on the meter installation. If any work is carried out that affects the installation configuration then the diagram must be amended to suit.

4. Other Notices

When gas compressors or engines are fitted on a gas supply, usually on non-domestic premises, notices are required:

● at the meter inlet, Fig. 32
● at the engine or compressor, Fig. 33.

For details of notices for industrial or commercial meters see Vol. 3, Chapter 2.

Bypassing Meters

Domestic meters are very rarely bypassed. However, in factories with continuous processes and in nursing homes and hospitals it may be essential for the gas supply to be maintained continuously.

A bypass is a pipe which links the service pipe with the internal installation. Two arrangements are shown in Figs 34 and 35. The bypass valve is normally closed and sealed by British Gas plc so that it will only be opened in an emergency.

The procedure for exchanging meters fitted with bypasses is given in Vol. 3, Chapter 2.

Regulators and Filters

Filters

Filters were fitted on domestic gas supplies to remove the very fine particles of rust which can be carried by the gas and would otherwise be deposited in regulators, meters and pipework. Where gas has been treated in the mains by oil fogging, filters are not usually necessary.

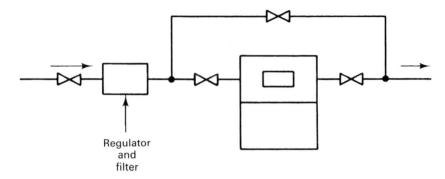

Regulator
and
filter

Fig. 34   Bypass to meter

A filter generally consists of a casing containing resin-bonded wool, glass fibre or other porous material through which gas can pass, the foreign matter being retained. The essential characteristics are:

- an acceptable pressure loss
- the ability to retain small particles
- a high dust holding capacity.

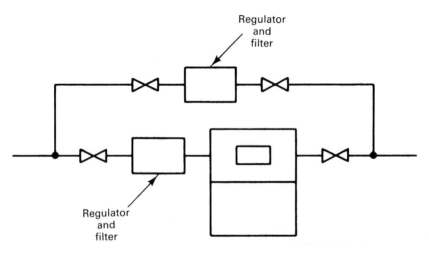

Regulator
and
filter

Regulator
and
filter

Fig. 35   Bypass to meter and regulator

Filters are incorporated as an integral part of some meter regulators and are also available as separate units. When separate, they should always be fitted upstream of the regulator. In all cases, the filter unit should face forward to allow easy access for cleaning or renewal.

Regulators

Details of the operation and installation of regulators generally are given in Vol. 1, Chapter 7.

Meters are normally fitted in association with either service or meter regulators. Although the two names were often interchanged, strictly there is a difference.

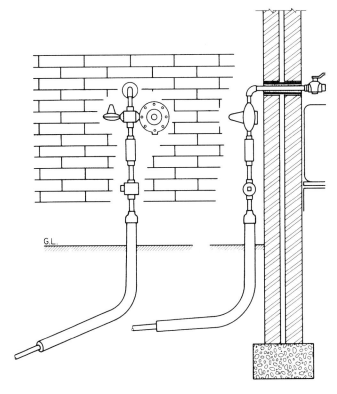

Fig. 36   Service valve, filter and regulator fitted externally; meter internally

A service regulator is fitted on a gas service pipe, that is, upstream of the meter control.

A meter regulator is fitted between the meter control and the meter, that is downstream of the meter control.

So a service regulator is on the service and the meter regulator is on the internal installation, as defined by the Gas Safety Regulations.

Meter Regulator

Most domestic meters are supplied through low pressure services and are fitted with meter regulators. The regulator usually connects directly on to the inlet of the meter and, in a number of Regions, used to incorporate a filter. The specification of the meter regulator is commonly:

- inlet and outlet connections 25 mm R1
- outlet pressure range 12 to 25 mbar
- maximum inlet pressure 75 mbar
- maximum gas rate 14 m^3/h (500 ft^3/h).

Service Regulators

These are generally used on medium or intermediate pressure services or on larger, non-domestic meter installations (see Vol. 3, Chapter 2). The type and size of regulator will depend on:

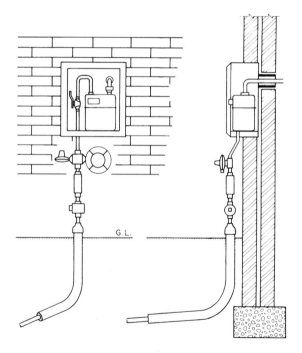

Fig. 37   Service valve, filter and regulator fitted externally, meter in a box

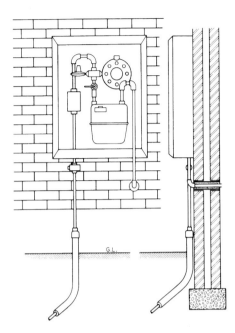

Fig. 38   Service valve fitted externally; filter, regulator and meter in a box

- the outlet pressure range required
- the maximum inlet pressure available
- the maximum gas rate required.

Regulators operating at inlet pressures above 75 mbar must be fitted with a relief valve and a vent pipe. The vent should be taken to open air at a safe height and the end turned down to prevent entry of rain or snow. A gauze may be used to keep out insects. The relief valve should be set to open at a pressure not above that used to test the installation on the outlet of the regulator.

A cock or valve must be fitted on the inlet to the regulator to allow it to be shut down in an emergency or for servicing.

After commissioning, all service regulators should be sealed.

The recommended control layouts for medium pressure installations are:

- service valve, filter and regulator fitted externally with the meter control and the meter within the premises, Fig. 36
- service valve, filter and regulator fitted externally with the meter control and the meter fitted in a meter box, Fig. 37

- service valve fitted externally with the filter, regulator, meter control and meter fitted in an external meter box; the outlet is taken through the base of the box before entering the premises, Fig. 38
- service valve, filter, regulator and meter are fitted externally with the meter control within the premises, Fig. 39.

In these installations the service valve used is a 'security' type, that is it may be closed by hand but requires a special tool to turn it back on.

On large installations, where the supply of gas must be maintained at all times, it may be advisable to provide a duplicate regulator stream (Vol. 3, Chapter 2).

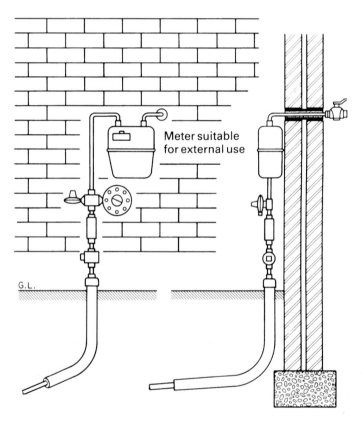

Fig. 39   Service valve, filter, regulator and meter fitted externally; meter control internally

Metered LPG Supplies

Metered LPG supplies should be installed to standards laid down in BS 5482 and the meter in accordance with BS 6400 where applicable.

Where a number of dwellings are supplied with LPG (propane) from a central storage tank, the first stage of the pressure regulation system is sited at the tank. This first stage regulator reduces the pressure to 0.75 bar.

Each individual dwelling will then have either a meter box, housing shut-off valves, the second stage pressure regulator and the meter, or the shut-off valves and second stage pressure regulator outside the dwelling with the meter fitted inside the property.

The second stage pressure regulator system also contains high and low pressure protection devices as described in Vol. 1, Chapter 7. It reduces the pressure to 37 mbar, the standard operating pressure for propane burning appliances.

Figure 40 illustrates a typical meter installation for LPG.

Testing of the High Pressure Stage of the Installation

Cylinder Installations

The following procedure shall be adopted for cylinder installations (see figure 42):

(a) Close off low pressure line at the emergency isolation valve.
(b) Break pigtail joint at cylinder 1 and couple gauge between pigtail and cylinder valve.
(c) Close off cylinder valve 2 and open cylinder valve 1 to charge system with gas.
(d) Close cylinder valve 1 and allow to stand for 5 minutes for equalization of temperature.
(e) Note pressure on gauge.
(f) If pressure drops, test individual joints with leak detection fluid.

For an installation using non-return valves the foregoing procedure shall be carried out on both cylinders.

Note: A suitable gauge would have a range from 0 bar to 10 bar and a dial not less than 100mm in diameter.

Bulk Installations

Where any pipework is subjected to full tank pressure, it shall be tested by the method described above. Where the first stage regulator

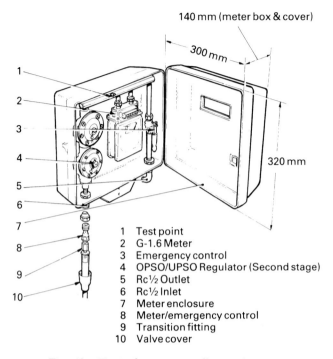

Fig. 40   Typical meter installation for LPG

is fitted directly to the tank valve, the connection should be tested for leakage by means of leak detection fluid.

Model Emergency Action Notice or Leaflet for Uses of Liquefied Petroleum Gas

The following is an example of an emergency action notice.

NOTE It is recommended that capital letters should be used where indicated and that the colour red should be used for words and phrases marked with an asterisk to reinforce the importance of these words.

EMERGENCY ACTION PROCEDURE

In the event of GAS LEAKAGE*

NEVER operate electrical switches
NEVER look for leaks with a naked flame
NEVER enter basements

CALL the gas supplier and consider calling the Fire Brigade. Wait outside.

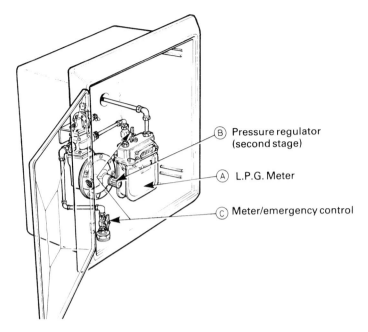

Fig. 41   Another LPG installation with a larger type of meter box

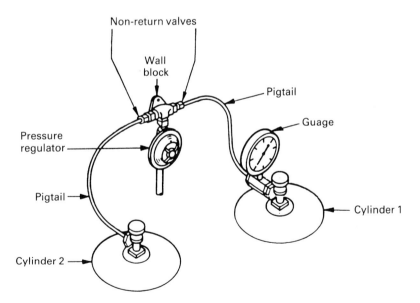

Fig. 42   High pressure leakage test

If safe to do so

- Extinguish all naked flames
- Turn off gas supply at cylinders
- Open doors and windows

DO NOT turn on the gas supply until it has been made safe to do so by a competent person.

In the event of FIRE*

- Call the Fire Brigade immediately and inform them that LPG cylinders are on the premises.
- Turn off the gas supply at cylinders if practical and safe to do so.
- Do not go near cylinder(s) in the vicinity of the fire.
- DO NOT turn on the gas supply until it has been made safe to do so by a competent person.

NOTE Refer to Chapter Three for low pressure testing.

CHAPTER 5

Flueing and Ventilation

Chapter 5 is based on an original draft by Messrs C. Bridgewater and E. A. Goreham

---

Introduction

This chapter provides a general guide to the flueing and ventilating of domestic gas appliances. Specific details of the flues for particular appliances may be found in chapters dealing with those appliances. Manufacturers' instruction sheets usually contain a wealth of information and should always be consulted.

The Gas Safety Regulations require that any person installing or servicing a gas appliance on any premises shall ensure that the:

- means of removal of the products of combustion
- availability of sufficient air for proper combustion
- means of ventilation of the room or space in which the appliance is fitted

are such that the appliance can be used without danger to any person or property.

Generally, if the following requirements have been complied with, the statutory regulations should be satisfied:

BS Code of Practice 5440: Installation of flues and ventilation for gas appliances of rated input not exceeding 60 kW (1st, 2nd and 3rd family gases).

Part 1: 1990: Specification for installation of flues.

Part 2: 1989: Specification for installation of ventilation for gas appliances.

Need for Flueing and Ventilation

Every gas burning appliance produces combustion products. If the appliance is clean and in good working order and there is an adequate supply of fresh air, the products will consist of non-toxic gases. If, however, these products are not removed and replaced by fresh air, then:

- the atmosphere will become vitiated (Vol. 1, Chapter 2)
- lack of oxygen will eventually cause incomplete combustion
- the appliance will cease to function properly
- the products of the incomplete combustion may contain toxic gases.

So, it is necessary to:

- remove the combustion products as soon as they are produced
- replace them with an equal quantity of fresh air.

This is normally achieved by:

- removing the products through a flue to the outside atmosphere
- supplying fresh air through a purpose-made ventilation opening.

Flues and ventilators are, therefore, necessary to ensure that appliances operate safely and effectively.

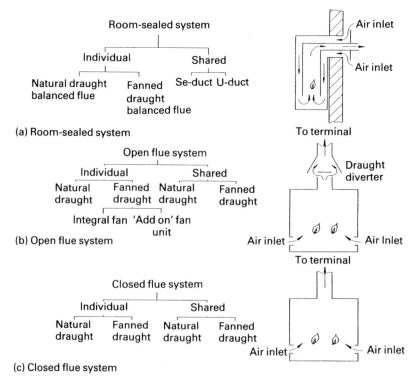

Fig. 1   Flue systems

Flues

Flues may be divided into a number of different categories (Fig. 1). They may be either:

- room sealed
- open
- closed.

Each of the categories can be:

- individual or shared
- natural or fanned draught (including 'condensing' appliances).

For example, the most common flue in older property is the 225 mm × 225 mm brick chimney, designed originally to vent a coal or wood fire. This is an individual flue for a single appliance. It is a form of open flue and it operates on natural draught.

Natural draught is the flue draught or 'pull' set up in the same way as a convection current (Vol. 1, Chapter 10). It is caused by the difference between the densities of the hot gases in the flue and the colder air outside, Fig. 2.

The flue draught or 'aeromotive force' will increase if:

- flue height is increased
- flue gas temperature is increased.

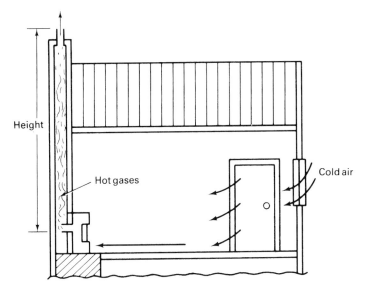

Fig. 2   Natural draught open flue

The flue draught will decrease if:

- the flue is cooled
- excessive resistance, due, for example, to horizontal fluepipe or 90° bends, are present.

Flue draught produced by natural means creates only a very small upward force in the flue. So mechanical aids can be used to provide a much more positive method of evacuating the products. Fanned draught can overcome flueing problems and permit a more flexible approach to flue design. It is dealt with later in the chapter.

Open Flue Systems

Also known as a natural draught conventional flue, the open flue system consists of four principal components:

- primary flue
- draught diverter
- secondary flue
- terminal.

The primary flue and draught diverter are normally integral parts of the appliance, whilst the secondary flue and the terminal are supplied to suit particular site requirements, Fig. 3.

Primary Flue

This creates the initial flue draught required to exhaust all the products of combustion from the appliance. It is a component part of the appliance and often part of the combustion chamber itself.

Draught Diverter

All open-flued appliances, with the exception of incinerators and open gas fires, should be fitted with a draught diverter or 'baffler' as it was once called.

The diverter is normally built into the appliance. However, when it is separate, it must always be fitted in the same room or enclosure as the appliance and generally directly on to the appliance flue outlet.

The draught diverter does three things:

- it diverts down-blow in the secondary flue
- allows flue products to be diluted
- breaks the pull of the secondary flue.

In adverse weather conditions, wind can blow down a flue, so reversing the flow of products, Fig. 4(b). If allowed to enter the

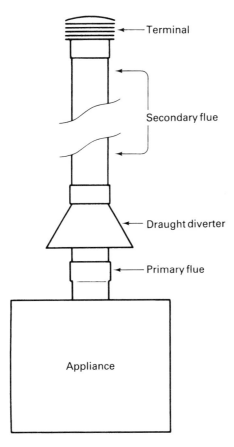

Fig. 3   Components of an open flue

combustion chamber, the down-blow would interfere with the combustion, so it is diverted away from the primary flue and into the room containing the appliance. Because it consists only of the products of complete combustion and fresh air, down-blow is harmless. Persistent down-blow should be investigated as it probably indicates a fault in the flue system.

Under normal operating conditions the draught diverter allows air to be drawn into the secondary flue from around the 'skirt', or flared opening, Fig. 4(a). This has the effect of increasing the volume of air available to carry the water vapour in the flue gases. So it helps to prevent condensation (see later section on condensation in open flues).

The flue pull in the secondary flue depends on the height of the flue and a number of other factors. This pull, if communicated to the

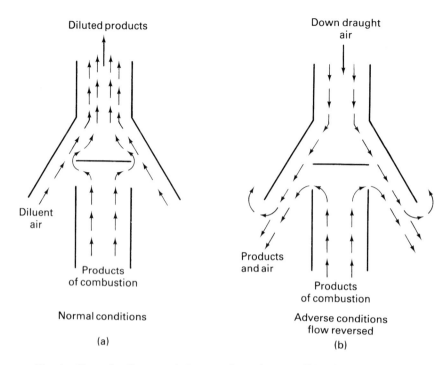

Fig. 4   Draught diverter: (a) normal conditions; (b) adverse conditions

combustion chamber, would have a direct effect on the appliance performance and efficiency. So different flues would have varying effects. The draught diverter acts as a natural flue break enabling the appliance, as designed, to have a standard performance and efficiency, irrespective of the flue length or location.

Secondary Flue

The purpose of the secondary flue is to pass the products of combustion and the diluent air from the draught diverter to the terminal.

Horizontal runs of flue make no contribution to the flue draught, but they do add to the resistance and tend to reduce the flow rate. Bends and other fittings have a similar effect. So, having due regard for its structural stability, appearance and termination, the flue should be as near vertical as possible with a minimum of bends, changes of section and other resistances. Horizontal or very shallow runs should particularly be avoided since, in addition to adding to the resistance, they encourage local cooling.

To maintain the temperature of the flue gases, the flue should be run inside the building for as much of its length as possible.

It is essential that the secondary flue immediately above the draught diverter is vertical because the flue gases are at their hottest at this point and can provide the greatest motive force.

It is sometimes necessary to allow for the flue to be disconnected so that the appliance may be serviced. For this purpose a slip socket, Fig. 5, or similar device is fitted at the base of the flue, usually just above the draught diverter.

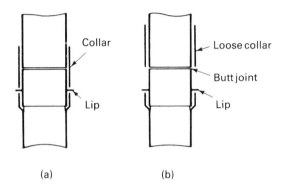

Fig. 5   Flue disconnecting fitting: (a) slip socket; (b) collar raised

Terminal

The terminal is fitted on the end of the secondary flue, outside the building. It should:

- assist in exhausting the flue gases from the secondary flue
- minimise down blow
- prevent the entry of rain and anything likely to block the flue, such as birds, leaves, or snow.

Approved terminals must pass a simple test for the entry of foreign matter.

This requires that any opening in a terminal should be such that it will admit a 6 mm diameter ball, but not a 16 mm diameter ball: with the exception that the opening in a terminal serving an incinerator should admit a ball of 25 mm diameter.

Typical open-flue terminals are shown in Fig. 6, 7 and 8.

Flue Sizes

Generally, a flue of a size equal to that of the appliance outlet will be adequate, provided that a certain minimum height is available. Flues for new or replacement gas fires shall have a minimum cross-sectional

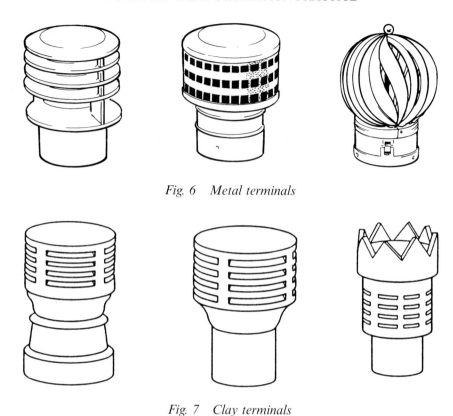

Fig. 6   Metal terminals

Fig. 7   Clay terminals

area of 12000 mm^2 unless the manufacturer's instructions specifically permit otherwise. Flues for decorative fuel effect gas appliances (which fall within the scope of BS 5871: Part 3: 1991) shall be so sized that they will contain a circle not less than 175 mm in diameter (BS 5440 Pt. 1 1990). Building regulations applicable from 1 April 1990 states that the flue size should be at least:

(a) in the case of a gas fire, a cross-section area of at least 12000 mm^2 if the flue is round or 16500 mm^2 if the flue is rectangular, and have a minimum dimension of 90 mm or

(b) for any other appliance a cross-sectional area of at least that of the outlet from the appliance.

While an undersized flue will cause spillage of products, an oversized flue is uneconomical and is more liable to condensation. So it is essential that the flue should not be undersized and desirable that it should not be oversized.

Fig. 8   Ridge terminal (J. Hinchliffe)

The method of determining the suitability of a proposed flue is to calculate its 'equivalent height', that is, the height of vertical flue to which it is equivalent.

The equivalent height may be calculated using the formula:

$$H_e = H_a \times \frac{(K_1 + K_o)_e}{(K_1 + K_o)_a - K_e H_a + K}$$

where

H_e is the height of the equivalent flue;
H_a is the vertical height of the actual or proposed flue;
K_1 is the inlet resistance of the flue;
K_o is the outlet resistance from the flue;
subscript e refers to the equivalent flue diameter;
subscript a refers to the actual or proposed flue diameter;
K_e is the resistance per unit length of the equivalent flue;
K is the resistance (other than the inlet and outlet resistances) of the actual or proposed flue.

Note: K_e and K are obtained from table 13(b) K_o and K_1 are obtained from table 13(a).

Example 1

Fig. 9 shows a 150 mm flue system fitted to a central heating boiler. Note: (the existing lean-to building has prevented the installation of a balanced-flue appliance.)

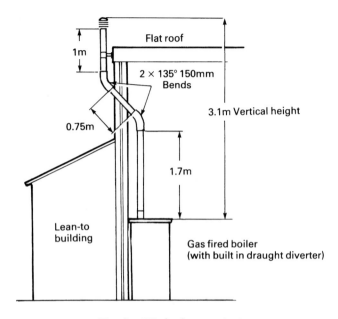

Fig. 9   Worked example 1

From Table 13(b)
    inlet resistance of actual flue   =   0.48
    outlet resistance of actual flue   =   0.48
    inlet resistance of equivalent flue   =   0.48
    outlet resistance of equivalent flue   =   0.48

From Table 13(a)
    other resistances of actual flue terminal (G.C.1. 150 mm)=   0.12
    pipe bend (2 × 135° 150 mm)   =   0.24
    3.45 m pipe 150 mm (3.45 × 0.12)   =   0.41
                                                0.77

$$\text{Equivalent height} = 3.1 \times \frac{(0.48 + 0.48)}{(0.48 + 0.48) - (0.12 \times 3.1) + 0.77}$$

$$= 3.1 \times \frac{0.96}{0.96 - 0.372 + 0.77}$$

$$= 3.1 \times \frac{0.96}{1.358}$$

$$= \mathbf{2.19 \ m}$$

This exceeds 1.0 m of flue pipe of the same diameter as the appliance flue spigot and is therefore satisfactory for this central heating boiler (see Table 14).

Example 2

Fig. 10 shows a precast block flue (231 mm × 65 mm) with a 125 mm round flue in the loft terminating with a ridge terminal. The flue is designed for a gas fire.

From Table 13(b)

inlet resistance of actual flue	=	3.0
outlet resistance of actual flue	=	1.0
inlet resistance of equivalent flue	=	3.0
outlet resistance of equivalent flue	=	1.0

From Table 13(a)
other resistances of actual flue:

ridge terminal (125 mm)	=	1.0
pipe bend (135° 125 mm)	=	0.25
3.4 m pipe 125 mm (3.4 × 0.25)	=	0.85

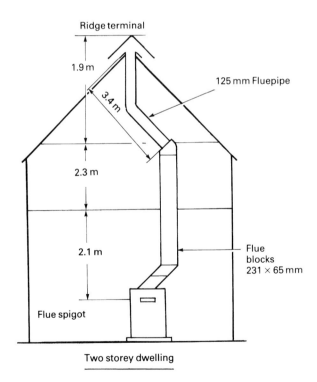

Fig. 10   Worked example 2

adaptor block (231 mm × 65 mm to 125 mm)   =   0.50
2 raking blocks (2 × 0.30)   =   0.60
4.4 m flue blocks 231 mm × 65 mm (4.4 × 0.65)   =   2.86
                                                     ‾‾‾‾
                                                     6.06
                                                     ‾‾‾‾

$$\text{Equivalent height} = 6.3 \times \frac{(3 + 1)}{(3 + 1) - (0.25 \times 6.3) + 6.06}$$

$$= 6.3 \times \frac{4}{4 - 1.575 + 6.06}$$

$$= 6.3 \times \frac{4}{8.485}$$

$$= \mathbf{2.97 \ m.}$$

This exceeds 2.0 m of 125 mm flue pipe and is therefore satisfactory for a gas fire connected to this precast block flue (see Table 14).

Example 3

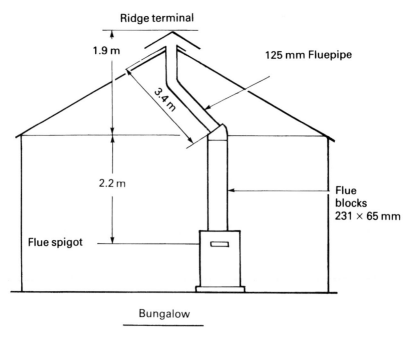

Fig. 11   Worked example 3

Fig. 11 shows a precast block flue (231 mm × 65 mm) with a 125 mm round flue in the loft, terminating with a ridge terminal. The flue is installed in a bungalow and is designed for a gas fire/back boiler unit.

From Table 13(b)

inlet resistance of actual flue	=	2.0
outlet resistance of actual flue	=	1.0
inlet resistance of equivalent flue	=	2.0
outlet resistance of equivalent flue	=	1.0

From Table 13(a)
other resistances of actual flue:

ridge terminal (125 mm)	=	1.0
pipe bend (135° 125 mm)	=	0.25
3.4 m pipe 125 mm (3.4 × 0.25)	=	0.85
adaptor block (231 mm × 65 mm to 125 mm)	=	0.50
2.2 m flue blocks 231 mm × 65 mm (2.2 × 0.65)	=	1.43
		4.03

$$\text{Equivalent height} = 4.1 \times \frac{(2 + 1)}{(2 + 1) - (0.25 \times 4.1) + 4.03}$$

$$= 4.1 \times \frac{3}{3 - 1.02 + 4.03}$$

$$= 4.1 \times \frac{3}{6.01}$$

$$= \mathbf{2.046 \ m.}$$

This is less than the 2.4 m required for a gas fire/back boiler unit fitted to a 125 mm flue and is therefore not suitable (see Table 14).

It must be noted that the factors given in Tables 13(a) and (b) and Table 14 do not apply to incinerators, decorative fuel effect gas appliances and inset appliances. The minimum equivalent height for these appliances will generally be included in the manufacturers data.

Table 13(a) Resistance factors for use in calculating equivalent heights

Component	Internal size	Resistance factor	
	mm		
	197 × 67	0.85	
	231 × 65	0.65	
Flue	317 × 63	0.35	per metre
blocks	140 × 102	0.60	run
	200 × 75	0.60	
	183 × 90	0.45	
Pipe	100	0.78	
	125	0.25	
	150	0.12	
Chimney	213 × 213	0.02	
	100 mm pipe	1.22	
90° bend	125 mm pipe	0.50	per fitting
	150 mm pipe	0.24	
	100 mm pipe	0.61	
	125 mm pipe	0.25	
135° bend	150 mm pipe	0.12	per fitting
	197 × 67	0.30	
	231 × 65	0.22	
	317 × 63	0.13	
Raking* block	Any	0.30	
Adaptor* block	Any	0.50	per block
	100 mm ridge	2.5	
	125 mm ridge	1.0	
Terminal	150 mm ridge	0.48	
	100 mm GC1	0.6	
	125 mm GC1	0.25	
	150 mm GC1	0.12	

* Raking block and Adaptor block are also known as Offset block and Transfer block respectively.

Table 13(b) Inlet and outlet resistances

Appliance	Inlet resistance K_i
Gas fire (12000 mm² equivalent flue size)	3.0
Gas fire/back boiler unit	2.0
Other appliances:	
100 mm spigot	2.5
125 mm spigot	1.0
150 mm spigot	0.48

Flue	Outlet resistance K_o
100 mm flue	2.5
125 mm flue	1.0
150 mm flue	0.48

Note: These factors do not apply to decorative fuel effect gas appliances nor to inset appliances.

When appliances are tested in the laboratories they are checked to ensure that they clear their combustion products when connected to a test flue. So, if in practice, an appliance is never connected to a less effective flue, it will always operate satisfactorily.

The minimum equivalent height of flues should therefore be the same as that of the current test flues. Minimum equivalent heights are given in Table 14:

Table 14 Minimum equivalent heights needed

Appliance	Minimum equivalent height
Gas fire (to be connected to precast block flues)	2.0 m of 125 mm flue pipe
Other gas fires	2.4 m of 125 mm flue pipe
Gas fire/back boiler unit	2.4 m of 125 mm flue pipe
Other appliances	1.0 m of flue pipe of the same diameter as the appliance flue spigot

Note: These factors do not apply to decorative fuel effect gas appliances nor to inset appliances.

British Standards are continually being changed and it is likely that test flue lengths may be altered at any time. Reference should always be made to current publications and bulletins for the latest information.

Precast Flue Blocks

Rectangular flue blocks, usually made from precast concrete or clay, can be used for some gas appliance flues. The blocks should comply with BS 1289 Flue blocks and masonry terminals for gas appliances. Part 1: 1986 Specification for precast concrete flue blocks and terminals and Part 2: 1989 Specification for clay flue blocks and terminals. They are usually built into the premises during construction, these are called bonded flue blocks and are cast with a bonding extension. Flue blocks can also be used to add a flue to an existing property, these are called non-bonded flue blocks and have no bonding extension. It should be noted that flue blocks are generally more resistive to the flow of products of combustion than are flue pipes of the same cross-sectional area. Not all appliances are therefore suitable for connection to flue blocks. Most of the blocks used in domestic properties have a minimum cross-sectional flue area of 12000 mm^2 and a minimum dimension of 63 mm. The 'aspect ratio', i.e. the ratio of the shorter dimension to the longer dimension, is usually 3:1 where the blocks are fully built in. Where the blocks are not fully built in the aspect ratio is generally around 1¾:1.

Bearing in mind the increased resistance of this type of flue, the following appliances must not be fitted to flue blocks with a cross-sectional area between 12000 mm^2 and 13000 mm^2 or with a minor dimension of 63 mm or less:

- incinerators and drying cabinets
- appliances with a flue outlet greater than 13000 mm^2
- decorative fuel effect appliance (within the scope of BS 5871: Part 3: 1991)

Gas fires and combined appliances incorporating a gas fire must only be connected to flue blocks if the manufacturers' instructions state that this is acceptable and the flue complies with building regulations.

Changes in the design of appliances sometimes means that a flue which has operated satisfactorily with an existing appliance cannot be assumed to be satisfactory for a replacement appliance of the same type.

Where it is necessary for a precast block flue to change direction, raking/offset blocks, Fig. 12(e), with a maximum offset of 30° must be used. When connecting flue pipes to flue blocks an adaptor/transfer block, (Fig. 12(d)) must be used. It is important to ensure that any flue pipe connected to an adaptor/transfer block does not project into the flue block and cause a restriction to the cross-sectional area of the flue. When connecting a gas fire or gas fire and back boiler or back circulator to a block flue, the manufacturers recess panel(s)/starter block(s) or fire recess panel/starter block for that particular appliance must be used*. Where gas fires are connected to the base of a flue block chimney there must be a debris collection space below the spigot of the fire. The collection space must be at least 75 mm deep with a minimum volume of 0.002 m^3. It is vital that blocks are correctly aligned during construction and that the block manufacturers recommended jointing material is used. Any material extrusions at the joints, that would reduce the effective cross-sectional area of the flue, must be removed during construction.

Using Tables 13 and 15 the minimum flue area of a concrete block flue or a combined block and pipe flue can be calculated. For example.

* A lintel/collector block should be installed above the fire recess panel/starter block.

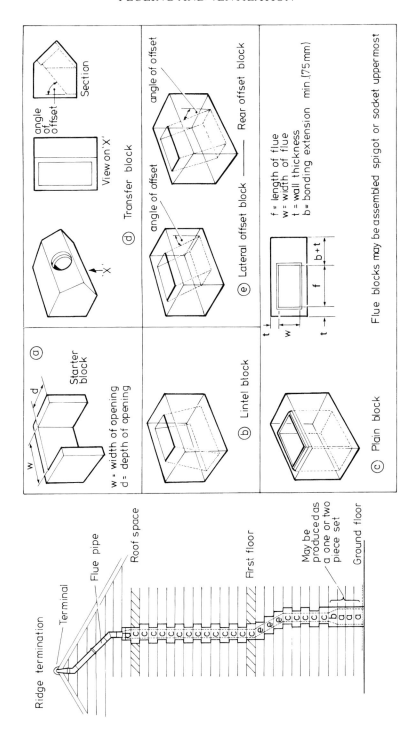

Fig. 12   Flue blocks for gas appliances

The flue has:

- 5 m concrete block flue   K = 1.5
- 1 off-set (2,135° blocks)   = 0.6
- 3 m flue pipe   = 0.0
- ridge terminal   = 1.0
- Total   K = 3.1

$$\text{Equivalent height} = (5 + 3) \times \frac{2}{(2 + 3.1)} \text{ m}$$

$$= 8 \times \frac{2}{5.1} \text{ m}$$

$$= 3.1 \text{ m}$$

The flue may be used for gas fires or fire/back boilers where manufacturer's instructions permit, or for any other appliance with an input rating of not more than 28 kW.

Termination of Open Flues

A flue terminal need not be made to any particular design, but it should comply with the requirements of BS 5440, BS 1289 and BS 715. Typical flue terminals are shown in Figs. 6, 7 and 8. The free area of the outlet openings should be twice the area of the flue.

The location of the terminal, rather than its particular design, has an appreciable effect on the performance of the flue system. It is absolutely essential that the terminal is positioned outside the building so that it is freely exposed to any wind and is not shielded by any roof, structure or object to such a degree that they create an undesirable pressure region around the terminal. To prevent the creation of such an undesirable pressure region BS 5440: Part 1: 1990 states that the terminal for an open flued natural draught flue system shall not be positioned within 1.5 m of a wall surface. Whenever an existing appliance is replaced, the terminal location must comply with the recommendations of BS 5440: Part 1: 1990.

Preferred positions are:

- at, or above the ridge of a pitched roof, by means of a ridge terminal, Fig. 13
- above a flat roof, Fig. 14
- above the intersection with a pitched roof, Fig. 14.

Table 15 gives the location for roof mounted terminals for individual natural draught open flue systems.

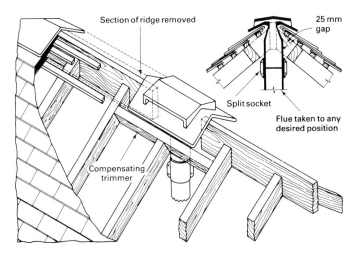

Section of ridge removed

25 mm gap

Split socket

Flue taken to any desired position

Compensating trimmer

Fig. 13   A typical ridge terminal installation

Table 15 Location of Roof Mounted Terminals for Natural Draught Open Flue Systems (Minimum Height of Base of Terminal)

Type of roof		Terminal location: Not within 1.5 m of the vertical side of a raised structure*			Terminal location: Within 1.5 m of a structure*
		Internal route Not on the ridge	*On the ridge*	*External route*	*Internal/external route*
Pitched	Pitch above 45°	1m above flue/ roof intersection (see Fig. 14(c)	At or above ridge level see Fig. 13	See Fig. 14(b) (1 m)	Terminal base 600 mm above the level of the top of the structure See Figs. 14(g) and 14(h)
	Pitch not above 45°	600 mm above flue roof intersection (see Fig. 14(d)		See Fig. 14(b) (600 mm)	
Flat	Without parapet	250 mm above flue roof intersection See Fig. 14(e)		The base of the terminal to be 600 mm above the level of the adjacent roof edge	
	With parapet	600 mm above flue roof intersection See Fig. 14(f)			

* A structure might be a tank room, chimney stack, dormer window or parapet.

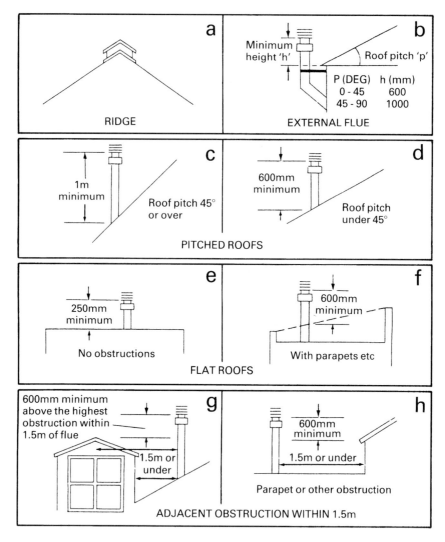

Fig. 14   Common open flue terminal positions

Condensation in Open Flues

New Installations

Condensation occurs in flues when the mean temperature of the flue gases is lower than their dew point. In practice a small amount of condensation is sometimes unavoidable, but is not normally a problem, particularly where the flue material is slightly absorbent.

The flue gas mean temperature can fall below the dew point because of excessive cooling caused by one or more of the following:

- excessive flue length
- a flue route flanked by exposed walls or passing through unheated parts of the building, for example, the roofspace
- a flue construction that cannot easily retain heat.

Now, you have seen that increasing the height of a flue will increase the flue pull. But increasing the height is also increasing the flue length and so may cause condensation. Obviously some compromise is necessary.

Table 16 indicates the maximum lengths of flues of various construction that will give freedom from condensation when used with a gas fire. The graphs in Fig. 15 deal with other types of appliances of known input ratings. Proposed flue installations can be checked against the tables and graphs to discover whether condensation may be expected.

If the data shows that condensation is likely to occur, the first consideration should be to insulate the flue. Proprietary insulated flues are generally constructed of double-walled pipe with a layer of insulating material between. A similar standard of insulation can be achieved by wrapping a single-walled pipe with mineral wool not less than 25 mm thick.

Table 16 Condensate-free lengths of individual open flue used with a gas fire (efficiency not greater than 70%)

Flue exposure	Condensate-free length		
	225 mm × 225 mm brick chimney: precast concrete block flue of area 13000 mm^2 and aspect ratio of up to 4:1 or area 20000 mm^2 and aspect ratio of up to 5:1	125 mm diameter flue pipe	
		Single-wall	Double-wall
Internal*	m 12	m 20	m 33
External	10	14	28

* An internal flue is one which has none of its surface exposed to external temperatures except for a length above the roof not exceeding the lesser of 1.5 m or one quarter of the flue height.

The proposed flue can then be rechecked in the table as double-walled pipe.

Sometimes, even with double-walled pipe it may not be possible to avoid condensation. In this case the next consideration should be the possibility of using an appliance with an alternative flue system, for example, a room-sealed appliance.

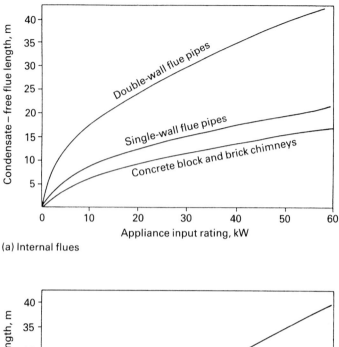

(a) Internal flues

(b) External flues

Fig. 15   Maximum lengths for condensate-free flues for appliances other than gas fires

If, however, a flue system which will create condensation must be installed, special care must be taken with the construction. The flue and jointing should be of non-permeable materials and adequate means must be provided for the removal of condensation.

A condense pipe of non-corrodible material and not less than 22 mm diameter should be run from the base of the flue to a suitable gully or hopper, outside the building. Figure 16 shows a method of fitting a condense pipe.

Existing Chimneys

When existing brick chimneys are to be used for flueing gas appliances, extra care should be taken to avoid causing condensation at all costs. The brickwork will have absorbed the sooty deposits from the previous solid fuel fires and any condensation which occurs will filter the deposits through to the finished plaster surface of any adjacent room. It makes a nasty, sticky, mottled mess and the customer won't be very pleased. Where an existing chimney or flue has given an unsatisfactory performance with a previously installed appliance or another fuel, it must be examined and any faults corrected before installing a replacement appliance.

Whenever a continuously burning gas appliance like a central heating boiler is installed, the chimney should be lined. A flexible, metallic flue liner of the same diameter as the appliance outlet should be used.

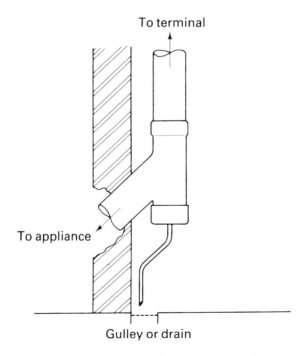

Fig. 16   Method of fitting condense pipe

Table 16 and Fig. 15 give the maximum flue lengths for which problems with condensation during normal winter conditions are unlikely. Table 17 also gives the appliance/chimney combinations that require the chimney to be lined.

TABLE 17 Appliance/chimney combinations which require the chimney to be lined

Application type	Flue length
Gas fire	More than 10 m (external wall) More than 12 m (internal wall)
Gas fire with back boiler unit	Any length
Gas fire with back circulator unit	More than 10 m (external wall) More than 12 m (external wall)
Circulator	More than 6 m (external wall) More than 1.5 m external length and total length more than 9 m
Other appliances	Flue lengths greater than those given in Table 16 and Fig. 15 (this Chapter).

Existing chimneys which have been used for solid fuel should be swept before being used for gas appliances. They should be checked for soundness and freedom from obstructions. Any dampers, register plates or restrictors must be removed or permanently secured to leave the main flueway unobstructed. Any under-grate air ducts must be sealed. Only one gas appliance may be connected to a chimney or flue and any other opening into the chimney or flue must be permanently sealed with materials no less substantial than those of which the chimney or flue is constructed.

Where an appliance is not fitted at the bottom of the chimney the flue must be sealed at a position not less than 250 mm nor more than 1 m below the point at which the appliance connects into the flue or chimney. A means of removing any debris that may collect in this space, must be provided. See Fig. 17(a). The method of connecting the appliance flue into the existing chimney or flue must be such as to prevent the entry of debris into the appliance flue spigot or flue connection. See Figs. 17(a) and (b).

Where a gas fire is fitted into a fireplace opening or any other type of appliance is connected to an existing flue or chimney that is not lined, a void below the point of the flue connection, or in the case of a gas fire, below the appliance flue spigot, must be provided. The void shall be as specified in Table 18. Provision must be provided for the inspection and removal of debris from this void.

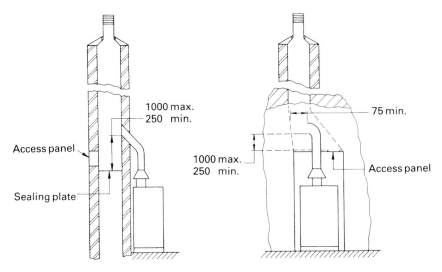

(a) Typical side view with appliance external to chimney
(b) Typical front view with appliance set in a recess

All dimensions are in millimetres.

Fig. 17   Connection to existing brick chimney: (a) Typical side view with appliance external to chimney; (b) Typical front view with appliance set in recess

TABLE 18 Voids below appliance connections

Circumstances	Minimum void dimensions	
	Depth	Vol.
	mm	m³
Any appliance fitted to brick chimney	250	0.012 (12 litres)
Any appliance fitted to a flue block chimney	75	0.002 (2 litres)

Flue Construction

Materials used for flue construction vary from brick and ceramics to stainless steel. All materials used should be:

- mechanically robust
- resistant to internal or external corrosion
- non-combustible
- durable under the conditions in which it is to be used

The materials most commonly used for open-flue installation in existing property are:

- Double-wall (often known as twin-wall) metal, conforming to BS 715: 1993
- Galvanised sheet iron or steel conforming to BS 715: 1993, generally confined to commercial or industrial equipment
- Vitreous enamelled steel, for internal flues only
- Flexible stainless steel flue liner, conforming to BS 715: 1993
- Precast concrete blocks for use in non-bonded flues and conforming to BS 1289: Part 1: 1986
- Clay blocks for use in non-bonded flues and conforming to BS 1289: part 2: 1989

When replacing an appliance, the existing flue system may be constructed of asbestos cement. This existing system may be re-used providing it conforms to BS 5440: Part 1: 1990. However, stringent precautions must be taken when working with asbestos. Vol. 1, Chapter 12 gives a list of points to be borne in mind when you come into contact or work with this material. Special fittings are available for connecting metal flues to asbestos-cement flues.

In property under construction, lined brick chimneys or precast concrete or clay flueblocks for use in bonded flues are generally installed.

Double-wall/Twin-wall Metal Flues

The pipes and fittings are in a variety of diameters and lengths conforming to BS 715: 1993. The sizes in general use are:

- diameters (mm) 50, 75, 100, 125, 150
- lengths (mm) 150, 300, 500, 1000, 1500. In addition adjustable/ telescopic lengths are available to eliminate the need for cutting to a required length.

All lengths of pipe and individual fittings have a male and female end.

The pipe and fittings are designed for the male end to be fitted uppermost and the joints are made by fitting the female end over the male. Where a bayonet type joint is used, the two components must be rotated until the full movement is taken up.

The insulating space between the inner and outer walls of the flue is generally about 8 mm. Where the pipe or fitting is to be used only for internal use, the outer pipe may be made of zinc-coated sheet steel or an aluminium coated steel. For external and internal use the outer pipe may be made from aluminium, aluminium alloy or stainless steel. Bends are usually adjustable and must not exceed 45° to the vertical, except for a section not exceeding 150 mm in length protruding from a rear outlet appliance. The pipe must be supported by brackets fitted throughout its length at intervals not exceeding 1.8 m or as stated in the manufacturer's instructions.

Where a flue pipe passes through a tiled or slated roof, the joint must be weatherproofed using a purpose-made plate with an upstand of 150 mm, Fig. 18. At the point where the flue passes through the ceiling, it must be encased in an insulating sleeve, Fig. 19.

The general criteria for protecting combustible material surrounding a flue pipe is that the material should not reach a temperature of more than 65° C. Proprietary methods may be used or a sleeve can be made of metal.

The sleeve must be large enough to provide an annular space around the flue pipe of not less than 25 mm. This space may be packed with non-combustible material and the ends may be covered by wall or ceiling plates.

If a flue pipe passes through a cupboard in which combustible material may be stored, it should be protected by a wire mesh guard at a distance of 25 mm around the pipe.

Vitreous Enamelled Steel

These pipes are made to BS 715, and are available in lengths from 150 mm so that little cutting is needed. When a pipe has to be cut, use the following method:

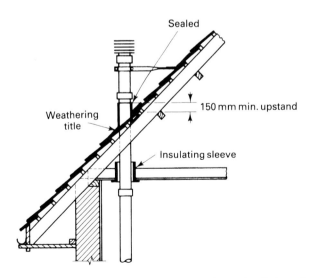

Fig. 18   Weatherproofing flue pipe through tiled roof

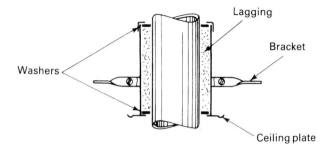

Fig. 19   Insulating sleeve

- bind a strip of adhesive tape (insulating tape) around the pipe at the point to be cut
- cut through the tape and the pipe using a hacksaw with a 32 t.p.i. (teeth per 25 mm) blade
- paint the cut end of the pipe to prevent rusting under the enamel, and paint any chipped areas.

The pipe is supplied with plain ends and is jointed by chromium plated connecting clips, Fig. 20.

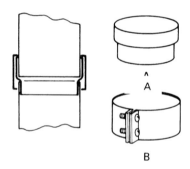

A, inner jointing ring; B, outer cover clip.

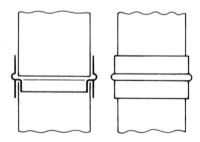

Fig. 20   Connecting clips for vitreous enamelled flue pipe

Brick Chimneys

It is now a requirement of the Building Regulations that chimneys should be lined. This lining, carried out during building construction, is normally by earthenware pipe with a low porosity.

Where it is necessary to line existing chimneys, a stainless steel flexible liner is generally used. The material required for this job is:

- adequate length of flexible, stainless steel flue liner of the same diameter as the appliance outlet
- two sealing/clamping plates
- an approved terminal of appropriate diameter.

The method of installation, Fig. 21, is as follows:

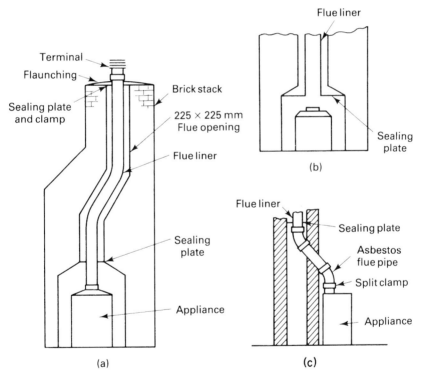

Appliances have integral draught diverters

*Fig. 21   Installing flue liner: (a) direct connection; (b) back boiler unit;
(c) appliance fitted externally*

- ensure that the chimney is clear of soot, brickdust or other debris
- remove the existing chimney pot and the flaunching
- draw the liner into the chimney from either the top or the bottom, whichever is the more convenient
- fit the clamping plate to the top of the liner with sufficient liner protruding to allow for the rebuilding of the flaunching and fitting to the terminal
- allow the clamping plate to rest on the top course of brickwork so as to close the chimney opening
- fit the terminal and rebuild the flaunching, making a weather-proof joint from the edge of the brickwork to the terminal socket so leading water away from the terminal and the liner
- fit a bottom clamping plate to close the base of the flue opening, after threading the liner through the plate
- if the appliance is situated in a fireplace opening, cut the liner to length to engage in the appliance outlet and make a suitable joint
- if the appliance has no direct flue connection, cut off the excess liner to give a flush finish.
- if the appliance is fitted outside the opening, extend from the appliance outlet into the side or base of the chimney with flue pipe, ending with a socket into which the liner is jointed.

Where it is necessary to bend a flue liner, the minimum bending radius, when measured to the inner side of the liner, shall be no tighter than that stated in the flue manufacturer's installation instructions. This bending radius must be no more than three times the diameter of the flue liner (BS 715: 1993).

The Standard also states that the manufacturer's instructions must include a warning that flue liners have to be installed in continuous lengths, without joiners, within the chimney stack.

Precast Concrete Blocks

These hollow concrete blocks are designed specifically for building into the walls of domestic properties during construction. The dimensions of the blocks vary but they are generally of a size so that normal brick laying techniques may be used for their installation. The flueway is formed by laying one block on top of another so that the hollow sections are in line.

Blocks that are built-in (bonded) during the construction of a building and contribute to the structural strength of the building are called bonded flue blocks (see Fig. 12). Non-bonded flue blocks are generally added to existing buildings and although they are likely to be supported by (tied-to) the building, they do not add to its structural strength (see Fig. 22).

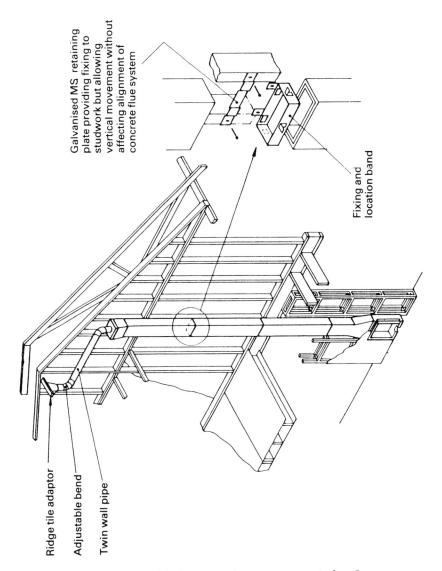

Fig. 22 Non-bonded flue block system for masonry or timber frame construction

The blocks are joined using the material recommended in the manufacturer's installation instructions. For gas fires, starter blocks or recess panels are used and these have openings similar in size to a standard fireplace opening and provide a void below the flue spigot for any debris to collect (see Fig. 12). Where appliances are being fitted into starter blocks or recess panels it is vital to ensure that they are of adequate size for the appliance to be fitted.

Care is required to ensure that no displacement takes place and that the joints between blocks are completely sealed. Any surplus material within the flueway must be removed.

The surface of the flue block is covered only by the cement rendering and plaster on the wall. Consequently the temperature of the plaster surface will be affected by the temperature of the flue gases. If the surface temperature rises much above 65° C, the plaster will crack and it will be difficult to make wallpaper paste adhere to it. In order to overcome this problem, direct rendering and plastering onto the flue blocks should be avoided when some appliance types or combinations of appliances are connected to the flue. A reduction in the surface temperature can be achieved by providing an air space of some other form of insulation between the flue block and the plaster finish.

This places a limitation on the type of appliance which may be fitted to a block flue. Gas fires are generally satisfactory but, if the installation of central heating appliances is being considered, it is advisable to check whether excessive surface temperatures may be expected and the flue may need to be insulated.

The design and construction of precast flue block systems should be in line with the recommendations made in BS 1289: Part 1: 1986 Specification for precast concrete flue blocks and terminals, and BS 1289: Part 2: 1989 Specification for clay flue blocks and terminals.

Flueing Incinerators

Incinerators are flued directly to outside atmosphere so that no spillage of products of combustion, smoke or particles of burnt refuse can occur within the premises. Omitting the draught diverter also ensures that the maximum possible flue draught is maintained to help complete the incineration.

Whilst the general rules for the other open flued appliances apply, special attention must be paid to the route of the flue, the material to be selected and the need for inspection and cleaning.

A completely vertical flue is the most desirable. The minimum vertical height above the appliance should be not less than 1.5 m with minimal diagonal lengths connected at an angle of more than 45° to the horizontal.

The higher temperature of the flue products from the incinerator prevents the use of aluminium or other flue materials likely to be short lived under these conditions. If an existing asbestos cement flue is to be used for an incinerator with an input of up to 45 kW the flue pipe should comply with BS 567. Where the incinerator has a capacity between 0.03 m³ and 0.09 m³ the flue pipe must be of heavy quality conforming to BS 835. Flue pipes made from asbestos cement can no longer be installed, flues made from substitute materials complying with BS 567 and BS 835 are acceptable. When connecting to, or replacing an existing asbestos cement flue pipe, the precautions given in Vol. 1, Chapter 11 must be followed.

The flue should be fitted with an inspection cover or means of disconnection for cleaning purposes.

Room Sealed Flue System

Balanced Flued Natural Draught Appliances

A room sealed or balanced flued appliance provides the solution to a number of the difficulties experienced with open flued appliances. It is a complete unit, meeting its needs for combustion air and products extraction without requiring the design of an additional, elaborate flue system. Room-sealed flues should, wherever practicable, be used instead of open flues.

With a natural draught balanced flue it is necessary for the appliance to be fitted on an external wall. Some models allow the appliance to be fitted to an internal wall with an external wall adjacent to one side.

Some fan assisted or fanned draught models can be fitted remote from external walls and are described later in this chapter.

The principle of the balanced flue can be applied to most space and water heating appliances. Commonly central heating units, convector heaters and water heaters use this system.

The appliance is sealed from the room in which it is installed and obtains air for combustion from outside the building, through a duct in the wall, Fig. 23. The products of combustion are vented by a smaller duct which passes through the air duct. The two ducts are often concentric and the flue duct usually terminates a short distance beyond the face of the combustion air inlet.

Because both the products outlet and the air inlet are at the same point, they should be affected equally by any wind pressure. The motive force in the flue should only be that supplied by the difference in densities of the hot products of combustion and the colder incoming air.

The terminal, which is an intrinsic part of the appliance, should be fitted in such a position that:

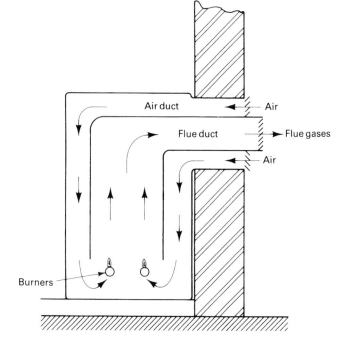

Fig. 23   Operation of natural draught balanced flue appliance

- the products of combustion cannot re-enter the building
- the terminal is exposed to free air movement
- adjacent or opposing obstacles do not affect the wind balance around the terminal.

A clear expanse of wall is preferable, with the terminal positioned in accordance with the manufacturer's installation instructions. If these instructions are not available, Fig. 24 and Table 19 illustrate the minimum dimensions of terminals from corners, obstacles etc.

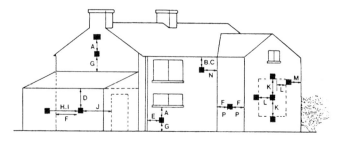

Fig. 24   Balanced flue terminal positions

TABLE 19 Minimum Distances of Balanced Flue Natural Draught Flue Terminals from Corners, Obstructions etc.

Dimension (see Fig. 24)	Terminal Position	Minimum Distance mm	Appliance Input Under 3 kW mm
A	Directly below an openable window, air brick etc.	300	300
B	Below gutters, soil or drain pipes	300	200
C	Below eaves	300	200
D	Below balcony or car port roof	600	400
E	From vertical soil or drain pipes	75	75
F	From an internal or external corner	600	See M, N or P
G	Above, the ground, a roof or balcony	300	300
H	From a surface facing a terminal	600	600
I	From a terminal discharging towards another terminal	600	600
J	From any opening in a car port (door, window etc.) into the dwelling	1200	1200
K	Vertically from a terminal on the same wall	1500	1500
L	Horizontally from a terminal on the same wall	300	300
M	From a single external corner		200
N	From a single internal corner		300
P	From a double corner (both sides of terminal)		500

The outlet part of the terminal is relatively hot compared to its open flued counterpart. This is due to the lack of air dilution in the

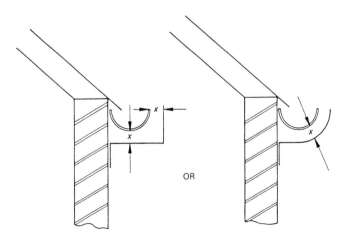

Note Air gap = 5 mm minimum

Fig. 25   Shielding of plastic gutters

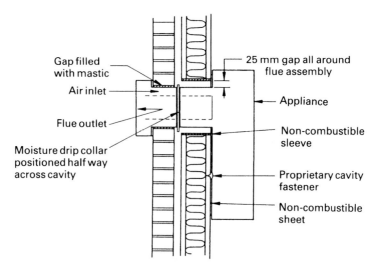

Fig. 26   Circular balanced flue fitted to a timber framed structure

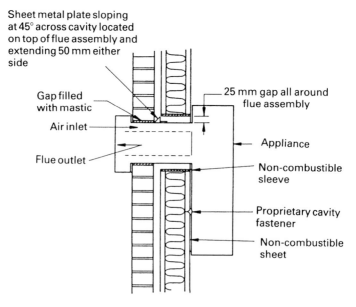

Fig. 27   Rectangular balanced flue fitted to a timber framed structure

products of room-sealed appliances. It is therefore necessary to fit a guard where the terminal is within 2 m of ground level and where the occupier or the public have access and could easily touch the terminal. Where the terminal is less than 1 m below plastic guttering or

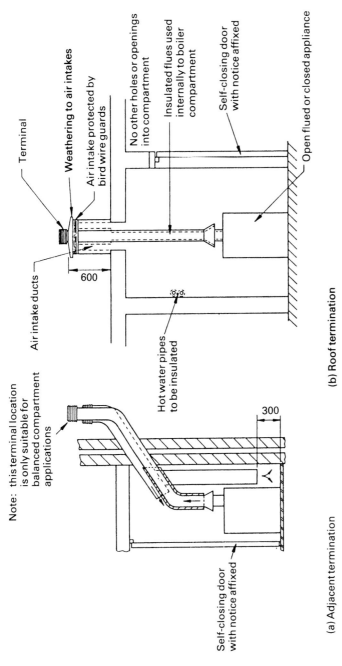

All dimensions are in millimetres

Fig. 28   Balanced compartment installation: (a) adjacent termination; (b) roof termination

0.5 m below painted eaves or any other painted surface, a shield, usually made from a suitable metal and at least 1 m long, should be fitted in order to protect the surface. There should be an air gap of at least 5 mm between the shield and the surface being protected (Fig. 25).

Where it is necessary for a flue assembly to pass through a wall, constructed of or containing combustible material, any specific installation instructions must be followed. If the manufacturers instructions are not available then the flue assembly passing through the combustible part of the wall must be fitted in a non-combustible protective sleeve. The sleeve must be sized to give at least a 25 mm air gap between the flue assembly and the protective sleeve. In addition similar protection is required where the appliance is fitted against an inner wall of combustible material and at the other side between the terminal backplate and any combustible outside wall. The dimensions and thickness of the material will depend on each individual appliance and it is therefore important to check the manufacturer's instructions.

The space between the flue assembly or protective sleeve and the surrounding structure must be sealed. It is also essential to prevent moisture from affecting the inner wall face. Methods of installing balanced flue appliances in properties of timber frame construction are illustrated in Figs. 26 and 27.

Balanced Compartments

A variation of the balanced flue concept is the balanced compartment (Fig. 28(a) and (b)).

The flueing and ventilation arrangements must ensure the complete combustion of the gas and full clearance of the products of combustion. They must also be installed following the flue and appliance manufacturers' recommendations.

The system consists of an open-flued appliance installed in a sealed, fire-resistant compartment or enclosure. The method is generally adopted for larger installations, heat input 35 kW to 60 kW, where room sealed appliances are not available. The air for combustion and the flue products are piped from and to outside atmosphere, terminating side by side with approved terminals. Where different terminals are fitted, the terminal with the lower resistance should be fitted to the flue pipe. A tee, fitted at high level in the air supply, allows for compartment ventilation.

The air inlet should discharge 300 mm above the floor near to the appliance. The air supply pipe/duct should be sized in accordance with Table 20.

TABLE 20 Balanced compartment air inlet pipe/duct sizing

Ducted to low level vent inside the compartment (see Fig. 28(a))	Ducted to high level only (see Fig. 28(b))
1.5 times the minimum air vent area for a high level direct to outside air opening as specified for an open-flued appliance, of similar input, in BS 5440: Part 2: 1989.	2.5 times the minimum air vent area for a high level direct to outside air opening as specified for an open-flued appliance, of similar input, in BS 5440: Part 2: 1989.

Example 1

An appliance with an input of 45 kW ducted to low level vent inside the compartment.

450 mm² × 1.5 (see Table 20) of free area required per kW
450 × 1.5 × 45 = 30375 mm² free area required
If square duct is required $\sqrt{30375}$ = 174.28
175 mm × 175 mm duct is required

If round duct is used 30375 = πr^2

$$r^2 = \frac{30375}{3.142}$$
$$r^2 = 9667.4$$
$$r = 98.3$$

diameter of round duct/pipe required is 196.6 or **200 mm**

Example 2

An appliance with an input of 55 kW is ducted to high level
450 mm² × 2.5 (see Table 20) of free area required per kW
450 × 2.5 × 55 = 61875 mm² of free area required
If square duct is used 248.74 or **250 mm × 250 mm duct required.**

If round duct is used 61875 = πr^2

$$r^2 = \frac{61875}{3.142}$$
$$r^2 = 19692.87$$
$$r = 140.33$$

diameter of round duct/pipe required is 280.66 or **280 mm.**

The products flue should be equal in area to the flue spigot on the appliance. The route of the products flue should follow the basic principles of individual open flues, with the air supply pipe/duct taking a similar route. The position of the terminal of a balanced compartment flue is not as critical as that of a normal open-flued appliance. However, when it is desirable to terminate the flue of a balanced compartment in a position not normally acceptable for an open-flued appliance, then confirmation that this position can be used must be sought from the appliance or flue manufacturers. The combustion air supply terminal should be below the height of the products flue terminal but by no more than 150 mm.

The compartment or enclosure must provide a room sealed effect when the appliance is in operation. To provide this effect the only openings into the compartment or enclosure, when the appliance is operating, must be through the combustion air and flue products openings. Any door into the compartment must be self-closing and fit flush onto a draught sealing strip fitted on the door frame. It is not permitted to have a door opening into a bathroom or any room containing a bath. There must be a notice on the door/access cover stating that it must be closed when the appliance is operating. The door should be fitted so that it operates an electrical isolating switch which shuts off the appliance when the door is open.

In order to keep the heat, transferred to the apartment, down to a minimum, the length of flue within the compartment and any pipework carrying hot water in the compartment must be insulated. Twin-wall flue, manufactured to BS 715 and insulating material not less than 19 mm thick should meet the requirements.

Balanced compartments are particularly suited to basement or cellar installations, or where it is otherwise difficult to obtain an adequate supply of combustion air. Figure 29 shows a combined flue and air inlet suitable for a sealed boiler house.

Fanned Draught Systems

Fans were first used for the removal of products of combustion from commercial and industrial appliances. Their use spread to shared flue systems in high rise flats where gas appliances were used for space heating and water heating. Finally, they were incorporated into individual domestic appliances and fitted to individual open flue systems.

A fanned draught system has the following advantages over a natural draught system.

- removal of the flue products is more positive and less dependent on wind conditions
- gives greater freedom in siting the appliance and its flue terminal
- flue outlet and air inlet ducts may be smaller
- can give higher dilution of products
- appliance heat exchangers may be made smaller and the overall weight and size of the appliance may be reduced

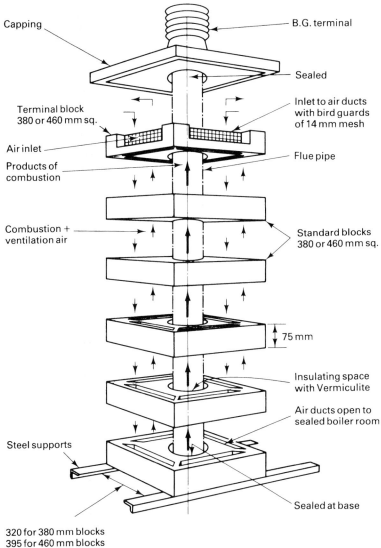

Capping

B.G. terminal

Sealed

Terminal block
380 or 460 mm sq.

Inlet to air ducts
with bird guards
of 14 mm mesh

Air inlet

Products of
combustion

Flue pipe

Combustion +
ventilation air

Standard blocks
380 or 460 mm sq.

75 mm

Insulating space
with Vermiculite

Air ducts open to
sealed boiler room

Steel supports

Sealed at base

320 for 380 mm blocks
395 for 460 mm blocks
These dimensions will not obstruct air ducts

(a) Exploded diagram

Fig. 29   Combined flue and air inlet: (a) exploded diagram; (b) Installation

- the thermal efficiency of the appliance can be improved by increasing the surface area of the heat exchanger and utilising the latent heat from the products of combustion as the water vapour condenses.

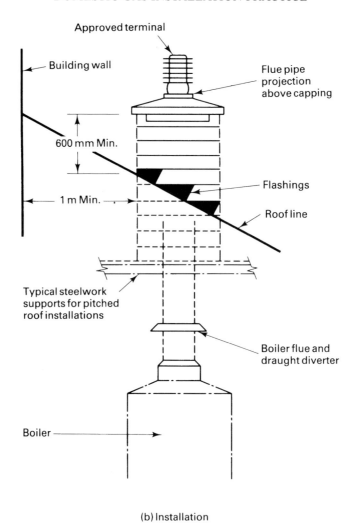

(b) Installation

The disadvantages are:

- the fan noise may be unacceptable
- additional safety devices may be required
- the system may be more expensive and can involve additional servicing operations
- when the thermal efficiency of the appliance has been increased by utilising the latent heat from the products of combustion, facilities must be provided to remove the resulting condensate.

Types of Fanned Draught Systems

Individual Fanned Draught Systems

There are three basic types of fanned draught flue systems:

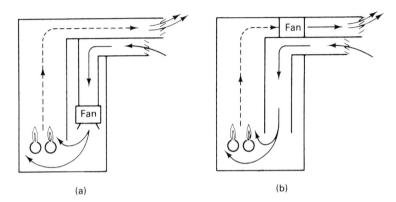

(a)   (b)

Fig. 30   Forced draught balanced-flue appliance: (a) fan at inlet;(b) fan at outlet

- balanced-flued appliance with fan in air inlet, Fig. 30(a)
- balanced-flued appliance with fan in flue outlet, Fig. 30(b)
- open or closed flued appliance with fan in secondary flue, Fig. 31.

Balanced-flued appliances with the fan in the inlet can be further sub-divided into two categories:

- fan delivering air only to the combustion chamber
- fan delivering air/gas mixture to the burner

Open-flued fanned draught systems can be divided into:

- normal dilution, with a CO_2 concentration in the discharged products of about 5%
- high dilution, with a CO_2 concentration of about 1%.

Balanced-flued Appliances

Every fanned draught system should, in principle, have two safety controls to monitor flame and combustion air or products flow failure respectively.

The siting of the terminal of the balanced-flued fanned draught appliance is not so critical as on a natural draught appliance and a clear expanse of wall around the terminal is not absolutely necessary.

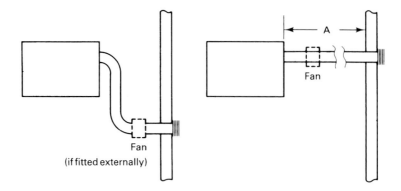

Possible flue paths using open flued appliances fitted with a fan in
the flue outlet or secondary flue. On some appliances distance 'A'
can be up to 6 metres

Fig. 31   Fanned draught open or closed flue appliance

However, positions where the products could create a nuisance should
be avoided. A terminal for a fanned draught balanced-flue boiler is
shown in Fig. 32.

With appliances having an inlet fan, the appliance casing should be
checked for leakage. Fans fitted on the outlet operate in the high
temperatures of the products and care must be taken over lubricating
the bearings.

Manufacturer's installation instructions must always be carefully
followed.

TABLE 21 Minimum dimensions of fanned draught balanced-flue terminal positions

Dimension (from Fig. 24)	Terminal Position	Distance mm
A	Directly below an opening, air brick, window etc.	300
B	Below gutters, soil or drain pipes	75
C	Below eaves	200
D	Below balconies or car port roof	200
E	From a vertical drain or soil pipe	75
F	From an internal or external corner	300
G	Above ground, roof or balcony level	300
H	From a surface facing a terminal	600
I	From a terminal facing the terminal	1200
J	From an opening in a car port (e.g. door or window)	1200
K	Vertically from a terminal on the same wall	1500
L	Horizontally from a terminal on the same wall	300

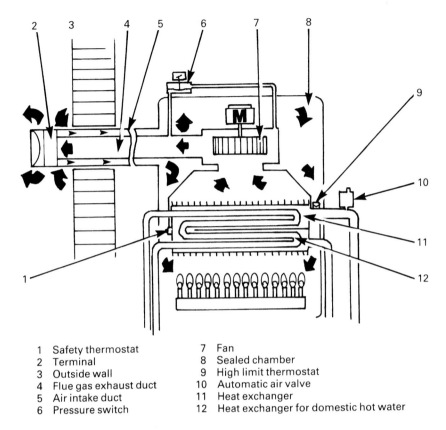

1	Safety thermostat	7	Fan
2	Terminal	8	Sealed chamber
3	Outside wall	9	High limit thermostat
4	Flue gas exhaust duct	10	Automatic air valve
5	Air intake duct	11	Heat exchanger
6	Pressure switch	12	Heat exchanger for domestic hot water

Fig. 32   Fanned draught balanced-flue and terminal

Open-flued fanned draught Appliances

A number of open-flued fanned draught appliances are currently available from certain manufacturers. The flue size, route etc. plus the size and type of fan and any additional controls, will have been determined by the appliance manufacturer. The manufacturer's instructions must be strictly adhered to when installing or servicing these appliances. In addition fanned draught units, designed for fitting to open flues, also give greater freedom in siting. It is possible to use wall terminals, except where very high winds are likely. The fan unit may be fitted anywhere in the secondary flue provided that it is accessible for servicing and not likely to be cooled so that condensation will occur.

The amount of air extracted mechanically from the room is usually less than that taken by a similarly rated appliance on a natural draught flue. If the room also contains other open-flued appliances, care must be taken to ensure that there is an adequate supply of ventilation air and that the fan pull does not have any adverse effect on the natural draught flues.

A fanned draught open-flue system must incorporate a safety control that will cut off the gas supply to the main burner if the flow in the secondary flue is insufficient to clear the products of combustion for a period of 6 seconds. The appliance will also incorporate a flame supervision device. When the safety control, monitoring the flow of products through the secondary flue, shuts off the supply of gas to the main burner a manual operation must be made to re-establish the supply.

The flow sensor is often a vane-operated switch but, alternatively, a thermostat may be used. The thermostat would be positioned near the lower rim of the draught diverter. When a reduced flow through the flue caused spillage at the diverter, the thermostat would heat up and switch off the gas supply.

Fig. 33   Fanned draught unit

As with the balanced-flued appliances, the maker's instructions must be followed closely. Permissible flue lengths vary with the appliance and fans are sized to give a particular performance under the conditions specified.

The size of the fan must take into consideration the volume of the products of combustion to be removed, plus the route and size of the flue. The products of combustion must be fully cleared even against adverse wind conditions. If the flue terminal is located in a position that is acceptable on a natural draught open flue (see Table 15) then the fan size is calculated against an adverse wind pressure of 0.15 mbar. When the terminal position is not as specified for a natural draught open flue then the fan size is calculated against an adverse wind pressure of 0.75 mbar.

The flow rate required to clear the products of combustion varies according to the type of appliance. Table 22 gives the concentration of carbon dioxide and the minimum flow rates that will normally give clearance of products of combustion from specific types of appliances, in all but exceptional circumstances.

TABLE 22 Minimum flow rate for fanned flues

Type of Appliance	Maximum % of CO_2 concentration	Minimum flue flow rate in m^3/h per kW input
Gas Fire	1	9.7
Gas Fire and Back Boiler Unit	2	4.9
Decorative Fuel Effect Gas Appliances	0.3	32.4
All other appliances (except incinerators)	4	2.4

High dilution systems or 'fan-diluted flue systems', as they are also called, are used principally for commercial installations. They are installed when it is impracticable to use a natural draught flue and when a large, high rated appliance must discharge its flue products at low level. The Local Authority will generally want to be involved in the design of the system to ensure that no nuisance is created.

The system is shown in Fig. 34. A short flue from the appliance rises to join a horizontal duct. Fresh air is drawn into the duct by a fan and mixed with the products of combustion from the flue. The mixture is then discharged to outside atmosphere. Both inlet and outlet should preferably be on the same wall.

A damper is usually fitted in the diluent air duct. It is adjusted so that no spillage occurs at the draught diverter and then locked in position.

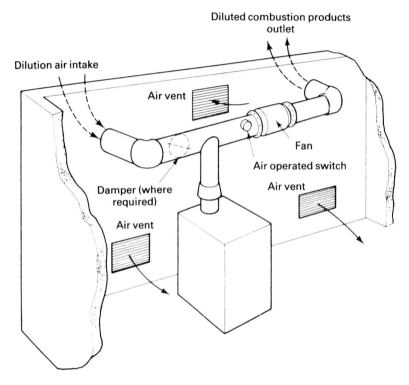

Fig. 34   Fan diluted system

The temperature of the mixture of air and products is relatively low, about 49° C.

As with the normal dilution systems, a flow monitoring device, usually an air flow switch, is required.

Condensation in a high dilution system is extremely unlikely. On normal dilution systems, however, the possibility is the same as for natural draught flues and tables of maximum condense-free lengths should be consulted.

The fan must be sized to deal with the total volume of fresh air and products of combustion. This 'design volume' flow rate can be easily calculated for any percentage of CO_2 required.

As a rough guide, the design volume flow rate (Q) in cubic metres per hour, is approximately 10 times the appliance rated input in kilowatts for every 1% CO_2 or

$$Q = \frac{10}{\%CO_2} \times kW \text{ appliance input } m^3/h$$

For example,

Boiler, 100 kW rating. 5% CO_2, required.

$$Q = \frac{10}{5} \times 100 \ m^3/h$$

$$= \ \mathbf{200 \ m^3/h}$$

The actual formula, if required, is:

$$Q = \frac{9.5}{\%CO_2} + 0.2 \quad \times \ \text{appliance rated input in kW}$$

For example,

Boiler rating 100 kW
1% CO_2 concentration required.

$$Q = \frac{9.5}{1} + 0.2 \quad \times \ 100 \ m^3/h$$

$$= 9.7 \times 100 \ m^3/h$$

$$= \ \mathbf{970 \ m^3/h}$$

When fitted in a secondary flue the fan should be capable of delivering the design volume flow rate against a static pressure of the total system pressure loss plus:

- 0.25 mbar for a roof termination
- 0.5 mbar for a wall termination.

Shared Fanned Draught Systems

There are three shared flue systems to which fanned draught may be applied. They are:

- common flue systems
- branched flue systems
- high dilution systems.

Shared flues are dealt with in the next section. The high dilution shared flue, Fig. 35, is a variation of the common flue system where two or more appliances in the same room share a common secondary flue. It operates in the same way as the individual fan-diluted system. The damper in the diluent air supply should be set so that no spillage occurs at the draught diverter of the appliance farthest from the fan.

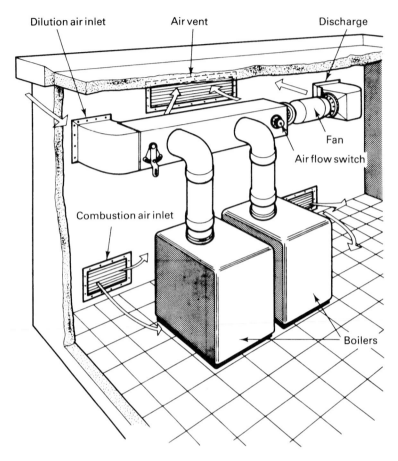

Fig. 35   High dilution shared flue

Shared Flue Systems

The types of shared or multiple flue systems are:

- open flued appliances   common flues
  branched flues
- room-sealed appliances   Se-duct
  U-duct

Common Flues

Where several open-flued appliances are installed in the same room, or enclosure, they may be connected into one main/common flue. Each appliance should have a flame supervision device and a draught diverter. There should be a minimum of 600 mm of vertical secondary flue height on each appliance, before it joins the common flue, Fig. 36.

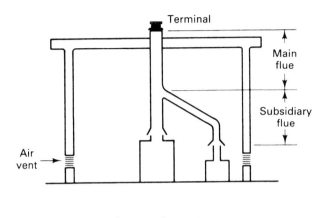

Common flue system

Fig. 36   Common flues

The common flue must be of sufficient size to accept the total sum of all the appliance outlets. In all other aspects, the considerations of components, route and construction are similar to those for individual open-flued systems.

Branched Flues (e.g. appliances installed on different floors of a building – flats, offices etc.)

These may be used where open-flued appliances are situated in different rooms, where these rooms all have the same aspect for windows and ventilation openings. For example, on successive floors of blocks of flats or offices, Fig. 37. Appliances must have flame supervision devices.

The subsidiary flue or limb provides the flue pull for each appliance to clear its products into the main duct. The minimum height of the subsidiary flue should be 3 m for a gas fire and 1.2 m for other appliances, except for incinerators, which need specialist consideration.

Subsidiary flues branch into the main flue which must be fitted internally. The main flue must run the total height of the installation and terminate in free air. Where the flue passes through a pitched roof the terminal must be fitted above the level of the ridge. In the case of the flue terminating above a flat roof the terminal must not be less

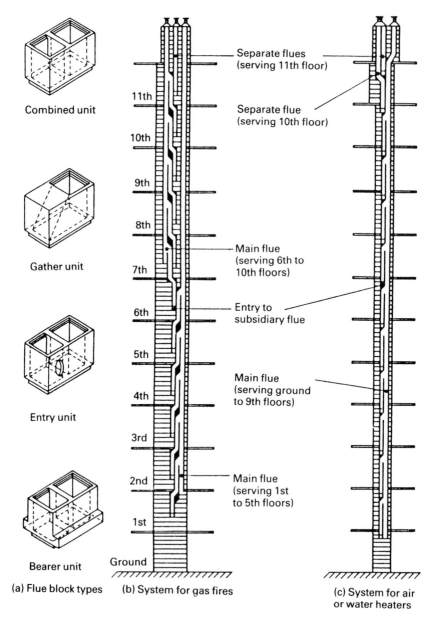

Combined unit

Gather unit

Entry unit

Bearer unit

(a) Flue block types

Separate flues
(serving 11th floor)

Separate flue
(serving 10th floor)

Main flue
(serving 6th to
10th floors)

Entry to
subsidiary flue

Main flue
(serving ground
to 9th floors)

Main flue
(serving 1st
to 5th floors)

11th
10th
9th
8th
7th
6th
5th
4th
3rd
2nd
1st
Ground

(b) System for gas fires

(c) System for air
or water heaters

Fig. 37   Branched/Shared flue system and flue block types

than 600 mm above the roof or any parapet that is within 1.5 m of it.
Where there is a structure within 10 m, measured horizontally, of the

terminal, then it must be fitted at a level corresponding to the height of the structure but reducing by one third of the difference between the measured distance and 1.5 m.

Where two or more appliances (other than incinerators) are installed in different storeys of a building and are served by a main flue, this flue must have a minimum cross-sectional area of 40,000 mm^2. All of the appliances being served by a main flue must be of the same type. The maximum number of appliances and their total input must not exceed those specified in Table 23 and Fig. 37.

TABLE 23 Appliances discharging into a shared main flue by way of subsidiary flues

Type of appliance	Nominal cross-sectional area of main flue			
	Between 40,000 mm^2 minimum and 62,000 mm^2		62,000 mm^2 and above	
	Maximum No. of appliances	Total input rating	Maximum No. of appliances	Total input rating
Gas fire	5	kW 30	7	kW 45
Instantaneous water heater	10	300	10	450
Storage water heater, central heating unit or air heater	10	120	10	180

There should be no offsets in the main flue and it must not be inclined at an angle greater than 10° to the vertical. Subsidiary flues may be angled to make the connection into the main flue but this angle must not exceed 45° to the horizontal.

To ensure its correct functioning, the top appliance should be at least 6 m below the terminal. This often means that it has to be vented separately.

The materials for construction are similar to those for other open flue systems. Pre-cast concrete blocks or metal flue pipes are preferred. Condensation is extremely unlikely.

The Se-duct and U-duct shared room-sealed, natural draught flue systems described in the next two sections are not suitable for appliances burning third family gases.

Se-duct

As its name implies, this was developed by the then South Eastern Gas Board. The system allows room-sealed appliances, in multi-storey dwellings, to receive air for combustion from a common duct and to discharge products of combustion to the same duct.

The arrangement is shown in Fig. 38. Two horizontal air inlets are taken from opposite sides of the building and connect into the base of a single, vertical duct rising through the centre of the building and discharging above roof level.

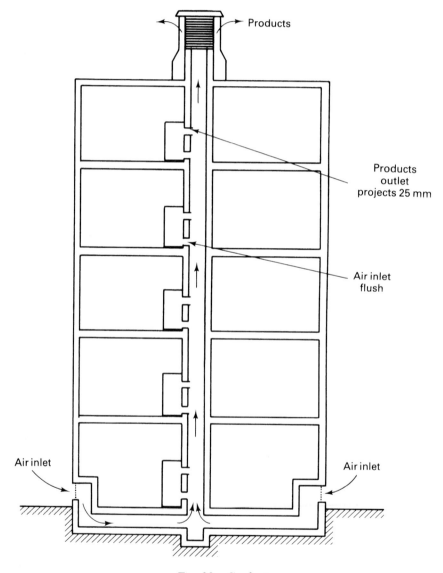

Fig. 38   Se-duct

The appliances, which are adapted versions of balanced-flued appliances, are connected into the vertical duct, on each floor.

The vertical duct is normally made of pre-cast concrete sections in standard sizes and heights. It may also be constructed from fire resistant board of suitable thickness and strength. Floor bearing units are used to spread the load imposed by the weight of the duct. The air inlet limbs may be constructed of similar materials, cast in situ, or formed from metal. Buildings supported on columns may not need horizontal air inlets.

The appliances are connected to the Se-duct through pre-formed holes or holes cut in situ. The appliance spigots fit into the holes and must be carefully sealed. The air inlet spigot should be flush with the inner face of the duct whilst the outlet should project the distance specified by the appliance manufacturer.

Terminal design has been standardised. It should be located above the line of any parapet or structure within 1.5 m. The base of the opening must always be at least 250 mm above the roof.

U-duct

This is a variation of the Se-duct and is used where it is not easily possible to obtain combustion air from the base of the building. Air is brought down from the top of the building in a vertical duct, adjacent to the duct which vents the appliance. The two ducts are joined at their base to form a 'U' shape, Fig. 39.

The size of the duct is slightly larger than that of a Se-duct, the area of each limb being 1.4 times that of a corresponding Se-duct.

Terminals for U-ducts have been standardised and their location should be considered as for Se-ducts.

General

Both Se-ducts and U-ducts may be used for a single appliance, where suitable.

Neither single, nor multiple appliance Se-ducts and U-ducts may be used with fanned draught operation.

Condensation is extremely unlikely to occur in Se-ducts or U-ducts and no precautions are necessary.

The sizing of branched flues, Se-ducts and U-ducts is critical and is normally a specialist activity, during the design stage. Tables of the sizes of branched flues and Se-ducts for various appliance combinations are given in BS 5440, Part 1: 1990.

Ventilation

Every gas burning appliance needs fresh air for combustion. Open-flued and flueless appliances take their air from the room in which

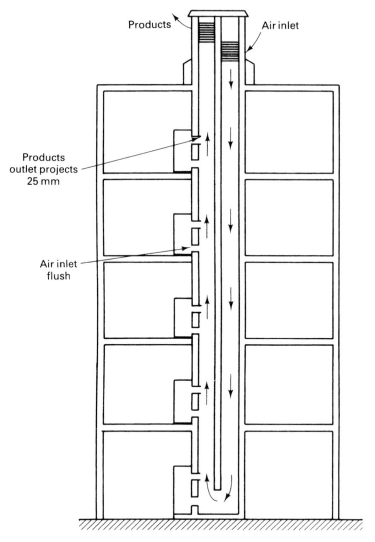

Products

Air inlet

Products
outlet projects
25 mm

Air inlet
flush

Fig. 39   U-duct

they are fitted, so an adequate supply of fresh air must be provided. In addition, where any appliance is fitted in a purpose-built compartment, allowance has to be made for ventilating the compartment.

As well as providing ventilation for appliances, fresh air is also needed for people. At rest an adult breathes in about a quarter of a litre of air per second (0.24 litres/s). On top of this, additional air change is required to prevent vitiation and noticeable body odour. Usually about 4 litres/s in total is needed per person.

Air enters any building through cracks around doors and windows. Suspended wood floors admit air and some even percolates through the fabric of the building. Recent work at Watson House has shown that even with up-to-date methods of insulation and weather stripping it is virtually impossible to reduce the area of air openings into habitable rooms to below 3500 mm^2. Air entering through this area is called 'adventitious ventilation'. The ventilation figures which follow make allowance for this adventitious ventilation as recommended by BS 5440: 1989. Installation of flues and ventilation for gas appliances of rated input not exceeding 60 kW (first, second and third family gases).

When air vents are fitted to communicate directly with outside air, the following points must be borne in mind:

(i)   When fitting air vents into a cavity wall a continuous duct of adequate size must be fitted across the cavity. Precautions must be taken to ensure that this duct does not transfer water to the inside wall.

(ii)   Where an air vent is fitted above a balanced flue terminal it must be at least 300 mm from the terminal. This applies to natural draught and fan assisted balanced-flue appliances.

(iii)   Where an air vent is fitted near an open flue terminal the air vent must be at least 600 mm from any part of the terminal on a natural draught appliance and at least 300 mm from any part of the terminal on a fan-assisted appliance.

When fitting air vents that communicate with the outside air they should not be installed in a position where they can become easily blocked e.g. by leaves or snow etc. or where they may draw in contaminated air e.g. from car exhausts etc. They should also be sited so that they do not produce draughts within the room as these will create uncomfortable conditions for the occupants of the room.

With air vents that do not communicate directly to outside air the following points must be borne in mind:

(i)   The room or internal space that the air vent communicates with must communicate with the outside air. Where two air vents are fitted in series the resistance to the flow of air is small enough to be acceptable without increasing the size of the air vent. However, when three or four air vents are fitted in series the free area of each air vent shall be increased by 50% in order to overcome the resistance to the flow of air.

(ii)   An air vent fitted in an internal wall (not a compartment wall) must be no more than 450 mm above the floor. This will help to reduce the spread of smoke in the event of fire.

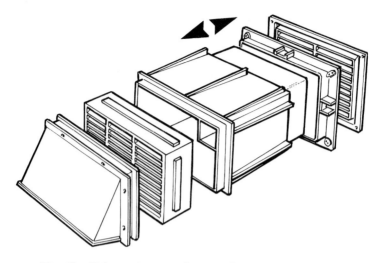

Fig. 40   Telescopic cavity liner and air vents (Stadium)

(iii)   No air vent must communicate with a roofspace or under-floor space if the space communicates with other premises. Where air vents do communicate with roof spaces due consideration should be taken regarding the problem of condensation and the vent must be located in a position that will prevent blockage by insulating material.

(iv)   No air vent should penetrate a protected shaft.

(v)   No air vent supplying air to an open-flue appliance must communicate with any room or internal space that contains a bath or shower.

Fig. 40 shows a ventilation set with a telescopic liner which enables the installer to make adjustments to fit wall and cavity widths and to accommodate building differences and irregularities.

Fig. 41 shows a ventilator set that can be installed with the help of a core drill, the interior and exterior wall back plates cover any inadvertent surrounding damage to wallpaper, plaster or brickwork.

Existing air vents should also be taken into consideration when assessing ventilation requirements.

Flueless Appliances

The following appliances do not require any specific provision for room ventilation:

- refrigerator
- gas heated towel rail

- airing cupboard heater
- boiling ring

More than one boiling ring in a room should be regarded as a hotplate, for which some ventilation may be necessary.

TABLE 24 Room air vent requirements. Flueless appliances

Minimum free area requirements (in mm²) of permanent openings direct to outside.

Type of appliance	Maximum rated input limit	room volume (m³)				Openable window or equivalent also required**
		< 5	5–10	10–20	> 20	
Domestic oven, hotplate, grill or any combination	none	10000	5000*	nil	nil	yes
Instantaneous water heater	12 kW	Installation not permitted	10000	5000	nil	yes
Tumble drier	none	10000	10000	nil	nil	yes
Fixed space heater in a room	50W/m³ of heated space	10000 plus 5000/kW by which the rated input exceeds 3 kW				yes
Fixed space heater in an internal space	100W/m³ of heated space	10000 plus 2500/kW by which the rated input exceeds 6 kW				yes
LPG space heaters complying with BS 5258: Pt. 10 or Pt. 11, in a room	50W/m³ of heated space	5000 plus 2500/kW by which the rated input exceeds 2 kW				yes
LPG space heaters complying with BS 5258: Pt. 10 or Pt. 11, in an internal space	100 W/m³ of heated space	5000 plus 1250/kW by which the rated input exceeds 4 kW				yes
Refrigerator	none	nil				no
Boiling ring (2 or more shall be treated as a hotplate)	none	nil				no

* If the room or internal space containing these appliances has a door which opens directly to outside, no air vent is required.

** Alternative acceptable forms of opening includes any adjustable louvre, hinged panel or other means of ventilation that opens directly to outside air.

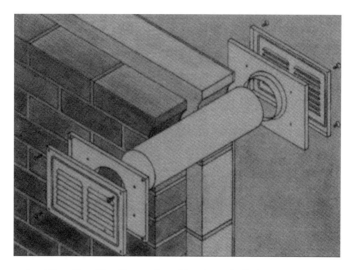

Fig. 41   Ventilator fitted by core drill (Stadium)

The remaining types of flueless appliances require at least an openable window. The area of additional ventilation required will depend on the input rating of the appliance and the volume of the room in which it is fitted. Table 24 gives the minimum areas required.

Open-Flued Appliances

An open-flued appliance with an input rating of not more than 7 kW does not require an air vent.

An appliance with an input rating of more than 7 kW requires a vent with a minimum effective area of 450 mm^2 for every 1 kW in excess of 7 kW, fitted in the room or space containing the appliance. The formula to find the ventilation requirements for an open-flued appliance is: (Heat input in kW − 7 kW) × 450 mm^2 = Effective ventilation area in mm^2. The vent should be either direct to outside air or to an adjoining room or space which is vented directly to outside.

When an open-flued appliance is fitted in a compartment, air vents are required at high and low levels, Fig. 42. The size of the vents is given in Table 25.

The compartment must also be labelled in order to warn against the dangers of blocking the air vents.

These vents allow for combustion air and also for cooling the compartment. They should preferably be to outside air, when their size may be reduced.

If the vents open into a room, that room must be provided with an air vent.

TABLE 25 Open-Flued Appliances, minimum air vent free area for compartments

| Position of vent | Compartment ventilated: | |
	to room or internal space	direct to outside air
	mm² per kW of appliance maximum rated input	mm² per kW of appliance maximum rated input
High level	900	450
Low level	1800	900

Room-Sealed Appliances

A room-sealed appliance does not require an air vent in the room or space containing the appliance.

When a room-sealed appliance is fitted in a compartment, air vents are required at high and low levels, Fig. 43. The size of the vents is given in Table 26.

The vents allow simply for cooling the compartment. They should preferably be to outside air, when their size may be reduced. If, however, they open into a room, there is no need for the room to be vented.

Compartment vents, whether for room-sealed or open-flued appliances should be either both into the same room or both on the same external wall.

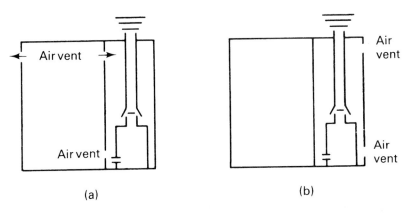

Fig. 42   Open flued appliance, air vents in compartment: (a) to inside; (b) to outside air

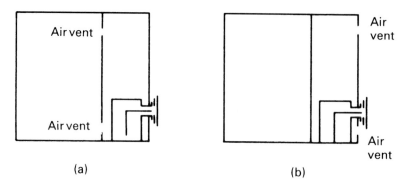

Fig. 43   Balanced flue appliance, air vents in compartment: (a) to inside; (b) to outside air

The vents should be sized on the maximum rating of a range-rated appliance or on the total rating of all the appliances when more than one is fitted in the compartment. A compartment which contains a balanced-flued appliance must be labelled to warn against the dangers of blocking the air vents.

TABLE 26 Room-Sealed Appliances, minimum air vent free area for compartments

Position of vent	Compartment ventilated:	
	to room or internal space	direct to outside air
	mm² per kW of appliance maximum rated input	mm² per kW of appliance maximum rated input
High level	900	450
Low level	900	450

Multiple Appliances

When more than one flued and/or flueless appliance is fitted in the same room, the air vent requirement is the largest of the following:

- the total rated input of flueless space heating requirements
- the total rated input of open-flued space heating appliances, except where any interconnecting wall between two rooms has been removed leaving two similar chimneys fitted with similar gas fires, then no air vent is required if the total rated input of the fires does not exceed 14 kW
- the largest individual rated input of any other type of appliance

If the room contains oil or solid fuel burning appliances they should be treated as gas appliances of similar type and rating. For a solid fuel open fire or small closed stove of unknown input, the free area of the required air vent shall be taken as 2500 mm^2. If an oil or solid fuel burning appliance only shows the rated output, then the input can be calculated by using the following formula:

$$\text{Input} = \frac{\text{Output} \times 10}{6}$$

Typical terracotta wall ventilator

The unobstructed fraction is about ⅓

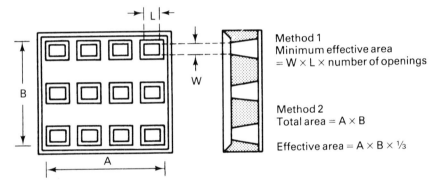

Method 1
Minimum effective area
= W × L × number of openings

Method 2
Total area = A × B

Effective area = A × B × ⅓

Typical sheet metal ventilation grille

Unobstructed fraction is about ⅔

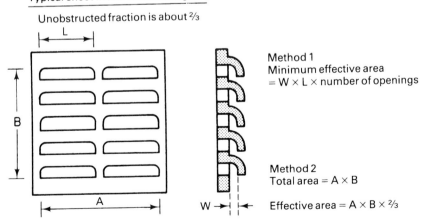

Method 1
Minimum effective area
= W × L × number of openings

Method 2
Total area = A × B

Effective area = A × B × ⅔

Fig. 44   Estimating effective area of vents

Construction of Air Vents

The materials which may be used to make air vents are specified in BS 493. Other materials may also be used provided that the vent is corrosion resistant and possesses adequate strength and durability. The most important factors are the size of the vent and its acceptability to the customer.

Proprietary vents are generally marked with their effective area by the maker. Alternatively, the maker should be able to supply this information. Failing this, a reasonable approximation can be made by:

- measuring the size of one of the openings, Fig. 44
- calculating its area
- multiplying by the number of openings.

As a guide, the area of the openings is approximately:

- $1/3$ of the total area of a terra cotta ventilator
- $2/3$ of the total area of a sheet metal ventilator.

The effective area may be estimated by multiplying the area of the ventilator by the appropriate fraction.

The openings in the air vent should permit the entry of a 5 mm diameter ball, but prevent entry of a 10 mm ball. They should disperse the flow of incoming air so as to avoid direct draughts in the room. Vents should not be fitted with any additional gauze or screen.

Installation of Air Vents

Air vents should be positioned so that air currents do not pass through areas of the room in which the occupants are normally sitting or standing. Air from outside should be entrained at a point near to the appliance and ceiling level vents can prevent draughts at head level. Figure 45 shows alternative methods of providing ventilation.

Direct Ventilation

An air vent in an outside wall should be located:

- not less than 600 mm from any part of a natural draught open flued terminal or not less than 300 mm from a fanned draught open flued terminal
- not less than 300 mm directly above any part of a balanced-flue terminal.

The joints between the vent and the wall must be adequately sealed to prevent water penetration.

Where a vent crosses a cavity wall it must be housed in a suitable duct which will not weaken the weather resistance of the cavity.

Indirect Ventilation

An air vent in an inside wall should be located:

- not more than 450 mm above floor level, so as to reduce the spread of smoke in case of fire
- not into the wall of any room containing a bath or shower.

Where two vents are in series the additional resistance is very small and there is no need for their area to be increased. However, when three or more vents are in series, the area of each vent should be increased to at least 50% more than the area required for a single vent.

Testing Open Flues

Inspection

When an appliance is installed or serviced, the flue should be visually inspected to ensure that it is:

- complete and unobstructed
- serving only one room or appliance
- fitted where necessary with an approved terminal that is correctly sited
- installed with a weathertight joint between the terminal and flue system
- free of any dampers or restrictor plates that should have been removed or permanently secured to leave the main part of the flue unobstructed.

Testing for Spillage

When an appliance is installed, or serviced, the flue should be tested for spillage. This shows whether the flue is operating satisfactorily and if the ventilation is adequate.

The test should be made using one of the following methods:

- hold a smoke match about 3 mm up inside the skirt of the draught diverter or gas fire canopy
- blow smoke from a smoke puffer just below the outer edge of the draught diverter or gas fire canopy.

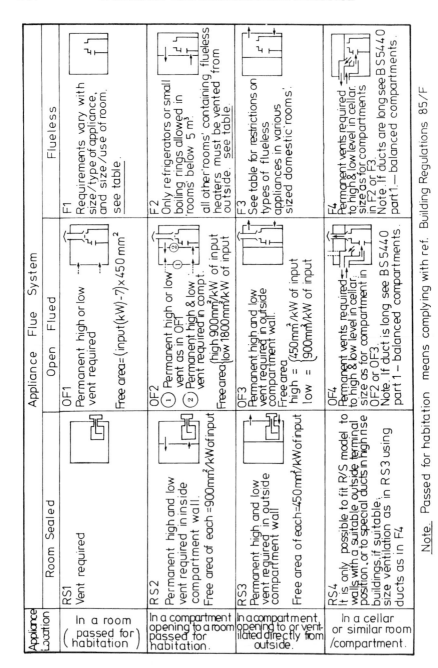

Fig. 45   Vent requirements on appliance flue systems

In all cases, spillage is indicated by the smoke being blown out-wards.

Where possible the test should be made all round the edge of the draught diverter or the canopy of a gas fire.

The procedure for carrying out the tests is as follows:

- close all doors and windows in the room containing the appliance
- light up the appliance
- after 5 minutes, test for spillage.

If spillage occurs, check to establish whether it is momentary, when it may be ignored, or due to the flue still being cold, when the test should be repeated after a further 10 minutes.

If spillage persists, check the flue and the air vent and correct any faults. An appliance should not be left connected to the gas supply unless it has passed the spillage test.

Testing the Effect of Extract Fans

Where fans are installed in the same, or an adjacent room of a premises which contains any type of open-flued appliance, their operation may affect the performance of the appliance flue.

Examples of fans that may affect the performance of a flue, are:

- room extractor fans
- fans in the flues of open-flued appliances
- fans in cooker hoods
- circulating fans of warm air heating systems (any type of fuel).

If the air inlet to the room in which the appliance is fitted is not adequate, spillage of products will occur at the appliance draught diverter or canopy.

Where this possibility exists, the following tests should be applied:

- carry out a test for spillage as detailed in the previous section.

If spillage occurs when the fan is not running or is isolated from the appliance, it indicates a fault in the flue or inadequate ventilation. The fault must be corrected before the test can be continued.

If there is no spillage with the fan not running, continue the tests as follows:

- open any doors connecting the room containing the appliance and the room containing the fan
- close all other doors and all windows
- switch on the fan

- test for spillage.

If spillage occurs, gradually open a window in the room containing either the appliance or the fan until spillage ceases. Estimate the area of the opening.

If the area is more than 6000 mm^2, specialist advice should be sought. When any additional vent has been installed the tests should be repeated.

As a general guide an extra air vent of up to 5000 mm^2 free area, fitted in either the same room as the appliance or the fan, will be sufficient to overcome the problem.

Comfort Ventilation

In Vol. 1, Chapters 5 and 10, comfort conditions were discussed in some detail. It became clear that comfort depended not only on heating, but also on ventilating. An adequate air change is required to:

- carry away excessive water vapour so that the relative humidity remains low
- provides fresh air for the occupants to breathe
- remove odours and air-borne bacteria
- provide air movement to give a feeling of freshness
- remove tobacco smoke and cooking vapours.

In domestic premises the recommended air change is two per hour in the habitable rooms. This is equivalent to about 5 to 6 litres/second per person for a family of four in an average room. It compares with the minimum of 6 litres/s recommended for persons with light sedentary occupations in a factory.

There is usually no difficulty in providing the required air change in domestic premises. Indeed, the problem is often that of preventing excessive ventilation with its attendant cold draughts and heat losses. This is particularly true in rooms heated by solid fuel fires discharging into unrestricted open brick chimneys. Air changes of eight per hour are easily obtained if doors and windows are not effectively weather stripped.

Heat loss from ventilation was referred to in Vol. 1, Chapter 10. It is further discussed in Vol. 2, Chapter 10.

Work to determine the full extent of adventitious ventilation and to devise methods by which it may be controlled, is continuing in research laboratories.

Whilst it may never be practicable to apply these methods completely effectively to older properties, new building methods and materials should result in a considerable saving in energy.

Gas Safety (Installation and Use) Regulations 1998

These new regulations were introduced on 31 October 1998.
Regulation 30 covers room-sealed appliances.

Reg. 30 (1) states that only a room-sealed appliance may be fitted in a room intended to be used as a bath or shower room.

Reg. 30 (2) states that in a room intended for use as sleeping accommodation, no appliance over 14 kW heat input may be installed unless it is a room-sealed appliance.

Reg. 30 (3) states that if an appliance of 14 kW heat input or less is fitted in a room intended to be used as sleeping accommodation it must

(a) be a room-sealed appliance; or

(b) incorporate a safety control designed to shut down the appliance before there is a build up of a dangerous quantity of products of combustion (an anti-vitiation device).

Note: Regulation 30 (2) and (3) came into force 1 January 1996.

CHAPTER 6

Cookers, Refrigerators and Laundry Appliances

Chapter 6 is based on an original draft by A. Pawsey

COOKERS

Introduction

Since its introduction about 150 years ago, the gas cooker has evolved to become one of the most important and most complex of domestic appliances. Originally a very simple contrivance, it gradually replaced the solid fuel cooker at a time when electricity was superseding gas for lighting. So, by developing the cooker, the Industry sustained and expanded the gas load and changed from being a supplier of lighting to become a supplier of heat. In recent years, increasing competition has accelerated the evolution of the cooker even more rapidly and this trend is likely to continue.

The gas cooker usually consists of three cooking units:

- hotplate
- grill
- oven.

These may be obtained separately or as one complete appliance. The trend in modern house building towards large, better appointed kitchens has encouraged the development of separate and built-in/slide-in units.

Developments in design have concentrated on:

Hotplates

- ease of cleaning
- improved automatic ignition
- improved simmering control.

Grills

- high level, under hotplate and situated in oven
- even heat distribution over large areas
- rotisseries, snack turners and kebabs
- self-supporting pans.

Ovens

- automatic ignition with flame supervision and re-ignition
- automatic cooking control
- oven lights and viewing windows
- catalytic oven linings (easy-clean).

Generally more attention is being paid to styling, additional features and electronic devices.

Cooking Processes

The basic cooking processes use heat to bring about changes in the food.

In meat, the connecting tissue between the fibres is broken down at temperatures above 72° C and the meat becomes tender. High quality meat may be roasted at high temperatures, about 220° C. The cheaper, tougher cuts require lower temperatures, about 160° C for longer periods.

Many vegetables can be eaten raw or lightly boiled, at 100° C to break down the cells. Root vegetables may need prolonged boiling or steaming under pressure for shorter periods.

Stews and casseroles are cooked for longer periods to soften the meat and vegetables and to transfer the flavours of the vegetables, herbs and wine into the meat.

When cakes are baked, the gluten in the flour liquefies and, at the same time, bubbles of steam and carbon dioxide form in the mixture. The cake rises. As cooking progresses the gluten sets and locks in the pockets of gas which give the cake its texture.

Frying temperatures range from 150 to 195° C.

The food is heated in different ways in each of the three cooking units and each unit is designed for particular cooking processes.

Hotplates

These are used for:

- boiling
- simmering (cooking at just below boiling point)

- frying
- pot roasting
- braising
- steaming.

Some hotplates incorporate a 'griddle', that is, a flat steel plate, heated from below. This is used for dry-frying of bacon, eggs, hamburgers or for drop-scones or pancakes.

Hotplate burners transfer their heat by conduction and some radiation from the flame. They have high primary aeration to ensure complete combustion and to allow the pan to be placed close to the burner head. The closer the pan to the burner, the greater the amount of heat that is supplied. However, heat is wasted if flames are allowed to lick up the side of a pan. To avoid this, use large pans on large burners. The present tendency is to provide burners with slightly lower heat inputs to minimise indoor pollution.

Various types of solid hotplates have been designed and some are used commercially. A new form, produced for domestic cookers on the Continent, uses a sheet of ceramic material heated from below by thermostatically controlled surface combustion plaques. Each burner has a flame supervision device and the combustion products are vented by a flue.

A recent innovation by a UK manufacturer has been the introduction of ceramic discs fitted above the burner. The heat passes through the disc and around its edges to heat the contents of the pan above. The manufacturers claim a 'streamlined' look and an easy-clean surface.

Grills

Grills are used for:

- toasting
- grilling
- browning off or glazing food already cooked elsewhere.

Browning food is called 'salamandering' in a commercial kitchen. Although carried out under a gas fired grill it was originally done by a 'salamander' which was a circular iron griddle made red-hot in the fire.

A grill cooks by radiation from a 'fret' above the flame on to the food below. Since heat is radiated in proportion to the fourth power of the absolute temperature of the source, the fret is heated to incandescence at as high a temperature as possible. Frets have been made from cast iron, ceramics or, more commonly, expanded heat-proof steel mesh.

Ovens

Ovens are used for:

- roasting
- baking
- casseroles or braising (after sealing off the meat on the hotplate)
- baked puddings and custards.

They may also be used for warming plates or keeping food hot at temperatures of 110 to 120° C.

Table 27 Cooking Guide

Cooking Process	Oven Heat	Gas Mark	Approximate Central Oven Temperature °C
Plate warming	very cool	$^1/_4$	105
Keeping food hot	very cool	$^1/_2$	120
Milk puddings, egg custard	cool	1	135
Rich fruit cakes, braising	cool	2	150
Low temperature roasting, shortbread	moderate	3	165
Victoria sandwich, plain fruit cake, baked fish	moderate	4	175
Small cakes, choux pastry	fairly hot	5	190
Short pastry, swiss rolls, souffles	fairly hot	6	205
High temperature roasting flaky pastry, scones	hot	7	220
Puff pastry, bread	very hot	8	230
Small puff pastries, browning cooked foods	very hot	9	245

Ovens cook by convection. The traditional British oven is directly fired and heats by natural convection. The burner, mounted at the rear of the base, creates a circulation and the flue removes a controlled amount of air and combustion products, Fig. 1. Gaps are left at the back of the oven shelves for the circulating gases. No cooking utensils larger than those supplied by the maker should be used, to avoid interference with the circulation.

Because the circulation is relatively slow, it creates zones of temperature. The bottom of the oven is typically one 'gas mark' lower in temperature than the centre and the top of the oven is one mark higher. In a British oven, foods requiring different cooking temperatures may be cooked at the same time.

Continental and American ovens are indirectly heated by burners below the base, Fig. 2. Heat enters at various points and the temperature is even throughout the oven. Only foods requiring the same cooking temperature may be cooked together.

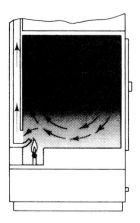

Fig. 1   Directly heated oven

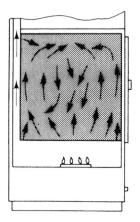

Fig. 2   Indirectly heated oven

Some ovens use forced convection to give more even cooking and increased efficiency. Early forms of 'fan convection ovens' simply had a paddle at the back of a normal oven to stir up the air. Later models have controlled air flow within the oven space.

The British practice of calibrating thermostats in 'gas marks' is unique. The mark is a more accurate setting than mere temperatures and is dependent on other factors as well. It gives an accurate guide to cooking rate so that food may be cooked in any gas oven at the same mark, even though the centre temperatures of different ovens may vary by as much as 56° C.

Although marks are not related to definite gas oven temperatures Table 27 gives an approximate guide to their central oven temperatures.

Design Standards

In the United Kingdom cookers and other appliances are designed to conform to certain British Standards e.g. BS 5258, 5386 and 6332 and are issued with a British Standard kitemark and/or safety mark. In other European countries appliances are designed to conform to the standards of that particular country. These different laws relating to product safety and other matters have caused trade barriers to be erected between member states of the European community. In 1985 Community Ministers agreed to tackle the problem of trade barriers by passing a series of Community Laws (Directives) to enable trade barriers to be lifted by the end of 1992. The law which applies to gas appliances is called 'The Gas Appliance Directive'.

Before the introduction of this Directive the main requirements for the safety of cookers sold in Britain had been covered generally by the 'Consumer Protection Act, 1987' and specifically by the 'Gas Cooking (Safety) Regulations, 1989'.

The Directive lays down the legal requirements that will apply throughout the European Community. Community members are required to amend or introduce new legislation to conform with the requirements of the Directive. In the United Kingdom the Directive will be covered by the 'Gas Appliance (Safety) Regulations, 1995'.

Gas Appliance (Safety) Regulations 1995 (GASR)

These Regulations cover the specific requirements for all gas appliances and must be followed before the appliance can be sold. There was a transitional period up to 1 January 1996 when appliances offered for sale in a Community country only needed to satisfy the requirements of that particular country.

The main provisions of the new Regulations are:

(i)   The appliance must be safe, it must comply with the requirements of the Directive. For cookers these requirements are detailed in the European standard EN 30, adopted as BS 5386: Part 4: 1991 Built-in domestic cooking appliances.
Part 3: 1980: (with amendments 1983/84/85/87/91/92:) Domestic cooking appliances burning gas, and
Part 6: 1991: Specification for domestic gas cooking appliances with forced-convection ovens.

(ii)   The appliance must be tested, it must be tested by a recognised authority (Notified Body). In the United Kingdom there are three such Notified Bodies, British Gas, Calor Gas and the British Standard Institution.

(iii)   The cooker must be quality guaranteed, the manufacturer must ensure that the cooker produced conforms to the tested design. Monitoring will be carried out by one of the Notified Bodies.

(iv)   The cooker must carry a CE mark issued under the authority of the Notified Bodies. The CE mark (Fig. 3) is followed by the last two figures of the year in which it was marked on the appliance together with the identification symbol of the relevant Notified Body, it will identify the cooker that conforms to the Regulations. Up to January 1996 there may be many cookers which meet the regulations and carry the BS kite or safety mark or even European certification marks. Even after 1 January 1996 the cooker could carry both marks denoting that it meets a standard beyond the Regulation requirements.

Fig. 3   CE mark

Domestic Cooking Appliances

There are three classes of appliances:

class 1 : Free-standing appliances

class 2 : Appliances for building in between two furniture units
        1 : Appliances made in one complete unit
        2 : Appliances consisting of one or more ovens or oven/grills, placed beneath the worktop and possibly, a hotplate built into the worktop

class 3 : Appliances for building into a kitchen unit or worktop.

Fig. 4   Free-standing cooker – High level grill (Stoves)

Fig. 5   Free-standing cooker – range (Cannon)

Fig. 6   Free-standing cooker – waist level grill (Stoves)

Free-standing

The ordinary cooker can be an oven with the hotplate above and a storage space or drawer below (Fig. 4). There is little difference in the sizes of hotplates and they generally have four burners. The grill is sometimes at high level and may be fitted with an electrically rotated rotisserie. Some models have the grill at waist level and this usually means that the storage space or drawer at the base of the cooker is omitted (Fig. 6). On other models, particularly continental, the grill is located inside, at the top of the oven.

The range shown in Fig. 5 is a free-standing cooker, it has one normal size oven and a smaller one, both with automatic control. This model also incorporates a griddle plate next to the hotplate burners.

Another model of a free-standing range has a six-burner hotplate and a single large oven.

Slide-in Appliances

These appliances are for fitting between two furniture units, they can be a complete cooker (Fig. 7) or an oven (with or without a grill), fitted under a working top, with a separate hotplate fitted above (Fig. 8). Some free-standing cookers can be fitted as slide-in cookers. (This information would be included in the manufacturer's instructions.)

Fig. 7   Slide-in cooker (Stoves)

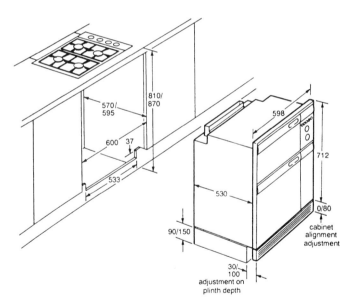

Fig. 8   Slide-in oven/grill with separate hotplate

Built-in Appliances

During the last decade these were the most rapidly expanding range of appliances; however the slide-in types are becoming just as popular. Up to about 1980, three units, the hotplate, the oven and the grill were produced independently, now more and more manufacturers are fitting the grill above or in the oven and few independent grills are available.

Hotplates are mainly four burner units although some two and three burner hobs are produced. Many incorporate a hinged lid manufactured from toughened glass (Fig. 9). Independent grills were designed to be fixed to a wall and could be obtained in open, fold-away or enclosed forms, Fig. 10 shows an open grill.

The greatest variety is in built-in ovens. Types available are:

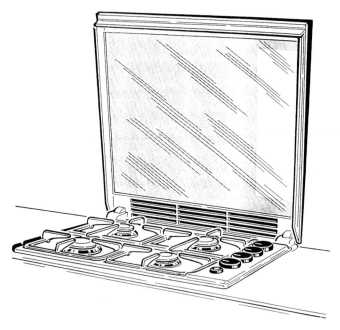

Fig. 9   Built-in hotplate (New World)

- single ovens
- double ovens
- ovens containing a grill
- ovens with a separate grill compartment (Fig. 11)
- fan convection ovens
- high level ovens
- ovens fitted with a microwave oven (electric) as a separate unit.

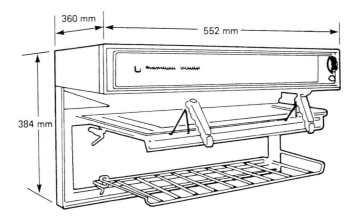

Fig. 10   Independent grill unit (Moffat)

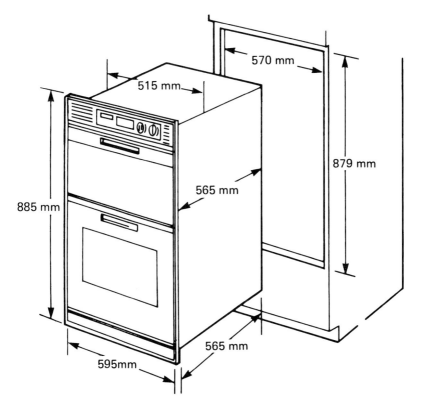

Fig. 11   Oven with separate grill compartment (Parkinson Cowan)

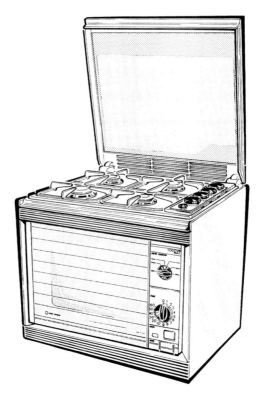

Fig. 12   Hob/microwave unit (New World)

Other Cooking Appliances

A few other appliances are made. These include a combined hotplate and microwave cooker (electric) installed as one unit (Fig. 12). The combined gas/electric cooker, usually a gas hotplate and electric oven and the gas/electric built-in hotplate is usually fitted with two gas burners and two electric hotplates.

A table cooker is an appliance intended to rest on a raised support or stand, it consists of:

- a hotplate having one or more burners and possibly a griddle
- an oven
- possibly a grill.

Another appliance, developed from the original solid fuel version, is the heat storage cooker, it is fired by a single burner and has two ovens, a solid hotplate and a means of producing hot water. Finally there is still available the older type of small hotplate which stands on a working surface.

Construction

The construction of cookers has changed as new production processes and materials have been developed. Continual development is essential to maintain the gas cooker in a competitive position with its electrical rival.

General Construction

Most cookers are made of mild steel panels which are welded, screwed or pop rivetted together. The steel is protected by vitreous or stoved enamel. Where appearance is not important the panels may be aluminium coated. A few free-standing cookers and most built-in ovens are built up on a chassis.

Hotplates

Hotplates generally use spreading flame burners; they can be all the same size or variable. Burners are sealed to the spillage tray to prevent spillage entering the underside of the hotplate. Some hotplates have a pair of pan supports, on others each burner has its own individual pan support. An example of a hotplate can be found on pages 118 and 119 in Vol. 1. The burners often have factory-set aeration and use retention flames for stability.

Pan supports vary considerably from the large, vitreous enamelled cast-iron type to the more popular small individual supports on each burner. These are often of thin section cast-iron or pressed steel.

Spillage trays have changed from the small individual lift-out bowls under each individual burner and are now a single tray with each burner sealed onto it.

Where hotplate taps are mounted on a traditional float rail, the injector for each burner is fitted onto its tap. With other arrangements the injectors may be mounted centrally in the hotplate.

Grills

The most common type of grill has a single burner at the rear, firing forward beneath an expanded metal fret. Combustion products are vented from slots near the front of the grill canopy, Fig. 13. Other designs include a double-sided burner, running centrally from front to back, with frets on both sides. Combustion products are passed back to a central point above the burner for venting.

The position of the burner in relation to the fret and the contour of the fret determine the effectiveness of the grill. Burner port sizes and the spacing of the ports are varied to achieve even heating.

An alternative form of grill is the surface combustion grill, using a natural draught radiant burner, Fig. 14. This has a metal gauze rather than the ceramic tiles used industrially. The gas burns on the surface of the gauze, producing an evenly heated radiant area. The burner heats up quickly and gives even heating when turned down.

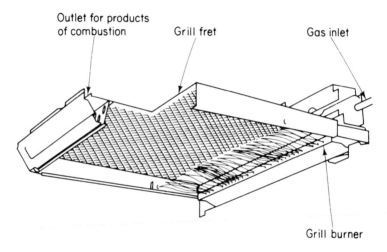

Outlet for products of combustion   Grill fret   Gas inlet

Grill burner

Fig. 13   Conventional grill

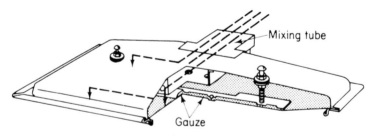

Mixing tube

Gauze

Fig. 14   Surface combustion grill

Ovens

Most ovens are insulated with aluminium foil or foil and glass fibre, held in place with wire. The oven itself is in the form of a shell, open at the bottom rear for the burner and its air supply and flued from the top. The oven flue has a break in it to prevent accidental blockage from affecting combustion. It vents above the hotplate on a free-standing cooker and above the oven on built-in types.

Oven doors must have a good seal. This is usually achieved by the door being held against a glass fibre or synthetic gasket by spring loading or an adjustable catch.

Some ovens have inner, self cleaning, catalytic linings. These are cleaned by heating at a special high setting, too high for cooking, at a temperature of 245° C.

Ovens may also be cleaned by 'pyrolitic cleaning', that is, heating to about 500° C. This technique has not yet been used on ordinary domestic gas ovens in this country.

Free-standing Cookers

The base of some free-standing cookers contain a storage compartment. It also gives access to the levelling feet, rollers or castors. The space often houses any batteries, fuses and ignition or control devices.

At the rear of the oven there is usually a cut-out at low level into which the 'stability bracket' locates (Fig. 20). This bracket prevents the cooker being tilted accidentally. The position of the gas connection is usually at hotplate level and to one side. It is normally $R^{1}/_{2}$. The oven flue is covered by a plate to prevent overheating the flexible connection.

Cookers which require an electrical supply are usually provided with a 2 m length of 3-core cable which enters the cooker behind the splash plate.

Built-in Cookers

Hotplates are generally developed from free-standing designs but the float rail and the taps are located at one side. The burners and controls are contained in a pan which projects through a hole in the worktop. The outer edge of the hotplate rests on the worktop with a silicone rubber seal to prevent spillage entering the cabinet. Clamps hold the hotplate firmly in position. Normal connections are $R^{1}/_{4}$.

Most independent grills are based on those fitted to free-standing cookers. They have mounting holes in the backplate and connections are $R^{1}/_{4}$ or 8 mm.

Built-in ovens normally rest on a shelf within the cabinet, on levelling feet. A full width fascia fronts the appliance and contains vents for combustion air and cooling air at the base and combustion products and cooling air at the top. Controls are usually located above the oven. The gas supply is usually $R^{1}/_{4}$ and located at the base.

A choice of left or right hand opening doors is common.

Controls

With the exception of automatic cooking controls most other devices are described and illustrated in Vol. 1, Chapter 11.

Taps

All hotplate and grill burners are controlled by safety taps which must be pushed in before they can be turned. Plug taps are most commonly used and are often fitted with moulded labyrinths or fine drillings and simmer position stops to provide easy control of turn down. All taps turn clockwise to close, including Continental designs. After opening, further turning on Continental taps reduces the gas rate.

Cookers with retractable grills have a further tap or cut-off device which prevents gas passing to the grill burner when the grill is retracted.

Hotplates with closable lids should be fitted with a device to prevent gas passing to the burners when the lid is closed. Fig. 15(a) shows the hotplate with the lid and valve open and (b) with them closed.

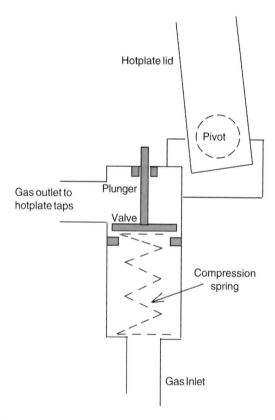

*Fig. 15(a)   Hotplate safety cut off valve, principle of operation
(hotplate raised, valve open)*

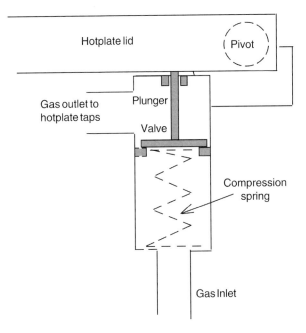

Fig. 15(b)   Hotplate safety cut off valve, princple of operation (hotplate lowered, valve closed)

A similar device is used when the grill is located in the oven. When the oven door is open, a valve controlling the gas supply to the grill is open and when the door is closed, the valve is closed. When testing for soundness on an appliance fitted with this type of device the hotplate lid or oven door should be open to test the appliance up to the burner control.

Regulators (Governors)

Regulators may exceptionally be found on some cooking appliances although the meter regulator is now the main pressure control. Cooker regulators were usually weight loaded, single diaphragm regulators with $R\frac{1}{2}$ connections.

Thermostats

The most common form of oven control is the combined tap and thermostat. This is usually a liquid expansion type. It is usually mounted on the float rail with the sensing phial located in the flueway, outside the oven. The thermostat gives modulating control, reducing the flame to a point where the gas input rate exactly matches the oven heat loss. A drilling, usually in a screw, provides a bypass so that the flame will not be extinguished if the valve is closed completely. It also gives the lowest oven temperature.

All thermostats have the standard settings from $\frac{1}{4}$ to 9 but some have additional positions at either end of the scale. Below $\frac{1}{4}$ may be a 'Low' or 'Hold' setting giving temperatures of about 65 to 75° C. This is for keeping food hot after cooking without it becoming 'dehydrated' or dried out. Above 9 may be a 'High' setting for cleaning oven linings.

Some thermostats incorporate a microswitch which is closed, when the tap spindle is pushed in or turned, to operate the spark ignition.

As well as being used on ovens, a similar theremostat has been used to control hotplate burners, the sensing element being held in contact with the pan by means of a spring.

Flame Supervision Devices

All modern ovens are fitted with a flame supervision device (FSD). These are sometimes thermoelectric but more often vapour pressure types are used, commonly containing mercury vapour. Flame rectification is also used.

The device is fitted in the supply to the oven burner only and does not protect the crosslighting pilots or heater flames. EN 30 allows gas rates of up to 600 W in the oven to be unprotected. Most devices operate independently of other controls but a few incorporate a microswitch to cut off the electric supply to the spark generator when the heater pilot is established.

Ignition

Many different types of igniters have been used on cookers but virtually all current models have repetitive spark generators. These have one outlet for use on a built-in oven or up to six outlets for a free-standing cooker. Some ovens, hotplates and grills have reignition.

Igniters have three types of generators; these are:

- piezo
- battery
- mains electricity.

Piezo igniters on cookers are usually the impact type. Because the energy output is low, these igniters generally serve one, or at most two, ignition outlets. A single igniter can be made to light four hotplate burners by means of flash tubes.

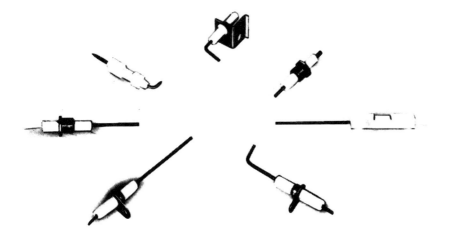

Fig. 16   Spark electrodes (Vernitron)

Battery powered generators are similar to mains generators. These are fitted on appliances which have no mains-operated electrical controls and so do not need a mains supply.

Mains spark generators are the most common. In these, an electric charge is built up in a capacitor until a trigger device, such as a neon or a thyristor allows it to discharge suddenly. The discharged current is stepped up by a transformer to the high voltage, 10 to 15 kV, needed to produce a spark. After each spark the capacitor recharges and the cycle is repeated. This type of generator can be made to supply any number of outlets.

All high tension spark systems have their efficiency reduced when subjected to high temperatures, humidity and spillage. Under these conditions the current leaks to earth by tracking or through poor insulation. Moisture on the electrodes reduces the voltage and the capacitive effects of the insulated leads increases the duration of the spark, so reducing its power.

To counteract these effects on some hotplate burners, the electrodes are located under the burner cap.

Reignition is controlled by flame conductance. A reigniter detects the presence of a flame by passing a minute current through it. This current is used to prevent the ignition capacitor from becoming charged, so stopping sparks from being generated.

As soon as the flame is extinguished, the detector current stops flowing, the capacitor charges up and sparking resumes until the flame is re-established or, on an oven, the flame supervision device shuts down.

There are many different designs of spark electrodes, used on all types of appliances, Fig. 16. Igniters normally use a single electrode, sparking on to the burner itself, with the current returning to the earth connection of the generator through the metalwork of the appliance. At times, a second electrode is used instead of the burner body and sometimes this electrode is connected back to the generator by a separate conductor. Spark gaps are typically 3 to 4 mm.

Reigniters have two electrodes, one for striking the spark and the other for detecting current. It is essential for correct operation that these electrodes are positioned exactly in accordance with the manufacturer's instructions. Failure to do this may result in the spark being produced at a point where the air/gas mixture is not readily ignitable or where the flame electrode cannot detect a flame.

Automatic Cooking

Automatic oven control is fitted to a number of cookers. It is designed to turn on and light the oven at a preset time and turn it off after a selected cooking period. The control originally consisted of a synchronous electric clock which operated micro-switches that controlled an igniter and a solenoid valve in the main burner supply. The pilot valve was sometimes mechanically operated and built into the device. The user could choose either manual or automatic control or switch the oven off at any time.

The control also housed an independent timer which rang at the end of any selected period.

A typical automatic cooker control of this era is shown in Fig. 17. This type of control is still used in a small number of new cookers today.

Since the early 1980s, more and more automatic cookers have utilised the latest microchip technology. The electronic automatic timer control system usually has a Light Emitting Diode (L.E.D.) display giving the time of day and generally using the 24 hour clock. There are also a number of controls operating the following functions:

- manual or automatic cooking
- duration of automatic cooking time
- start time for automatic cooking
- stop time for automatic cooking
- displayed time adjuster
- minute minder control.

Many of the modern automatic cooker controls allow the customer to operate a fully automatic system, e.g. where the oven will turn on

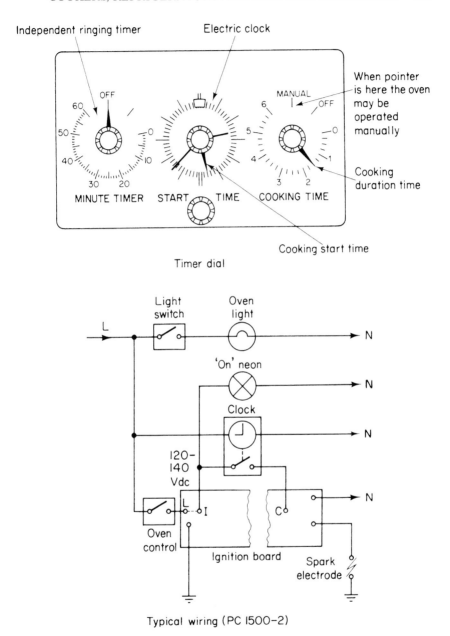

Timer dial

Typical wiring (PC 1500-2)

Fig. 17   Automatic oven control (see Ch. 13 for symbols)

and off at the pre-selected time or automatic off only, e.g. where the cooking period is started manually by the customer and the timer is set to turn off automatically at a pre-selected time. Figs. 18 and 19 illustrate the modern type of display facia layout and the wiring and functional flow diagrams associated with the system.

Electric Controls

Electric switches are used to control a number of devices, including:

- ignition devices
- motors driving various rotary spits
- oven interior lights
- 'Manual/Hold' oven settings

Single spark generators use a changeover switch which charges the capacitor and then discharges it. Most switches are single pole on/off switches.

Solenoid valves are used for automatic cooking and also on the 'Manual/Hold' setting of ovens. In this case the valve is mounted in a bypass across the oven control and, when open, allows a low rated supply to pass to the main burner through a restrictor.

Electrical test points are provided on some new appliances. These enable a service engineer to diagnose electrical faults with a multimeter from a central terminal block or test block, mounted in a readily accessible position.

Installing Gas Cookers

The installation of gas cookers should be carried out in accordance with the recommendations of the current BS Code of Practice 6172

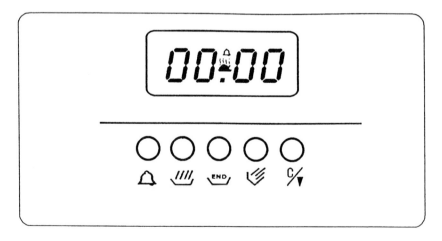

Fig. 18(a)   Automatic oven control (Programmer) (New World)

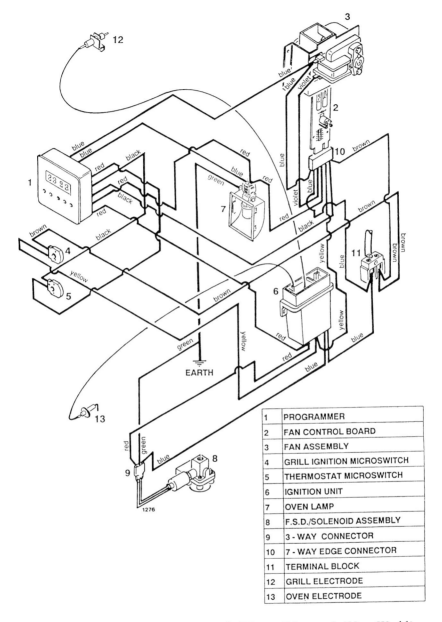

1	PROGRAMMER
2	FAN CONTROL BOARD
3	FAN ASSEMBLY
4	GRILL IGNITION MICROSWITCH
5	THERMOSTAT MICROSWITCH
6	IGNITION UNIT
7	OVEN LAMP
8	F.S.D./SOLENOID ASSEMBLY
9	3 - WAY CONNECTOR
10	7 - WAY EDGE CONNECTOR
11	TERMINAL BLOCK
12	GRILL ELECTRODE
13	OVEN ELECTRODE

Fig. 18(b)   Automatic oven control (Wiring Diagram) (New World)

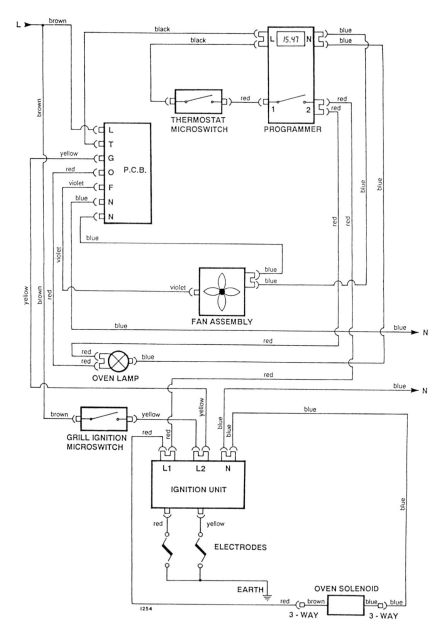

Fig. 18(c)   Automatic oven control (Functional flow diagram)
(New World)

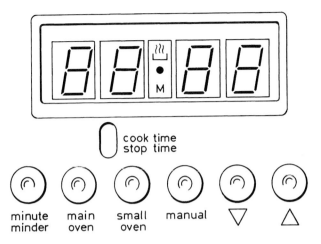

Fig. 19(a) *Automatic oven control (Automatic Clock controls) (Cannon)*

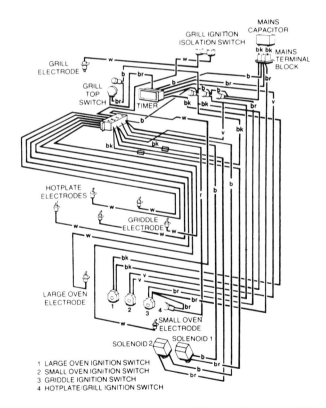

Fig. 19(b) *Automatic oven control (Wiring Diagram) (Cannon)*

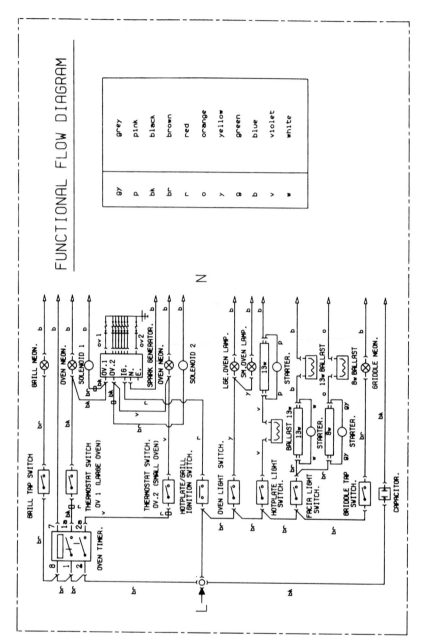

Fig. 19(c)   Automatic oven control (Functional flow diagram) (Cannon)

and the Gas Safety (Installation and Use) Regulations 1998. Because of the numerous variations between different models, the relevant manufacturer's instructions should be consulted.

Location

The cooker should be sited in a well lit, draught-free position. It should be conveniently placed in relation to the sink, working space, refrigerator etc., but be separated from them by working surfaces.

A hotplate or a free-standing cooker should not be fixed in a corner so that the saucepan handles foul the side of the wall. It should not be possible for spillage to become trapped between the sides of the cooker and adjacent units.

Fire Prevention

There should be a space of at least 900 mm between the top of a hotplate and any combustible surface above it. The minimum space above a grill should be 450 mm unless a greater distance is specified by the maker. Walls and work surfaces may require protection adjacent to a cooker, particularly if the hotplate is below worktop level. In some cases, a plinth may be used to raise the cooker.

Gas Supply

The recommendations for the fitting of the gas installation pipes are given in BS 6891.

The supply pipework should be sized in relation to its length and maker's recommendations. Usually 15 mm (DN 15) nominal bore is adequate for a normal free-standing cooker and 8 mm (DN 8) for a built-in unit.

The cooker should normally be connected to the supply by a length of flexible tubing into a bayonet-fitting, plug-in connector, as shown in Chapter 1, or with an angled plug-in socket and a straight plug-in adaptor. This form of connection was proposed by the King Report. It prevents strain on the supply pipes whilst allowing easy and safe connection or disconnection for cleaning or decorating.

The appliance should be made stable by the fitting of a stabilising bracket, Fig. 20, or, if desired, a restraining chain. The cooker should normally not be pulled right out with the flexible tubing connected. It should only be moved far enough to unplug the connector, so the connection should be conveniently sited for access.

Electric Supply

The electricity supply should be taken from a correctly wired and earthed 13A socket outlet by means of a suitably fused plug fitted to

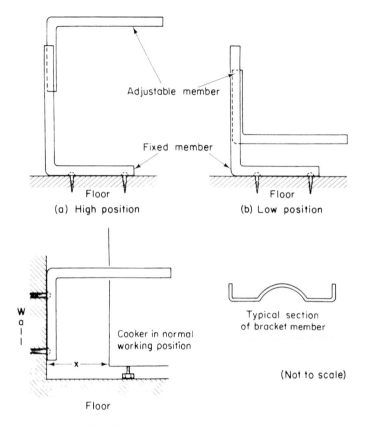

Fig. 20   Cooker stabilising bracket

the end of the cooker cable. The socket should be conveniently located not more than 1.5 m from the cooker. The cable should be shortened, if necessary, and not left slack or in loops. The cooker cable and the plug and socket should not be subjected to excessive heat by exposure to products and hot surfaces.

Ventilation

The requirements for ventilation are outlined in Chapter 5. The main points are as follows:

- Rooms up to a volume of 5 m^3 require 10,000 mm^2 of permanent ventilation to outside air plus an openable window or equivalent.

- Rooms of volumes between 5 m^3 and 10 m^3 require 5,000 mm^2 of permanent ventilation to outside air unless they have a door which opens directly to outside, when no purpose-made ventilation opening is required.
- All rooms containing cooking appliances must have an openable window or equivalent
- Larger rooms require no purpose provided ventilation
- If an extract hood is fitted it should be mounted with its lowest edge at least 2 m above floor level.

Built-in Units

These may require special consideration.

Care must be taken to ensure that hotplates are securely clamped into position and that the seal between the unit and the worktop is sound.

Grills must be securely mounted on walls by means of fixing devices which are unaffected by heat.

Ovens require a circulation of cooling air between the oven and the inside of the cabinet. Deflector plates above the flue outlet must be correctly positioned.

Built-in ovens, drop-in hotplates and other individual units should be connected to the installation pipe by means of rigid or semi-rigid connections. Flexible connections used with free-standing cookers should not be used as the appliance is not intended to be moved by the customer.

General

It is essential to the correct operation of the cooker that it should be accurately levelled. Normally, adjustable feet or rollers are incorporated for the purpose.

Commissioning

The manufacturer's instructions should be complied with when putting an appliance into operation.

After connecting the appliance, carry out a check on all aspects of its operation. This should include:

- test the installation for soundness and purge
- check ventilation
- check that all removable parts have been correctly assembled or positioned
- check the ignition on all burners, ensure spark gaps are correct

- check burner pressure is correct
- check flame height and aeration on all burners: hotplate burners should be well aerated, grill and oven burners should have slack flames but without yellow tips
- check oven flame supervision device operation
- check thermostat and bypass by heating up and turning down to a low setting
- check oven door seal by checking its grip on a 50 mm strip of paper approximately 0.125 mm thick, shut in the door
- check automatic cooking control and timer
- check ancillary electrical devices, lights, rotating spits or snack turners
- instruct the customer on lighting and operating the appliance.

Servicing

Although it is possible for the customer to carry out most of the cleaning operations on a gas cooker, there is still the need for regular periodic servicing. The operations should include:

- examine the appliance and discuss its performance with the customer
- check ventilation
- check electric cable undamaged and connections to plug and appliance are secure
- check electric supply earthed and correctly connected
- check gas flexible tubing sound and undamaged, plug-in connector operates satisfactorily
- if fitted, check stabilising bracket or chain are secure and effective
- check cooker level, if fitted check moveable base operates satisfactorily
- check taps, ease and grease if required
- check burner pressure correct: governor clean and level (if fitted)
- check ignition devices, electrodes, clean ceramics and insulation, reignition satisfactory, flash tubes and any pilot jets clear, battery voltage adequate
- clean burners, clear main ports, retention slots and crosslighting ports
- check flame picture on all burners, clear injectors and adjust aeration where necessary
- check flame supervision device phial clean and correctly located, heater flames correctly aligned
- check operation of FSD

- check oven tap and thermostat, ease and grease tap if necessary
- check any broken or disturbed joints for soundness
- check operation of thermostat and bypass rate correct
- check oven flueways clear, thermostat phial or rod correctly located
- check oven door seal and catch
- check automatic cooking devices
- check electric lights, switches, motors and solenoids
- clear up work area and wipe over appliance exterior.

Fault Diagnosis and Remedy

The faults which may occur on burners, regulators, igniters and control devices are dealt with in the appropriate chapters of Vol. 1. Electrical fault diagnosis is also covered in Chapter 13 of this volume. Other common faults are as follows:

Blockages

Many of the faults on gas cookers are caused by the blockage or partial blockage of jets or burner ports. The blockage may be due to:

- spillage
- corrosion
- deposits from cooking vapours
- burnt-on oils and fats deposited by splashing
- moisture.

These blockages affect combustion by reducing primary aeration. This may in turn cause:

- ignition problems with flash tubes or sparks
- poor combustion, unstable flames and smell from hotplate burners.

Blocked heater jets will cause the oven flame supervision device to fail to safety.

Blockage in the oven flue will reduce the supply of combustion air and the flame may become extinguished.

Spillage may also block spark gaps or bridge flame conductance circuits.

Blockage of grill burner ports may cause reduced grilling area or uneven heating.

Burner Pressure and Aeration

Incorrect burner pressure, which may be due to a faulty meter governor, usually causes ignition problems, particularly on the oven burner. High pressures can cause unstable flames. Over aeration or a higher gas rate in an oven burner can affect cooking. Although the thermostat should compensate by reducing the heat input, the increased rate of circulation around the oven can cause food to burn underneath.

Over aeration of a burner may also affect its ignition.

Cooking Problems

Faults on hotplates and grills are usually self-evident. Oven faults may be more difficult to diagnose.

Uneven cooking may be caused by the cooker being out of level. Misplaced oven linings or distorted shelves may have a similar effect.

When problems occur, check the following points:

On the Appliance

- flame picture satisfactory, aeration and pressure correct
- flame supervision device is fully open and operative
- thermostat operates, valve closes correctly and bypass is clear
- oven flue is clear and thermostat rod or phial is correctly positioned
- door seals properly
- oven furniture is correctly assembled, shelves right way round and level.

Cooking Method

- recipe followed faithfully
- correct size of tins used
- correct shelf height used
- thermostat mark correct
- correct cooking time allowed.

If the oven appears to be operating normally, but the customer complains of poor results, it is usually advisable to call in a home economist to carry out cooking tests which are sensitive to subtle faults in performance.

It may be possible to alter the thermostat calibration by plus or minus 1 mark. However, it is not generally possible to recalibrate a thermostat accurately against a thermometer placed in the oven. To be of any use the thermometer would have to be set to give BS oven temperature readings in an identical oven, known to give satisfactory results under the same conditions. Thermostats are usually designed for replacement rather than recalibration.

Automatic Cooking Problems

Whilst it is possible for the control to fail while the oven is on, generally a mechanical or electrical fault results in the oven remaining unlit.

Failure of the solenoid valve gives a similar result. On single oven cookers the oven can still be used on the manual setting, but on some double ovens the solenoid is required even for manual operation.

Cooking problems with automatic controls may be due to faulty operation of the control by the user. It is not always easy to interpret and understand the manufacturer's written instructions for use. Always ensure that the customer is completely familiar with the device and does indeed set it correctly.

REFRIGERATORS

Introduction

Gas refrigerators were originally introduced into this country from Sweden. For many years they were the product of a single manufacturer, with other makers joining the field as the gas refrigerator gained popularity. The appliance itself is efficient and reliable, virtually silent in operation and requiring very little attention. In spite of this, it is now not as popular as it once was. This is not for any technical reason. The gas refrigerator has simply been pushed out of the public view by an enormous increase in the number of electrical models produced.

The few gas refrigerators currently marketed are Family 3 models with capacities of around 0.1 to 0.17 m^3. Some recent models had a small separate freezer compartment mounted on the top with a capacity of about 0.03 m^3.

Refrigeration

Bacteria and mould-producing organisms are carried in the air and attack food as soon as it is picked, killed or prepared. If these organisms are allowed to grow and multiply they cause the food to decay. At normal room temperatures food can 'go bad' quite quickly and in hot weather the process is speeded up. However, when the weather is cold, food keeps longer. The colder food is kept, the slower the bacteria grow.

This becomes particularly pronounced at temperatures below 10° C, and, at 0° C the bacteria which cause food poisoning are dormant. So it is sensible to have a refrigerator for storing food for any length of time, most refrigerators operate at 0° to 7° C.

Refrigerators usually contain a frozen food storage compartment. This is colder than the main cabinet. It provides storage for pre-frozen food and produces a supply of ice cubes. The compartments of different models are run at different temperatures and are 'star rated' in accordance with BS EN 28187: 1992 as follows:

     * short-term storage, about 1 week, temperature −6° C
    ** medium-term storage, about 1 month, temperature −12° C
  *** long-term storage, about 3 months, temperature −18° C

These storage times are only a rough guide, some foods keep fresh longer than others. The temperatures given are those that must be achieved under specified test conditions. In practice conditions and temperatures may vary.

Domestic freezers are now quite common. The difference between a freezer and a frozen food compartment, apart from its size, is in the greater cooling ability of the freezer. It can freeze fresh food from room temperature to −18° C in 24 hours or less, as well as being able to maintain it at that temperature. If left on maximum setting, a freezer may reach temperatures of −25 to −30° C. Manufacturers quote the weight of food which can be frozen in 24 hours by their appliances.

The symbol for a freezer is shown in Fig. 21.

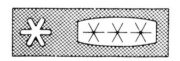

Fig. 21   Freezer symbol

Cooling Units

There are two types of cooling unit used in domestic refrigerators:

- compression type
- absorption type.

Although all gas refrigerators are the absorption type and nearly all electric refrigerators are the compression type, it helps to study both.

Both types cool the food by means of an evaporator situated inside the cabinet. Within the evaporator, heat from the food is absorbed by the evaporation of a liquid. This heat is rejected into the room air by the condenser, mounted behind the cabinet.

Absorption and compression units simply use different means of arriving at the same situation.

Both types rely on the fact that:

Most gases will condense into liquid if:

- the pressure is increased
- the temperature is reduced.

Most liquids will evaporate into gases if:

- the pressure is reduced
- the temperature is increased.

Compression Unit

The compression unit, Fig. 22 is the type fitted to most modern electric refrigerators. In it, a gas such as sulphur dioxide or 'Freon', which is the brand name for a range of refrigerants, is compressed by a motor-driven compressor. The hot, compressed gas is cooled in a heat exchanger or 'condenser' where, at room temperature, it condenses into a liquid. This liquid collects in a receiver from which it passes to an 'expansion valve' or orifice.

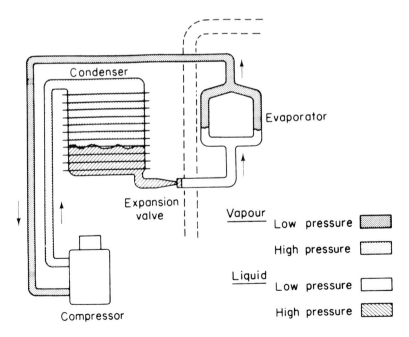

Fig. 22   Compression unit

The liquid is squirted through the expansion valve into the evaporator so experiencing a considerable reduction in pressure. This brings about evaporation which absorbs heat from the cabinet. The gas is drawn back to the compressor and recycled.

The compression unit is very efficient but it is noisier than the absorption type. It has moving parts which may wear and it can cause radio or TV interference if not adequately suppressed.

Thermostatic control operates by starting and stopping the motor.

Both sulphur dioxide and the range of 'Freon' gases, when released into the atmosphere are 'environmentally unfriendly'. Substitute gases which do not contain CFCs (chloro fluoro carbons) have been developed to replace them.

Absorption Unit

All gas refrigerators have this type of unit which may also be heated by electricity, l.p.g, paraffin, or by multi-fuel. It relies on gravity and the effect of partial pressures to operate. There is no change in total pressure around the sealed unit. The refrigerating fluid is ammonia, under pressure, and the unit also contains some hydrogen and water.

The relationship between the partial pressures in a mixture of gases was established by John Dalton, one of the greatest of English chemists. Dalton's Law of Partial Pressures states:

'The total pressure of a mixture of gases is the sum of the partial pressures of the gases composing it. That is, the sum of the pressures which each gas would exert if it was confined alone in the space occupied by the mixture'.

The absorption unit makes use of this fact and the unit operates as shown in Fig. 23.

The ammonia is dissolved in the water and the mixture collects in the boiler, where it is heated by a small gas flame. Hot ammonia gas is driven off. This rises up into the condenser where it is cooled and condenses into liquid ammonia at room temperature. The liquid ammonia runs down from the condenser and into the evaporator where it evaporates into the hydrogen which is present in this part of the unit.

The ammonia is able to evaporate without any overall reduction in pressure because the hydrogen supplies so large a proportion of the total pressure in the evaporator that the pressure of the ammonia vapour on the surface of the ammonia liquid is low enough to allow evaporation to take place. As the ammonia liquid evaporates it absorbs heat from the cabinet. The ammonia gas produced by the evaporation becomes mixed with some of the hydrogen and flows down into the absorber.

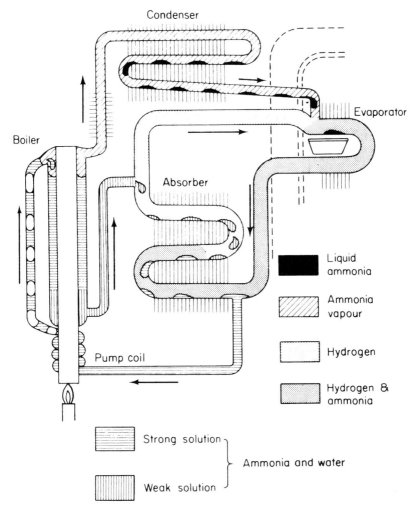

Fig. 23   Absorption unit

The absorber is a series of tubes constantly fed with a trickle of water (or very weak ammonia and water) from the bottom of the boiler. Ammonia has an affinity for water and is quickly dissolved into it leaving the hydrogen free to rise back into the evaporator.

The now strong ammonia solution flows down from the absorber to the lowest point in the system, the bubble pump or pump coil. This is a coil of tube around the combustion chamber in which the liquid is heated so that bubbles form. As the liquid boils, the bubbles carry the solution back into the top of the boiler.

From here, ammonia gas is driven off to begin another cycle and the water overflows into the absorber.

So there are three cycles which together make up the absorption system. They are:

- the ammonia cycle
  - ammonia is driven off in the boiler, is condensed, evaporated and reabsorbed into water
- the water cycle
  - water is left behind in the boiler, trickles down through the absorber, dissolves the ammonia gas and carries it back to the boiler via the pump coil
- the hydrogen cycle
  - hydrogen contributes to the evaporation, becomes mixed with ammonia gas, leaves the ammonia behind in the absorber and returns to the evaporator.

The absorption unit is less efficient than the compression type but it has no moving parts and near silent operation.

Thermostatic control modulates the gas rate to the burner.

GAS REFRIGERATORS

Construction

Gas refrigerators are manufactured to comply with the recommendations of BS 5258, Part 6: 1988 and BS EN 28187: 1992.

The refrigerator consists essentially of:

- cooling unit
- cabinet
- gas equipment.

Cooling Unit

Most of the unit is mounted behind the cabinet with only the evaporator situated inside the cabinet, at the top. Figure 24 shows a rear view of a typical refrigerator. The boiler is on the lower left-hand side. The absorber is beside it, about mid-way up and the condenser is at the top. The unit is made of steel tubing and usually enamelled black. It may be pressurised up to 23 bar. As well as containing the refrigerants it is usual for a rust inhibitor to be included when the unit is charged.

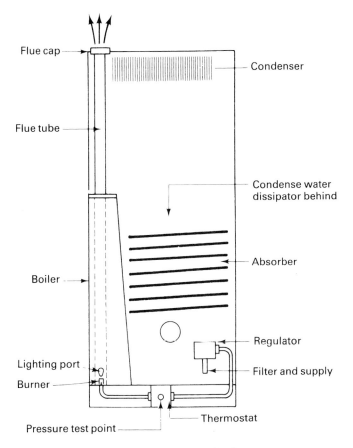

Fig. 24   Rear view of cabinet

Cabinet

This is an insulated cupboard. The outer case is painted mild steel sheet with an inner lining usually of plastic or aluminium. The space between is insulated with polyurethane foam or a similar material. Near to the boiler the insulation is glass fibre. The door is hinged at one side and the door liner is formed into bottle racks, egg shelf, dairy compartment and so on. The door has a synthetic rubber seal and is held closed by magnetic catches. The cabinet may have levelling feet and the top may form a working surface.

Gas Equipment

The gas equipment, Fig. 25, consists of:

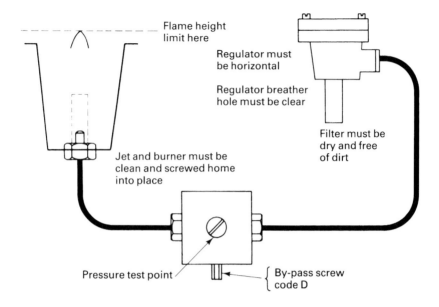

Flame height limit here

Regulator must be horizontal

Regulator breather hole must be clear

Filter must be dry and free of dirt

Jet and burner must be clean and screwed home into place

Pressure test point

By-pass screw code D

Fig. 25   Gas equipment (1st family gas)

- regulator (1st family gases)
- filter
- thermostat
- burner
- igniter – on some models
- flame supervision device (3rd family gases).

Regulator

An appliance burning 1st family gases should include a gas regulator, it is a small weight loaded regulator, usually set at about 5 mbar.

An appliance burning 2nd family gases is usually regulated at the meter and 3rd family gases by a regulator fitted to the container and complying with BS 3016.

Filter

A small, cigarette type filter is included. It may be fitted in a separate housing or be located in the inlet to either the governor or the thermostat.

Thermostat

A liquid expansion thermostat (Vol. 1, Chapter 11) controls the gas rate to the burner. It has variable settings and may be adjusted by the user to compensate for hot or cold weather. The bypass is usually a drilling in a screw which may be changed to suit the family of the gas supplied. The temperature sensitive element may be simply the end of the capillary tube. It is passed through the cabinet casing and located in a socket in the evaporator or clipped to one of the fins. Pressure test points are fitted on the thermostat to enable its operation to be checked. Typical thermostat settings are from 1 to 5 with a D for defrosting at the lower end of the scale and a C for continuous at the top. The thermostat is on bypass rate at D and full on at C. The usual setting for most of the year is 3 to 4.

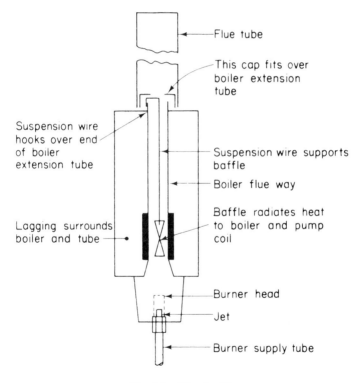

Fig. 26   Flue baffle

Burner

The burner is aerated and consists of a jet and a burner head. It is located centrally below the boiler flueway and the flame height should be checked with the maker's instructions. A baffle consisting of a length of twisted stainless steel, is suspended by a wire in the flueway at a critical height above the burner. It radiates heat to both the pump coil and the boiler, Fig. 26.

Igniter

Various types of igniters have been fitted to refrigerators and the piezo electric is currently used, where an igniter is fitted. Otherwise the appliance has a manual lighting port beside the burner.

Care and Use

Cleaning

The materials used in its construction make for easy cleaning and hygienic operation. The painted case may be wiped down with a mild detergent, abrasive cleaners should be avoided. The inner lining should be cleaned only with water and bicarbonate of soda. No other substances should be used, or food may be tainted. All internal surfaces should be dried after cleaning to prevent frost building up.

Food Storage

Cooling air circulates around the cabinet by natural convection. The cold air from the evaporator, being more dense, falls to the bottom of the cabinet. As it absorbs heat from the contents and the sides the air warms, becomes less dense and rises back up to the evaporator, Fig. 27.

Air contains water vapour. The hotter the air the more water it can hold. As air is cooled below its dewpoint the water condenses (Vol. 1, Chapter 5). This happens inside a refrigerator. Each time the door is opened, warm, moist air enters the cabinet. When this air makes contact with the cold evaporator it is condensed and frozen, forming ice crystals or 'frost' on the evaporator. Wet food or dishes containing water should be covered to prevent the water evaporating and to minimise frost formation.

The frost acts as a deodoriser, absorbing taints carried by the air circulation. So it is usually recommended that dairy products, like milk and cream which are susceptible to tainting, should be stored at the beginning of the air circulation. Strong smelling foods should be placed near to the end, Fig. 28. Covering or wrapping food guards against tainting and the manufacturer's instructions should be followed carefully.

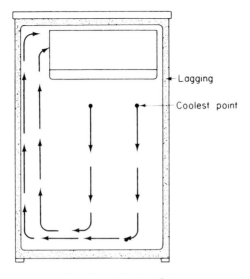

Lagging

Coolest point

Fig. 27   Air circulation

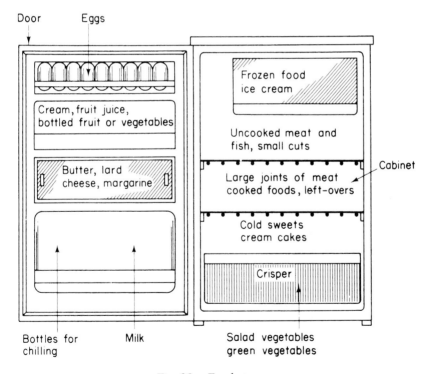

Door

Eggs

Cream, fruit juice,
bottled fruit or vegetables

Butter, lard
cheese, margarine

Frozen food
ice cream

Uncooked meat and
fish, small cuts

Cabinet

Large joints of meat
cooked foods, left-overs

Cold sweets
cream cakes

Crisper

Bottles for
chilling

Milk

Salad vegetables
green vegetables

Fig. 28   Food storage

Defrosting

A small amount of frost may not affect the operation of the refrigerator significantly. However, when it reaches a thickness of about 6 mm, it acts as a layer of insulation on the evaporator and reduces the rate of cooling.

To keep the refrigerator operating effectively the frost must be removed periodically. This may need to be done every 2 to 4 weeks depending on usage. Defrosting may be done, manually, by carrying out the following procedure:

- turn the refrigerator off, or turn the thermostat to a 'Defrost' setting if provided
- remove all contents and leave the door open
- to speed up defrosting, put containers of hot, but not boiling, water into the cabinet and the evaporator
- remove ice and water; never use metal implements to scrape the ice off the evaporator
- when all frost has melted, wash out and dry the cabinet thoroughly
- turn back on again, turn to normal thermostat setting.

Many refrigerators now have automatic defrosting which relieves the customer of this chore.

Automatic Defrosting

The automatic defrosting unit is a short circuit between the boiler and the evaporator, sealed by a condensation trap attached to a syphon, Fig. 29. Periodically the syphon empties the trap and hot ammonia gas is allowed to pass into the evaporator for about 1 hour. This is long enough to melt the frost but not so long that it affects the cooling of the cabinet.

In more detail the operation is as follows.

During normal operation of the cooling unit, ammonia gas condenses and a small amount collects in the trap of the automatic defrost unit. This condensation seals off the short circuit between the boiler and the evaporator, Fig. 29(a).

As the condensation builds up it reaches a point where the syphon is filled, Fig. 29(b). Now the condensed ammonia is syphoned back into the boiler, Fig. 29(c), leaving the trap clear and allowing hot ammonia gas to flow from the boiler, up into the evaporator, Fig. 29(d).

Defrosting now takes place. While this is continuing the condensed ammonia is beginning to collect again in the trap until it reaches a

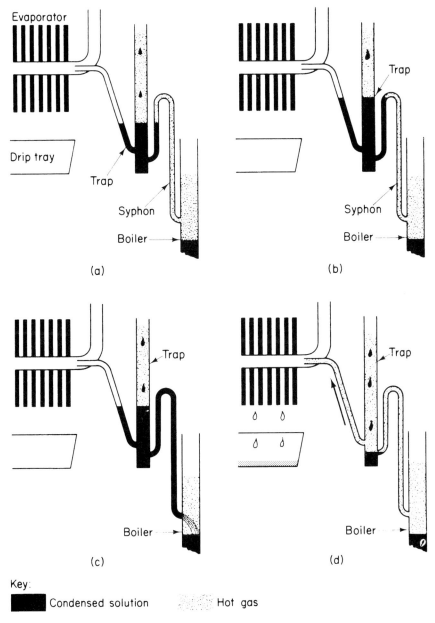

Key:
■ Condensed solution   ▨ Hot gas

Fig. 29   (a, b, c & d) Automatic defrost

level where the passage for the hot gas is once more sealed, Fig. 29(a). Defrosting stops, normal cooling is resumed and the cycle begins again.

Where refrigerators are fitted with a separate frozen food compartment, the auto defrost operates only on the main cabinet. The frozen food compartment should be manually defrosted periodically at about 8 to 10 week intervals.

Installation

This should be carried out in accordance with manufacturer's instructions and should comply with the recommendations of the current BS 5258: Part 6: 1988.

Location

The refrigerator should be located in a cool, dry, draught-free position, convenient for access and reasonable light. It should not be fitted in a larder or under a cupboard used to store perishable food.

There must be adequate clearance to allow a flow of air over the cooling unit and allowances must be made for lighting and cleaning. Usually a space of about 200 mm must be left above the flue outlet. If required by the maker, adjacent surfaces should be fireproof.

When the refrigerator is installed next to a cooker or other heat-producing appliance, it must be insulated from it by an air space of at least 25 mm or by thermal insulation.

Gas Supply

For appliances burning 1st and 2nd family gases, the inlet connection is $R_c\frac{1}{8}$ and 6 mm tubing should be adequate for the very low gas rate. The final connection to the appliance must be of flexible tubing if the appliance is mounted on rollers. Otherwise rigid connections may be used if required. Nylon tubing has commonly been used as a flexible connection. A control cock must be provided which must be fitted upstream of any flexible tubing. Alternatively a plug-in connection can be used.

For appliances burning 3rd family gases the inlet connection is threaded $R_c\frac{1}{8}$ or $R_p\frac{1}{8}$. Alternatively, the connection may be provided with compression fittings to BS 864 for 6 mm copper tube to BS 2871 or with steel tube to BS 1387. For portable appliances burning 3rd family gases, flexible hose to BS 3212 should be used.

Commissioning

When the supply is connected the following operations should be carried out:

- check that the appliance is level to an accuracy stated in the manufacturer's instructions
- if necessary level the appliance by levelling screws or packing
- check that the regulator, if fitted, is level
- check that the baffle is correctly located and that the flue is clear and correctly assembled
- test the installation for soundness and purge
- check the igniter, if fitted, and light the burner
- check the setting pressure and bypass pressure
- check the appearance of the flame
- check the operation of the flame supervision device (3rd family gases)
- check that all ancillary containers and equipment are correctly positioned
- set the thermostat to an average setting
- instruct the customer on
  - lighting and using
  - star-rating
  - defrosting.

Servicing

Like other gas appliances, refrigerators should be serviced regularly and at least once a year. The operations should include:

- examine the appliance and discuss its performance with the customer
- turn off the gas and disconnect the appliance if necessary
- remove the burner and plug or cover the hole
- remove flue extension and baffle
- clean flueway
- clean and replace flue extension and baffle
- check baffle location
- clean and replace burner
- check gas control and grease if necessary
- check that the filter is clean and dry, renew if necessary
- check that the regulator is clean and level, (if fitted)
- reconnect and turn on the gas, test for soundness
- check the igniter and light the burner
- check the flame picture and adjust to correct height
- check burner pressure and check thermostat operation and bypass pressure
- check flame supervision device (3rd family gases)
- set thermostat to recommended setting

- examine door fit, seal and catch
- check that appliance is level and stable
- check for adequate ventilation around the unit
- wipe down the appliance
- clear up the work area and leave the appliance operating correctly.

Fault Diagnosis and Remedy

1. Cooling Unit

The cooling unit is a completely sealed unit and cannot be repaired. If it becomes faulty it must be replaced by another unit. The two major unit faults are:

- internal blockage
- loss of refrigerant due to leak.

Internal blockage. Corrosion building up over a long time in an old unit may clog up one of the circulating pipes and stop the cycle from operating. Various remedies have been proposed including 'tilting'. This entails tipping the appliance, usually upside down, to make the liquids flow through the piping in the hope that they may clear the blockage. It sometimes works!
Leak of refrigerant. Leaks may be caused by corrosion or welding faults. They often occur where tubes are joined. There may be a smell of ammonia and usually yellow crystals form around the leak. Yellow marks may appear on the tubing.

2. Door Seal

It is important that the door should seal the cabinet effectively. If it does not, the performance will be impaired and excessive frosting will occur. To check the seal use a piece of stout paper about the size of a postcard (75 mm wide), Fig. 30.
　　Place the paper on the door frame and shut the door. Pull the paper gently, the gasket should grip the paper. If it does not, follow the maker's instructions for adjusting or renewing the seal. This may entail packing washers under the hinges or softening the gasket with water at about 60° C so that it may be reshaped.

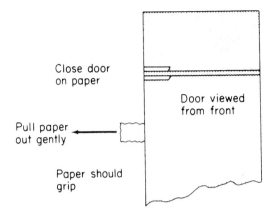

Close door
on paper

Door viewed
from front

Pull paper
out gently

Paper should
grip

Fig. 30   Testing door seal

3. Common Faults and their Remedies

These may be summarised as follows:

Symptom	**Action**
Cabinet does not get cold at all	*Check that gas is on and burner is lit* *If burner is out, check that gas supply and filter are clear, relight, turn thermostat to maximum setting. Recheck after one hour* *If not cooling, unit is faulty*
Burner does not stay alight	*Check for:* *• excessive draughts* *• incorrect setting pressure or faulty regulator, (if fitted)* *• choked or wrong size thermostat bypass* *• blocked or linted burner* *• blocked or obstructed flue*
Cabinet not cold enough	*Check that burner pressure is correct, filter and burner clear* *Check baffle location, if necessary clean baffle and flue* *Check condenser fins clean and ventilation unobstructed* *Check thermostat setting not too low, calibration correct* *Check for excessive frost on evaporator or cabinet overloaded, instruct the user*

Symptom	Action
Cabinet too cold	*Check thermostat:* • *end of capillary tube not in good contact with the evaporator* • *setting too high* • *wrong size, overlarge bypass* • *check setting pressure*
Cabinet above room temperature	*Check for excessive gas pressure; if correct, cooling unit is faulty*
Uneven cooling in evaporator	*Check cabinet level and readjust*
Excessive frosting	*Check door seal, reshape or replace as necessary* *Check gas pressure or thermostat setting too high*
Smell of 'gas'	*Check for flame impingement, flame height* *Check baffle location, baffle and flue clean* *Check burner pressure correct, burner clean*
Smell of ammonia	*Check for leak on cooling unit*
Food tainting	*Check for spilt liquid on or behind linings, cabinet cleaned with strong detergent* *Check for uncovered strong foods or bad food in cabinet* *Defrost and clean with bicarbonate of soda*

LAUNDRY APPLIANCES

Introduction

The market for gas laundry appliances is a fairly small and traditional one. At one time gas contributed to all home laundry operations with wash boilers, washing machines, tumbler dryers, drying cupboards and cabinets, airing cabinets and irons.

However, with the advent of new synthetic fabrics which were washed at lower temperatures and required little or no ironing and no airing, the automatic washing machine and the launderette began to take over the load. The only articles which may be washed in water at temperatures between 95° C and boiling point are white cotton or linen articles without special finishes. Most modern bedlinen is nylon or polyester-cotton which should not exceed 60° C, if white, and 50° C if coloured.

Boiling clothes is a thing of the past.
Appliances manufactured were:

- wash boilers
- drying cabinets
- drying cupboards

Although intended for domestic use, the drying cupboards and cabinets were mostly installed in commercial laundry rooms by local authorities. These appliances are no longer manufactured but are still to be found on the district.

A recent development has been the introduction of the gas heated tumble dryer.

Wash Boilers

Wash boilers were constructed of a copper or aluminium pan heated from below by an aerated ring burner, Fig. 31.

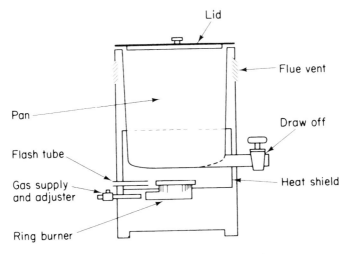

Fig. 31   Section through wash boiler

The pan had a draw-off or drain cock and a lift-off or hinged lid.

Products of combustion were vented from the top of the outer casing, normally into the room. Ignition was manual, sometimes by means of a flash tube. A screw restrictor to adjust the burner pressure was commonly provided in the gas connection which was screwed $R_c^{1/8}$.

The water capacity of the pan was usually 23 to 41 litres and the burner input rating was 4 to 6 kW.

Copper pans were usually coated on the inside with a tin-lead alloy, aluminium pans could be anodised. Pans should be drained and dried after use. Washing soda and abrasive or caustic cleaners must not be used. Only soap, or soapless detergent, should be used for washing. Manufacturers' instructions must be followed carefully otherwise the pan may become badly stained or even corroded.

Installing Wash Boilers

Recommendations for wash boiler installations were given in BS Code of Practice 335: Part 1 (obsolescent)*. Ventilation requirements are also detailed in BS 5440: Part 2.

Most boilers manufactured had heat input rates of 6 kW or less and were fitted in rooms, other than bathrooms, larger than 6 m^3. The room should have an openable window and permanent ventilation to outside air of at least 9 500 mm^2.

The wash boiler should be located close to a water supply and a drain. If a separate, permanent water supply is installed it must comply with local water authority regulations.

Although the final gas connection on the appliance is R$_c$⅛, the supply should be 8 mm to carry the 0.6 m^3/h gas rate required. A control cock must be supplied. The supply may be rigid but is usually flexible with a plug-in-connector to allow the appliance to be portable.

Before lighting the burner to set the pressure and check the flame appearance, always ensure that there is sufficient water in the pan to cover the joints around the bottom and on the draw-off. After adjusting, drain off the water and dry the pan. A new pan should be cleaned in accordance with the manufacturer's instructions before being put into use.

Servicing Wash Boilers

Because of its simplicity, the wash boiler requires very little attention. It will be necessary, occasionally, to remove and clean the burner and to clean out the flueways between the outside of the pan and the inside of the case. After cleaning, check the gas rate and the flame picture. Clean the flash tube, if fitted.

The drain tap may also need cleaning and unblocking. Where the tap has a washer, this may need renewing occasionally.

* Obsolescent indicates that the standard is not recommended for use in new equipment but needs to be retained to provide for the servicing of existing equipment that is expected to have a long working life.

If the customer has used only recommended washing powders in the pan, it should require no attention. It is, however, inevitable that some staining will occur in use. This should be accepted as normal.

Drying Cabinets and Cupboards

Appliances conform to the safety standards specified in BS 5258: Part 3 (now withdrawn)*. Units may however be found on the district which do not comply with the requirements. These appliances are ventilated and heated cupboards for drying washing and lower input units for airing only.

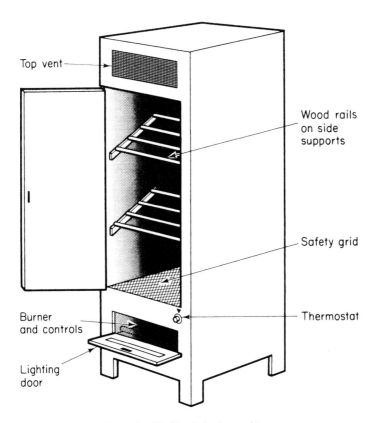

Fig. 32   Unflued drying cabinet

A drying cabinet is a self-contained appliance, usually unflued, Fig. 32. A drying cupboard is a purpose-built compartment with a gas

*Withdrawn indicates that the standard is no longer current.

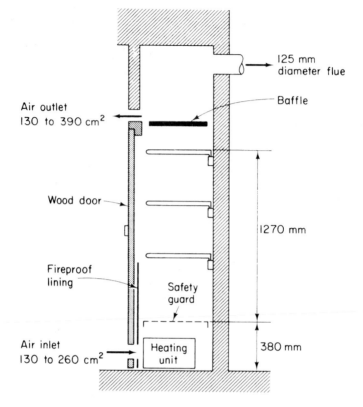

Fig. 33   Flued drying cabinet

heated unit at the base. It is usually flued to outside air, Fig. 33. In both appliances, air for combustion and for drying is drawn in through a low level ventilation opening. The air is heated by the burner and then circulates around the washing. The moisture-laden air is then exhausted either through high-level louvres into the room or to the outside air through a flue. Some cupboards are fitted to Se-ducts and are room-sealed. They have a linkage which closes shutters on the Se-duct inlet and outlet as the door is opened. The door itself has an effective seal and a latch.

The burner may be either a jetted bar or a box type and was usually controlled by a small governor and a variable rod type thermostat. Working temperatures in the centre of the cabinet were up to 104° C.

A protection plate above the burner spreads the heat around the cabinet and a safety guard prevents any accidentally fallen washing from creating a fire hazard.

The baffle in the top of the flued cupboard serves to control the rate of air flow through the cabinet and also acts, with the upper ventilation opening, as a draught diverter.

The manufacturer's instructions for loading the cupboard must be followed. If the cabinet is to dry the clothes satisfactorily, they must be spread out, not all bunched together.

Installing Drying Cabinets and Cupboards

British Standard 5440: Part 2: 1976, before its revision in 1989 gave recommendations for the installation of these appliances.

The outlet of a flueless drying cabinet should not be located where it is likely to cause condensation problems, that is, immediately below a cold surface. Absorbent surfaces and surface finishes can minimise condensation.

Drying cupboards must have their internal surfaces protected by non-combustible material up to, at least, the level of the safety guard.

Flueless cabinets must have heat inputs of not more than 2 kW. Higher rated units must be flued, the minimum flue size being 125 mm diameter, or its equivalent. A flueless cabinet requires a permanent ventilator to outside air of at least 9600 mm^2 effective area. There must also be an openable window in the room.

Flued drying cupboards require a ventilator of 450 mm^2/kW input. Where more than one appliance is fitted in the room, the ventilation area should be sized on the larger appliance. In communal drying rooms the ventilation area may be reduced to 6500 mm^2 per appliance and still further reduced if a fan-assisted common flue is used.

If several cupboards are connected to a common flue, of any type, flame supervision devices must be fitted to the burner of each appliance.

If gas is supplied to the burner of the cabinets through separate secondary prepayment meters, then the pilots must be supplied through a separate, permanent supply.

Servicing Drying Cabinets

There is very little to go wrong with a drying cabinet but the burner may become choked with lint and dust.

The main operations when servicing the appliance are:

- remove and clean the burner
- clean spreader plate and baffle plate if necessary
- replace and light the burner
- check gas pressure and flame picture

- check the thermostat operation
- check that ventilation louvres and flueways are clear and correctly sized
- check the door and shutter seals on room-sealed appliances
- wipe down the appliance or the cupboard door and clear up the work area.

Gas-heated Tumble Dryers

A recent innovation into this country has been the gas-heated tumble dryer – shown in Fig. 34. Prior to the introduction of this gas-heated appliance, electricity was used to heat the air for drying the clothes in tumble dryers. The gas-heated machine is claimed to have certain economical advantages over its all-electric counterpart. An electrical supply is, however still required to operate the following components:

- fan, to supply the combustion air (primary and secondary), plus a cooling flow of air around the clothes drum and motor
- motor, to rotate the clothes drum (forward and reverse)
- full sequence control unit.

The dryers are manufactured to BS 5258: Part 17: 1992 Specification for direct gas-fired tumble dryers (2nd and 3rd family gases).

Fig. 34   Gas heated tumble dryer (Crosslee)

They are installed to BS 7624: 1993 Specification for installation of domestic direct gas-fired tumble dryers up to 3kW heat input (2nd and 3rd family gases).

Generally domestic tumble dryers are designed to fit into the 600 mm wide openings of modern kitchen cabinetry. Most are fitted with castors at the rear and have feet at the front. Some dryers are designed to be fitted (stacked) on top of a front-loading automatic washing machine.

Figure 35 shows the general arrangement of components and the air flow in a gas-heated tumble dryer. The dryers are operated by an electro-mechanical timer fitted on the control panel. The electric supply to the full sequence control unit is via the following:

- switched electrical supply
- door microswitch (made when the door is closed)
- timer switch
- exhaust air thermostat (usually two are fitted, one to operate at 60° C, for normal fabrics, the other to operate at 50° C, for delicate fabrics, the required temperature being operated by a switch on the control panel)
- inlet air thermostats (fitted in the rear banjo to protect against overheating. One is set to operate at approximately 110° C and the other at around 120° C).

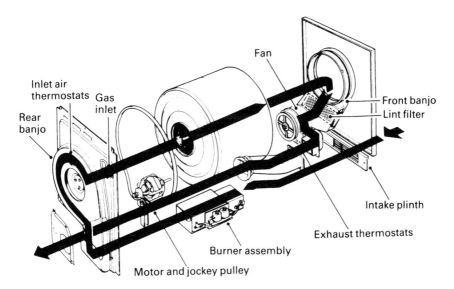

Fig. 35   General arrangements of major components and air flow

The sequence of operation of the control unit cannot commence until contacts are made in all the above-mentioned components. If any of the switch contacts are opened (by opening the appliance door, switching off the timer etc.) then the electrical supply to the control unit is removed and the control goes to a 'lock-out' position until the supply is re-instated by opening and closing the door, switching the timer off and then on again etc.

In normal operation when power is supplied to the control unit there is a pre-purge period of approximately 10 seconds created by the fan and motor running. At the end of this pre-purge period the following actions take place simultaneously:

- electricity is fed to the gas solenoid valves (two valves fitted in series in the gas supply – for additional safety)
- spark ignition commences at the ignition electrode close to the burner
- a neon light on the control panel switches on (this neon is connected in parallel with the gas valves and when lit indicates that they are open)
- the flame supervision electrode, in the flame path, starts to look for the presence of a flame by the partial a.c. rectification method.

If a flame is sensed before the end of the 10 second ignition period the ignition spark will stop and the control will continue in an operative condition until the electrical supply to the control is switched off. When this happens the control goes into 'lock-out' and one of the actions previously described must be taken to re-start the appliance.

If a flame is not sensed during the initial 10 seconds ignition period then the control will go to a lock-out position. If the flame failure electrode probe fails to sense a current of more than $1\mu A$ – this can be caused by starvation of gas, deviation of the gas flame, air flow failure etc. Then the control unit will immediately go to 'lock-out' position.

It is important to note that the flame failure electrode circuit will only be effective if the polarity of the electrical supply is correct and there is an efficient earth connection to the appliance.

Summary of Operation

With the electricity switched on, door switch closed and the timer in an open position power flows to the motor, fan, timer motor and flame control unit (through the selected exhaust air thermostat and inlet air 110° C thermostat). The fan and motor will both run during

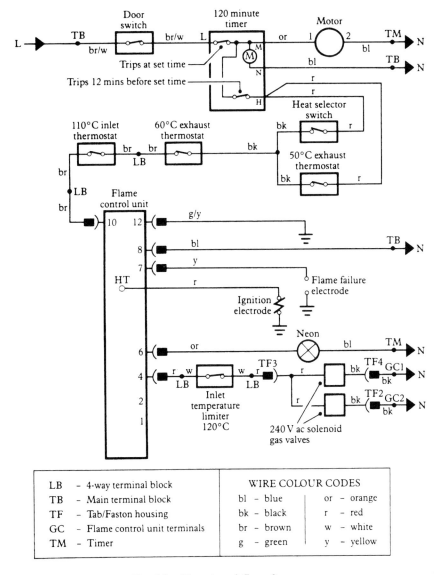

Fig. 36   Functional flow diagram

the 10 second pre-purge period. At the end of this period the spark generator is energised, the two gas valves will open and the neon indicator will light. As soon as a flame is sensed, by the flame supervision electrode, the spark generator will cease to operate and the appliance will operate normally. During this normal operation the selected exhaust air thermostat will protect the clothes against over-drying by switching the power on and off to the flame control unit. Twelve minutes before the drying period ends the timer will shut off the power supply to the flame control unit but the fan and motor will continue to give a cooling period until the timer shuts off.

A restriction to the flow of heated air can be caused by an overloaded drum or blocked air inlet/filter or blocked/restricted exhaust. This will cause overheating and is protected by the two air inlet thermostats. If the 110° C thermostat operates, power to the flame control unit will be cut off and the burner will shut off but the fan and motor will continue to run as long as the timer is on. When the thermostat cools to approximately 95° C the circuit to the flame control unit will be restored and if the timer is still on, the burner will re-ignite. The 120° C thermostat is a back-up to the 110° C thermostat. If the 120° C thermostat shuts off, power is cut off to the two solenoid valves and the neon indicator. The fan and motor will continue to run but absence of a flame at the flame supervision electrode will cause the spark generator to spark for approximately 10

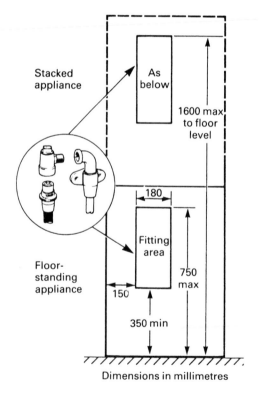

Fig. 37   Wall fitting location for gas supply pipe

seconds. As the thermostat cannot cool down to 90° C in that period the control unit will go into the 'lock-out' position. It will require a manual operation to clear this condition, switch the power on and off or open and close the door etc., when the thermostat has cooled. Until the fault has been rectified the occurrence is likely to be repeated and because of the high temperature involved, the flame control unit and 110° C thermostat must be changed.

Figure 36 shows a functional flow diagram of a typical gas-heated tumble dryer.

Installing Gas-heated Tumble Dryers

Gas-heated tumble dryers must not be fitted in a bedroom, bathroom or shower room. The room in which they are installed must have an openable window or other means of ventilation. At the times the dryer is in use the window must be kept open. When the appliance is fitted in a room of less than 10 m^3 volume a permanent ventilation area of 10 000 mm^2 is also required.

Tumble dryers are generally designed to be fitted to a flexible gas connection tube to BS 669. Fig. 37 illustrates the piping arrangements

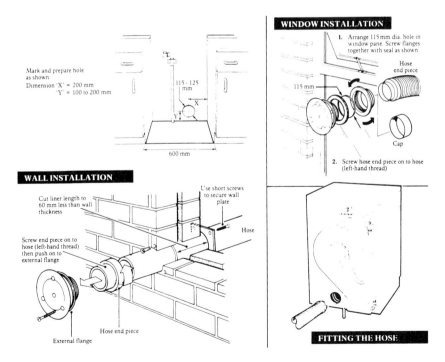

Fig. 38   Venting options, using kit

prior to connecting the flexible hose which should hang down from the bayonet connector. A 13 amp socket with an efficient earth connection must be fitted in close proximity to the dryer.

The appliance should have a vent hose terminating to outside atmosphere. Fig. 38 shows the installation of the vent hose kit either through a wall or a window.

It should be remembered that the fan on the tumble dryer could have an adverse effect on the flue of an appliance fitted in the same or an adjacent room. Spillage tests, as described in Chapter 5, should be carried out on appliances that might be adversely affected by the operation of the tumble dryer fan.

Servicing Tumble Dryers

The following operations should be carried out when servicing tumble dryers:

- remove and clean the burner
- check that any filters fitted in the air ducts are clean. If it is obvious that any filter is not being cleaned as per manufacturer's instructions then instruct the customer of the need to do this
- check the working pressure and flame picture
- check the operation of the flame supervision device
- check that all ventilation grills are correctly sized and clear.

Hot Fill Dishwasher

In the past the trend has been to have domestic dishwashers made for cold fill only, with an integral heating element.

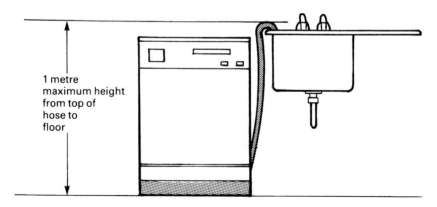

1 metre maximum height from top of hose to floor

Fig. 39   Discharging water into sink

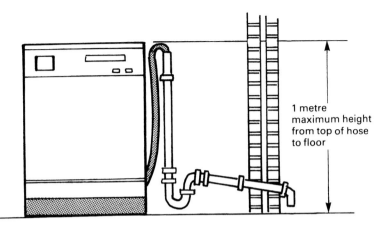

Fig. 40   Discharging by a fixed drain outlet

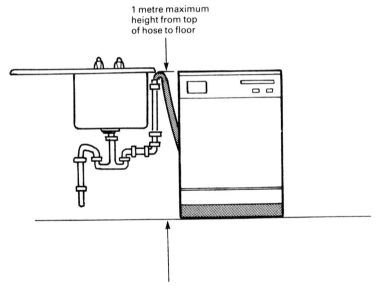

Fig. 41   Discharging into a universal waste trap

Now a manufacturer of gas appliances is marketing a 'hot fill' dishwasher, fitted with a lower rated element to top up the water temperature, so promoting a non-gas appliance supplied from a gas heated hot water system.

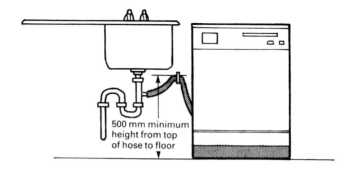

Fig. 42(a)   Discharging into an existing sink trap and waste

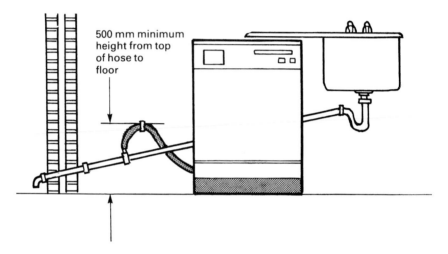

*Fig. 42(b)   Discharging into an existing sink trap and waste
              (alternative method)*

Installing Dishwashers

There should be a positive head of water, the lower the water head, the longer the machine will take to fill. The maximum recommended hot water storage supply temperature is 70° C.

The water hose connection has a $R_c\,3/4$ union connector at its free end. It is fitted by the manufacturers and should not be removed or replaced with an alternative fitting. The hose should be connected to the hot water supply through a separate tap or stop cock which serves only the appliance. The inlet hose must not be connected to a mixer tap.

Both the hot water connection and method used to discharge water into the drains should satisfy the local water authority bylaws. Multi-storey buildings might require expert advice.

The methods of discharging into drains are as follows:

Directly into a sink; Fig. 39 by hanging the hooked end over the edge of the sink, the top of the hooked end must not be higher than 1 m above the floor.

By a fixed drain outlet (Fig. 40). To prevent the water draining out of the machine a water break is essential and should not be lower than 500 mm above the floor. As with Fig. 39 the maximum permissible height from the floor to the top of the hose is 1 m. The drain stand pipe should have a minimum internal diameter of 38 mm and a trap should be provided between the stand pipe and drain to prevent odours from the drain being transmitted into the property.

Figure 41 shows a method of discharging into a universal waste trap and Fig. 42(a) and (b) discharging the water into an existing sink trap and waste.

CHAPTER 7

Space Heating

Chapter 7 is based on an original draft by B. Gosling

Introduction

In this country it is usual for us to need some heating in our homes for at least part of the day during eight out of the twelve months of the year.

Outside temperatures can vary from 0° C in the early morning frost to 20° C in the afternoon sunshine, during the same day. And we can certainly save energy by not heating rooms when they are not being used. So we need flexible, responsive forms of heating.

All of which explains the continuing popularity of the individual space heaters. They offer a simple, easily controlled means of obtaining heat where and when it is needed. They can be used economically and additional heaters may easily be added if more rooms are required to have a higher standard of heating. So the need of a growing child for a study-bedroom or an elderly relative for a bed-sitter can be met without difficulty.

Of course, it is desirable that the whole dwelling should be maintained at a minimum temperature of about 13° C. This is to:

- prevent water pipes and cisterns from freezing
- keep the fabric of the building and the furnishings warm and dry
- prevent condensation occurring on walls and ceilings
- reduce the time needed to raise the temperature in any room to full comfort level
- enable the occupants to move from room to room without discomfort.

Individual space heaters can provide the minimum temperature or 'background' heating, as well as the full heating in the separate rooms. The types of space heaters available are:

- Gas fires: these are either radiant or radiant/convector, hearth or wall mounted. They can be open or balanced flued, natural or

312

fanned draught and produce either natural or fanned convected heat.

Some modern gas fires have fanned flues plus fanned convected heat and when operating at their top efficiency they fall into the category of 'condensing' appliances.

The radiant heat from modern gas fires may be emitted from a traditional white fireclay radiant panel or a ceramic base and a fuel effect area.

- Convectors: these may be flueless, although these are not very common today as they can give a problem with condensation, open or balanced flued, natural or fanned draught. They may also produce natural or fanned convected heat.
- Cabinet (Mobile) Heaters: these are flueless LPG appliances usually burning butane and giving off mainly radiant heat.

Generally gas fires are fitted in living rooms, dining rooms and bedrooms where there are existing brick chimneys or concrete flues (flue blocks). Their output of radiant heat quickly provides a feeling of warmth and the combined radiant and convected heating elements soon heat up the whole room.

Since 1st January 1996 the Gas Safety (Installation and Use) Regulations have prohibited the installation of gas fires or other gas space heaters in any room used or intended to be used for sleeping accommodation unless they are room sealed or, fitted with a device which turns off the flow of gas to the burner before a dangerous level of fumes can build up (Regulation 30(3)). Such a device would operate using an oxygen depletion system incorporated into the thermo-electric flame failure system. These units are commonly called anti-vitiation devices or oxy-pilots. See the Gas Safety (Installation and Use) Regulations 1998 at the end of Chapter 5.

Convectors are commonly fitted in halls, often to supply background heat to the surrounding area and sometimes to the bedrooms etc. in the floor above. They may be used in any room and can effectively provide additional heating in a large living room in association with a radiant/convector fire.

Cabinet (Mobile) heaters burning LPG are often used as a temporary source of heat, sometimes to supplement the existing heating system, in very cold weather conditions.

In addition to gas fires, convector heaters and cabinet heaters used to improve the comfort conditions in a room, decorative live fuel effect appliances, giving off some heat but mainly installed for decorative purposes, are becoming increasingly popular. These are generally installed to provide a focal point in a living room or other already heated area.

The manufacture of gas fires, convectors etc. comes under BS 5258 Safety of Domestic Gas Appliances. It is covered by the following:

Part 5 Gas fires.
Part 8 Combined appliances: gas fire/back boiler.
Part 10 Flueless space heaters (excluding catalytic combustion heaters), 3rd family gases.
Part 11 Flueless catalytic combustion heaters (3rd family gases)
Part 12 Decorative gas log and other fuel effect appliances (2nd and 3rd family gases).
Part 13 Specification for convector heaters.
Part 16 Inset live fuel effect fires (2nd and 3rd family gases).

No European Standards (EN) are envisaged for the UK type of gas fire, therefore in Europe the British Standard is recognised.

Sizing Individual Heaters

The ideal method of deciding on the size of heater required is to calculate the heat losses from the room (Vol. 1, Chapter 10). This is the only accurate method since it takes into consideration the main factors which affect the heat loss. These are:

- the exposure and the aspect of the building; this varies from being on a hill facing N or NE to sheltered, facing S or SW
- house or bungalow; the bungalow has a roof over every room and the ground beneath every room with higher heat losses from both
- size of windows and type of glazing
- the structure and its capacity for transmitting heat
- the amount of heat lost by air infiltration.

However, there are times when it is necessary to estimate the size of heater required for a room, as when checking that a heater is, in fact, capable of producing the output required.

As a rough guide, experience has shown that the following figures may be used. These are for a house, with an average exposure and with average sized windows.

S.I. Units

The quantity of heat required to produce comfort conditions in a room is approximately:

Heat required = Q × volume of room in cubic metres

Where Q = 0.05 kW for heating by radiation
0.06 kW for radiant heat and convection
0.07 kW for convection

The formula therefore gives the output required from the heater.

To find the input rating, the efficiency must be taken into consideration. This is approximately:

- 50% for a radiant fire
- 60%*for a radiant convector
- 70% for a convector heater.

So, the inputs required are:

$$0.05 \times \text{volume of room} \times \frac{100}{50} = \text{volume} \times 0.1$$

$$0.06 \times \text{volume} \qquad \times \frac{100}{60} = \text{volume} \times 0.1$$

$$0.07 \times \text{volume} \qquad \times \frac{100}{70} = \text{volume} \times 0.1$$

But, $\text{volume} \times 0.1 = \dfrac{\text{volume}}{10}$

Then, for any room heater the input rating in kW is the volume of the room in m^3, divided by 10.

For example:

A room is 4 m × 3 m × 2.5 m

Therefore volume = 30 m^3

$$\text{Input rating of heater required} = \frac{30}{10} \text{ kW}$$

$$= 3 \text{ kW}$$

Imperial Units

Heat required = Q × volume of room in cubic feet
Where Q   = 5 Btu/h for radiant heating
              6 Btu/h for radiation and convection
              7 Btu/h for convection

*On average this is usually 65–75% but in the case of a condensing appliance may be 90%, so the appliance will sometimes be oversized. However, this will give quicker heating up and it may then be turned down to a lower gas rate to maintain the room temperature.

Inputs required are:

$$5 \times \text{volume of room} \times \frac{100}{50} = \text{volume} \times 10$$

$$6 \times \text{volume of room} \times \frac{100}{60} = \text{volume} \times 10$$

$$7 \times \text{volume of room} \times \frac{100}{70} = \text{volume} \times 10$$

So the input rating in Btu/h is 10 times the volume of the room in cubic feet.

GAS FIRES

Radiant Fires

A typical radiant gas fire is shown in Fig.1.

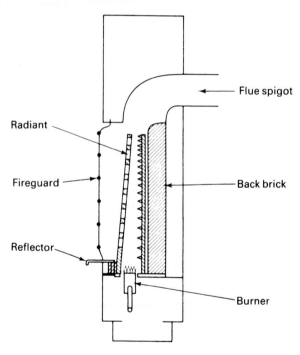

Fig. 1   Radiant gas fire

The fire consists of the following components that also form the radiant section of the traditional radiant/convector fire:

Outer Case

Usually pressed steel with a decorative finish of stoved enamel. It may have a wooden top and sides.

Firebox

This is an aluminised steel box which contains the radiants and the firebrick. The top is formed into a canopy which leads the products of combustion to the flue spigot. This projects about 50 mm from the back plate into the flue or fireplace opening.

Note: The radiant/convector fire has a convector heat exchanger between the top of the firebox and the flue spigot.

Firebrick

A refractory brick is normally fitted behind the radiants to prevent heat being lost into the chimney and to maintain the radiants at the highest possible temperature. The brick also prevents the firebox becoming overheated.

Radiants

Radiants have been made in many different shapes and sizes. They are produced from fireclay and are fragile, particularly after they have been in use for a time. The shape in common use is shown in Fig. 2. The flame from the burner burns inside the radiant and the hot gases raise the temperature to about 850° C to 950° C. Thorns or ribs are moulded to the internal surface to increase its area and improve the heat exchange. The radiant is tapered towards the top and the thorns are spaced and sized so as to distribute the heat from the gases and achieve an even luminous appearance.

To complete the combustion of the gas within the radiant, air is drawn in through the front slots. Radiants may be supported by hooks, moulded on to the back as in Fig. 2, or by a radiant support plate, Fig. 3. The support plate also acts as an air guide plate and controls the amount of secondary air passing to the flame. This limits the volume of excess air which would otherwise cool the radiant.

Since radiant output is proportional to the fourth power of the absolute temperature of the radiant (Vol. 1, Chapter 10), cooling would reduce the thermal efficiency of the fire, which is about 47%.

On radiant fires the radiant area is usually larger and the radiants themselves are sometimes taller than those on other fires which are

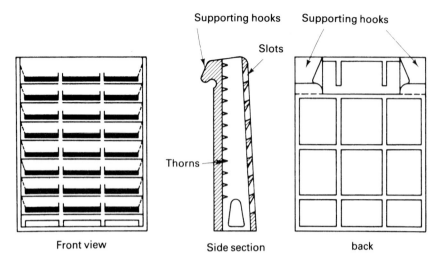

Fig. 2   Gas fire radiant

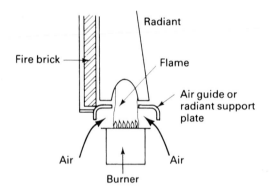

Fig. 3   Radiant support and air guide plate

designed to give a significant proportion of heat by convection. The overall output consists essentially of radiation with, perhaps, about 5% convection.

Reflectors

The radiant panel is usually surrounded by chromium-plated mild steel or stainless steel reflectors. This assembly is partly decorative and partly functional. It makes the radiant section appear larger and it reflects the heat waves away from the surrounding case. The bottom reflector, sometimes called a 'fender' helps to reduce floor temperatures immediately in front of the fire, Fig. 4.

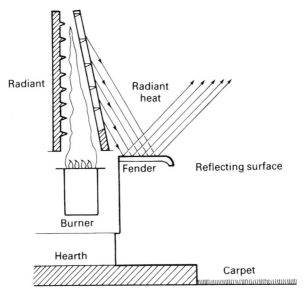

Fig. 4   Reflectors

Fireguard

Any fire fitted in a dwelling should have a fireguard to comply with the Heating Appliances (Fireguards) Regulations 1991. This law applies to all gas fires and normally a guard is incorporated in the design of the fire. From 1 January 1996 manufacturers must comply with the provisions of the Gas Appliances (Safety) Regulations 1995 (GASR).

BS 6778 covers the specification for fireguards for use with portable, free-standing or wall-mounted heating appliances.

Burner

Gas fire burners were described and illustrated in Vol. 1, Chapter 4. The burner ports are set in groups so that each radiant is filled with a flame.

Burners may be:

- simplex
- duplex.

On a simplex burner, all the flames are supplied by a single injector which is controlled by a simple tap, Fig. 5.

Duplex burners allow radiants or pairs of radiants to be controlled separately, Figs 6 and 7. Usually the outer pair of radiants may be

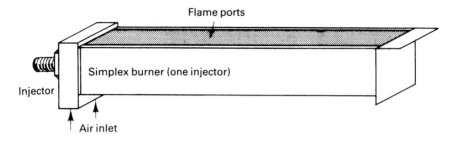

Fig. 5   *Simplex burner*

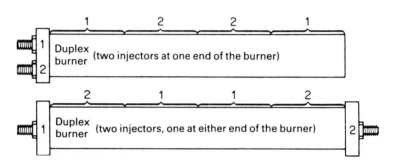

Fig. 6   *Duplex burners: (a) injectors at one end; (b) injectors at both ends*

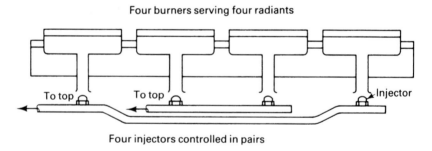

Fig. 7   *Duplex burner, injectors controlled in pairs*

turned off leaving the centre radiant/s at full-on rate. This enables the heat output of the fire to be reduced while still maintaining the highest possible radiant efficiency.

Fig. 8 illustrates a typical burner arrangement for a modern radiant/convector gas fire with live flame effect. There are in fact two burners. The main burner, which can have a variable gas rate, supplying the heat for the radiant/convector section of the fire. The

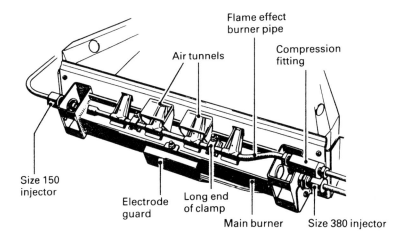

Fig. 8   Typical main burner and flame effect burner for the radiant heat/live flame section of a modern gas fire

flame effect burner, as its name implies, gives the effect of a flame thus providing an attractive focal point.

Igniter

Gas fires may usually be lit manually by a match or taper, but they all incorporate some form of automatic ignition. The types of igniter fitted are:

- filament, battery or mains powered (usually found on older type fires)
- permanent pilot (usually found on older type fires)
- spark, piezo electric.

The principles of these igniters were dealt with in Vol. 1, Chapters 2 and 11. Practical details are as follows:

Filaments nominally require 3 volts supplied either by two $1\frac{1}{2}$ volt HP batteries or from a mains transformer. Turning the appliance control knob past the full-on position, against a spring, closes the switch and opens the gas supply to the pilot jet. The filament glows and ignites the jet, so lighting the fire. When the tap knob is released, the spring turns the spindle back, shutting off the filament and its gas supply. A typical filament with leads, pilot tube and banjo connection is shown in Fig. 9.

If a filament is used in the horizontal or angled position it is important that the coloured dot stamped on the body is on top. This type of filament may be rotated on the gas supply tube.

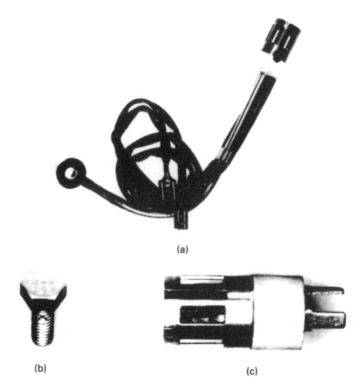

(a)

(b)

(c)

Fig. 9   Filament igniters: (a) igniter, complete with leads, pilot supply and banjo; (b) banjo securing screw; (c) igniter head

Permanent pilots are usually left alight when the fire is in frequent use, during the winter. The gas supply is generally taken from the main supply before the appliance burner control tap. A pilot tap, filter and adjuster are usually fitted in the supply.

Piezo spark generators may be operated by a cam in the appliance tap mechanism so that they function when the fire is turned on. Alternatively the impact type may be mounted separately, near the tap knob and operated by a push button. The spark gap may be between two electrodes or between a positive electrode and the burner, Fig. 10. The gap is usually 3 to 5 mm and the electrodes should not be bent with pliers as this may crack the ceramic insulation.

Radiant Convector Fires

This is the most popular type of fire and the open-flued model is the most common. The convected heat is at least 25% of the total output and usually nearer 50%. The efficiency is higher than that of a radiant fire, being about 65 to 75%.

Radiant/convector fires with a balanced flue, Fig. 16(a) and 16(b), are approximately 80% efficient while fires with fanned draught flues and fanned convection, Fig. 17, are about 90% when working at their highest efficiency.

The fire heats the room in which it is situated by both radiant and convected heat, Fig. 11. Typical radiant/convector fires are shown in Figs 12 and 13.

The convected heat is obtained by allowing air from the room to circulate around the firebox and become heated. This also reduces the heat loss through the back of the fire and helps keep the case cool. Because the case is relatively cool, decorative finishes and materials including wood, leather or vinyl can be used. In addition, the modern fire has a gas-to-air heat exchanger, Fig. 14.

The heat exchanger is a flat metal box made from aluminised steel or very occasionally, cast iron. It is flat to give maximum surface area for efficient heat transfer from the products of combustion to the air surrounding the heat exchanger. As soon as the air inside the fire case becomes heated it rises upwards and is discharged through louvres at the top. Cooler air from the room, which displaces the warm air, enters the fire at hearth level.

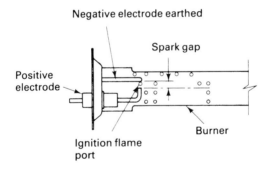

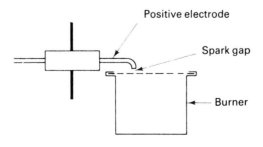

Fig. 10   Spark ignition

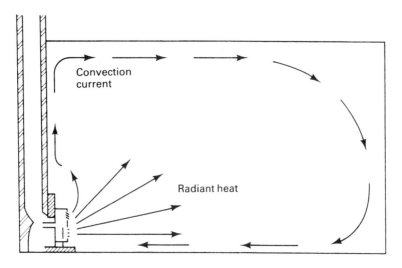

Fig. 11   Room heated by radiant convector fire

Fig. 12   Traditional type of radiant/convector gas fire (Robinson Willey)

Some radiant convector fires are enclosed with a glass panel covering the radiants, Fig. 15. This slightly reduces the radiant output but increases the overall efficiency.

Fig. 13   Live fuel effect type of radiant/convector fire (Valor)

With a conventional open flued fire any downdraught from the chimney can pass out into the room through the canopy. On a glass-fronted fire this opening is covered and a draught diverter is necessary to prevent the appliance being vitiated or smothered by its own products of combustion.

The principle of the balanced flue was covered in Chapter 5 and it has been applied to the radiant/convector fires, Figs. 16(a) and 16(b). The fire is similar to most other balanced flued heaters with the radiants visible through a 'glass' panel made from a transparent ceramic material. The panel has to be removed to exchange the radiants or carry out some servicing operations. It should be replaced strictly in accordance with the maker's instructions so that there are no leaks and combustion is not disturbed.

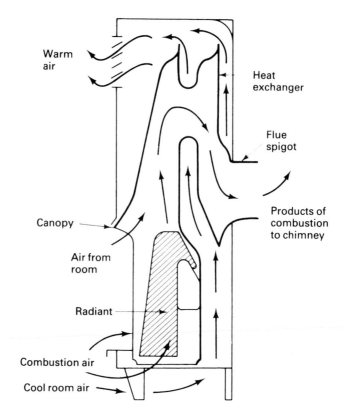

Warm air

Heat exchanger

Flue spigot

Canopy

Products of combustion to chimney

Air from room

Radiant

Combustion air

Cool room air

Fig. 14   Heat exchanger

A schematic drawing of a condensing gas fire is shown in Fig. 17. This appliance relies on the action of a toroidal fan, fitted in the flue, to entrain combustion air from the room in which the appliance is fitted and to create the flow of heated products through the twin section convected air heat exchanger and out through the CVPC (chlorinated polyvinyl chloride) flue. The action of the toroidal fan allows the appliance to be fitted to a flue pipe of approximately 28 mm diameter. Convected air is entrained from the room and passed over the twin section convected air heat exchanger then returned, heated, to the room by the action of the convection fan. In order to satisfy the requirements of the Gas Safety (Installation and Use) Regulations 1998, appliances of this type have to be fitted with controls that will shut down the appliance in the event of fanned flue draught failure. The flue can be up to 6 m long (depending upon the number of elbows used) and can run horizontally. Fig. 18 shows the

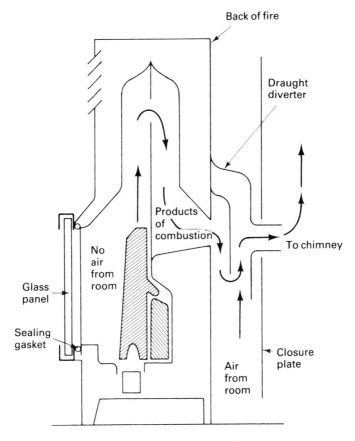

Fig. 15   Glass-fronted fire

general layout of a typical flue system on a condensing gas fire, including the arrangement for disposing of the resulting condensation.

Some radiant/convector fires were fitted with thermostats. These were usually liquid expansion types with the temperature-sensitive element in the form of a phial or coil of tubing. This sensing element was situated below and to the side of the burner in the path of the air passing up to the convection heat exchanger, Fig. 19.

The body of the thermostat, with the control knob and bellows were either mounted separately or formed part of the control tap assembly. Thermostats were often a source of complaint on radiant/convector fires and most modern appliances rely on manual control by the occupiers of the room in which the fire is fitted.

Some fires are available with imitation coal or log effects operated by electricity. The flicker is produced by spinners mounted on pivots

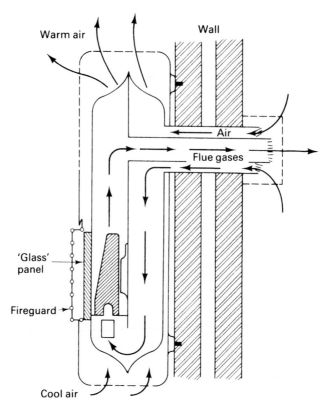

Fig. 16(a)   Balanced flue fire (traditional radiant type)

over the light bulbs and rotated by the current of warm air. The bearing at the centre of the spinner should be kept clean and dry. It may be lubricated with graphite by rotating it on the tip of a soft lead pencil. Do not use oil, it will become clogged.

There are three basic types of gas fires that have a flame effect. Fig. 13 illustrates one of these where the traditional radiant heat producing section of the fire is replaced by a flame effect. This appliance also has a heat exchanger for heating convected air; its overall efficiency is approximately 60–70%. The appliance also has a glass screen in front of the live flame effect, which controls the amount of air drawn through the fire and consequently helps to increase its efficiency. The glass screen does however mean that the appliance needs an integral downdraught diverter and checks for spillage have to be carried out at the skirt of this downdraught diverter.

Fig. 20 illustrates an inset live flame effect fire. This type of appliance also produces convected heat from a convector heat

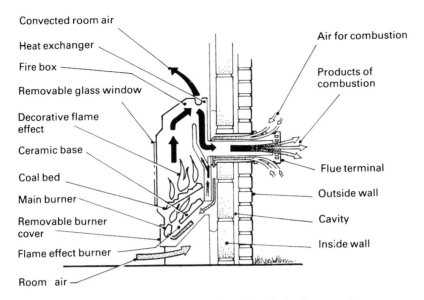

Fig. 16(b)   Balanced flue fire (live fuel effect type)

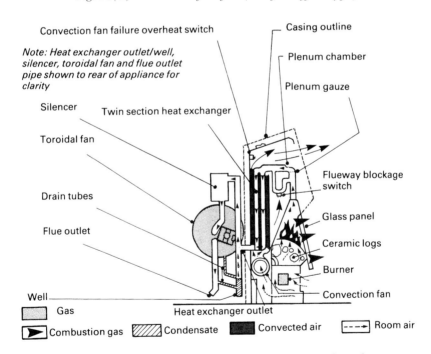

Fig. 17   A condensing gas fire, showing combustion, air and condensation flows

exchanger. The flame effect section is open and this usually limits the efficiency of the fire to around 40 to 50%. The appliance is fitted into a fireplace opening. Fig. 21 shows the main assemblies and component parts of an inset live flame effect gas fire.

A third decorative fuel effect fire is illustrated in Fig. 22. This appliance, which only produces radiant heat, has a very low efficiency and is generally used as a focal point in already heated rooms or spaces.

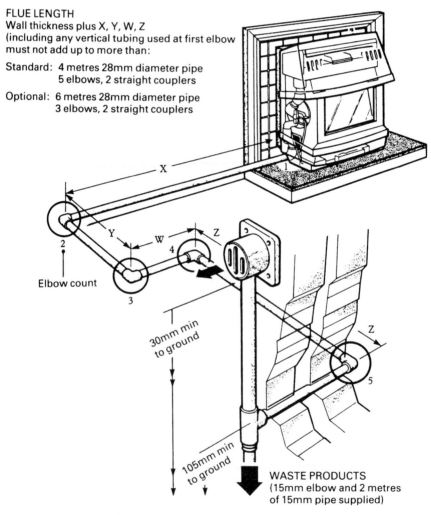

FLUE LENGTH
Wall thickness plus X, Y, W, Z
(including any vertical tubing used at first elbow
must not add up to more than:

Standard: 4 metres 28mm diameter pipe
5 elbows, 2 straight couplers

Optional: 6 metres 28mm diameter pipe
3 elbows, 2 straight couplers

Elbow count

30mm min to ground

105mm min to ground

WASTE PRODUCTS
(15mm elbow and 2 metres
of 15mm pipe supplied)

Fig. 18   The general layout of a condensing gas fire flue system

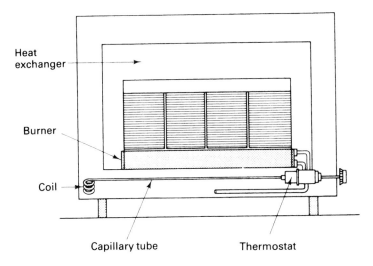

Fig. 19   Location of thermostat

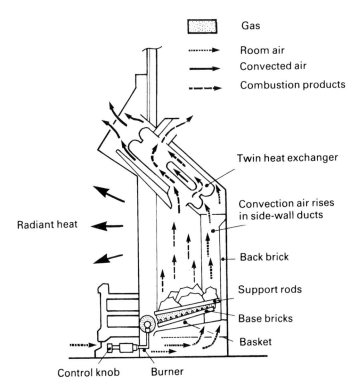

*Fig. 20   Inset live fuel effect fire, showing room air, convected air and
products of combustion flow*

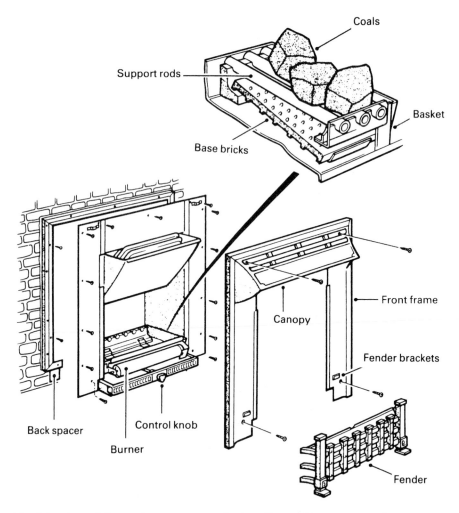

Fig. 21   Assemblies and components of a live flame effect inset gas fire (Cannon)

Another innovation is a glass-fronted live fuel effect radiant/convector gas fire, fitted with openable glass doors, Fig. 23.

With the doors closed it is a conventional radiant/convector fire, operating at its highest efficiency. With the doors open the radiant heat is increased at the expense of the convected heat and overall efficiency.

Note: It is vital that the logs/coal etc. of all fuel effect appliances are located in the appliance strictly to the manufacturers' instructions.

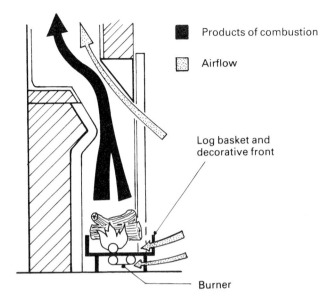

Products of combustion

Airflow

Log basket and
decorative front

Burner

*Fig. 22   A decorative fuel effect fire, where most of the heat produced goes
up the flue*

Fig. 23   Inset live fuel effect gas fire (Valor)

Installation of Gas Fires

To satisfy the requirements of the Gas Safety (Installation and Use) Regulations 1998, the installer of a gas fire must ensure that the flue, to which the appliance is to be fitted, is suitably constructed and in a proper condition for the safe operation of the appliance.

The installer must also ensure that there is a means of shutting off the supply of gas at the inlet of the appliance.

The procedure for fitting a fire where there is an existing gas installation is as follows:

- ensure that the flue and ventilation are suitable for the fire being fitted
- turn off all other appliances
- test the existing installation for soundness
- run the gas supply and fix the fire
- test the completed installation for soundness
- if sound, purge the supply and turn on other appliances (as necessary)
- check the appliance working pressure and adjust if required
- check the flue for spillage
- check the operation of any safety controls (FSD etc.)
- inform the customer on the safe and correct operation of the fire
- leave any installation and user's instructions with the customer.

The installation of gas fires etc. is currently covered by the following specifications:

BS 5440: Parts 1 and 2 – Installation of flues and ventilation for gas appliances of rated input not exceeding 60 kW (1st, 2nd and 3rd family gases).

BS 5871: Specification for installation of gas fires, convector heaters, fire/back boilers and decorative fuel effect gas appliances.

Part 1: 1991 Gas fires, convector heaters and fire/back boilers (1st, 2nd and 3rd family gases).

Part 2: 1991 Inset live fuel effect gas fires of heat input not exceeding 15 kW (2nd and 3rd family gases).

Part 3: 1991 Decorative fuel effect gas appliances of heat input not exceeding 15 kW (2nd and 3rd family gases).

BS 5482: Domestic butane- and propane-gas-burning installations.

Part 1: 1994 Specification for installations at permanent dwellings.

BS 6891:   1998 Specification for installation of low pressure gas
            pipework of up to 28 mm (R1) in domestic premises
            (2nd family gas).

The type of gas fires covered by Part 1 of the BS 5871 is shown in
Figs 12 and 13. They are manufactured to BS 5258 : Part 5 and
normally fit in front of a closure plate that covers a fireplace opening.
No purpose-provided ventilation is normally required for fires of this
type up to 7 kW input. A flue of 125 mm minimum across its axis is
normally required.

Part 2 covers inset live fuel effect gas fires manufactured to
BS 5258: Part 16. As the name implies these are either fully or
partially inset into a fireplace recess or opening. Figs 20, 21 and 23
illustrate this type of appliance which normally requires a flue with a
minimum axis of 175 mm and no purpose-provided ventilation for
inputs of up to 7 kW.

Decorative fuel effect gas appliances, manufactured to BS 5258:
Part 12 are covered by BS 5871 : Part 3. Fig. 22, shows an example of
this type of fire. They are either fitted in a fireplace/builder's opening
or under a purpose-made flue canopy. A flue with a minimum axis of
175 mm is normally required for this type of appliance plus purpose-
provided ventilation of at least 10 000 mm^2 for appliances up to
15 kW input.

Flueless space heating appliances must not be installed in a bedroom, bed-sitting room or any room containing a bath or shower.

Open flued appliances must not be installed in any room or internal
space containing a bath or a shower.

Open flued combination fire/back boilers or fire/back circulators
must not be installed in a bedroom or bed-sitting room.

Any appliance burning 3rd family gas must not be installed in a
room or internal space below ground level. Note the limitations on
fitting appliances in bedrooms, bath or shower rooms (Gas Safety
(Installation and Use) Regulations 1998, Regulation 30) at the end of
Chapter 5.

Gas Supply, Hearth Fixing

Most fires have an inlet gas connection situated below the burner in
the middle of the fire about 25 mm above the hearth level. It is
usually in the form of a union elbow with an $R_c\frac{1}{4}$ thread. The
fittings required to install the fire are shown in Fig. 24.

Gas Supply, Concealed Fixing

Figures 13 and 14 of Chapter 3 – Internal Installation, illustrate
methods used to connect concealed gas supplies to gas fires. The

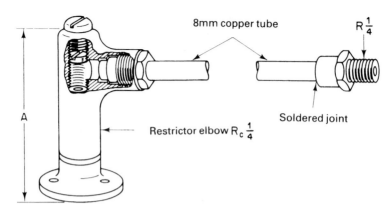

8mm copper tube

$R \frac{1}{4}$

Soldered joint

Restrictor elbow $R_c \frac{1}{4}$

A

Dimension A. 50mm without extension piece.
199mm with extension piece.

Fig. 24   Gas fire fixing kit

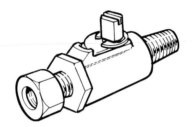

Fig. 25   R¼ × 8 mm Cu ball valve

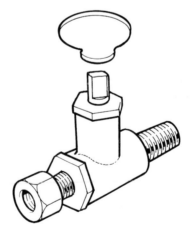

Fig. 26   Isolating valve with a loose key

installer must remember that any gas supply pipe buried in the structure or running within a chimney recess or fireplace opening, must be suitably sleeved or protected, e.g. coated or wrapped with PVC tape.

Fig. 25 shows a R¼ × 8 mm Cu ball valve that can be fitted to the inlet of the appliance as a means of shutting off the gas supply.

Where fires are fitted in rooms where the occupants may tamper with the means of shutting off the gas supply, e.g. nurseries or old people's homes etc., an isolating device with a loose key, Fig. 26, may be preferred.

Closure Plate and Spigot Restrictor

A 'closure plate' is a sheet of metal, usually 22 swg half-hard aluminium or 27 swg aluminised steel. It is sealed to the fireplace or flue opening with heat-proof tape which may be adhesive on one or both sides. This tape may have to withstand temperatures of up to 100° C and British Gas plc have produced a standard, PRS 10, for the tape used by their employees when fitting closure plates.

Closure plates are designed specifically for each individual fire. They control the rate at which air is drawn from the room into the chimney, so that the room is kept at a comfortable level without draughts. The plate incorporates relief air openings to give a ventilation rate for a typical fire of between 56 m³/h and 70 m³/h, Fig. 27. This is approximately two air changes per hour.

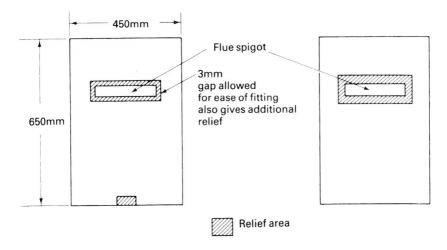

Fig. 27   Closure plate

The open area around the spigot plus the relief area at the base gives the total relief area. Sometimes this is all located around the spigot. It is important that the closure plate is correctly sealed on all four edges so that air can only be drawn into the chimney through the permitted openings. If too much air is allowed to pass under the plate it may starve the burner of combustion air when there is a strong flue pull. This will cause the flames to distort and reach downwards as in Fig. 28.

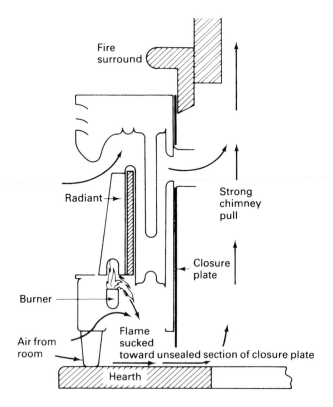

Fig. 28   Effect of unsatisfactory seal

On a normal radiant convector fire, the products of combustion have a choice of two ways out of the fire. One is via the canopy, which is open and unrestricted, the other is through the heat exchanger to the flue spigot. The resistance to flow is much greater through the tortuous path of the heat exchanger so, unless there is sufficient pull from the flue, the products will take the route of least resistance and escape to the room through the canopy opening. This is called 'spillage'.

A 250 × 250 mm brick chimney in an average 2-storey house has a height of 8.5 m and creates a flue pull of about 0.1 mbar, with a typical fire. Fires usually clear their products of combustion with a flue draught of between 0.025 and 0.03 mbar.

Increasing chimney height increases the flue draught and pulls more air from the room into the chimney. This air passes through the heat exchanger, so cooling it down and reducing the overall efficiency of the fire, Fig. 29.

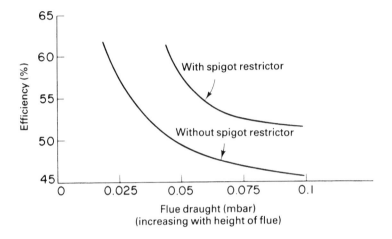

Fig. 29   Effect of flue draught on efficiency of fire

To offset the effect of a tall chimney, manufacturers usually recommend fitting a 'spigot restrictor' where a flue is more than 4 m high. Spigot restrictors are designed for particular fires to increase the resistance to flow through the heat exchanger and the size of the restrictor should not be altered. Follow the manufacturer's instructions carefully.

Testing Flues

Before fitting the fire the flue must be examined to ensure that:

- it serves only one room
- it is not blocked with paper or rubble etc.
- where solid fuel has been used, it has been swept
- any restrictors, such as dampers or register plates, have been removed

- there is adequate 'catchment' space, that is, volume behind the fire so that falling debris will not obstruct the products from the flue spigot
- any other openings (e.g. holes for circulating pipes, air supply to 'Baxi' fire etc.) into the recess are sealed.

Fit the closure plate and seal it to the fireplace opening on all four edges.

Apply a smoke match to the spigot opening. Check for adverse conditions. If there is a definite flow into the spigot opening, proceed to fit the fire. Any persistent down draught should be corrected, possibly by fitting an approved terminal to the top of the flue.

If no flow is indicated, introduce some heat into the flue by means of a blow torch. If there is still no flow, the flue may be blocked and the fire should not be fitted.

If the flue is satisfactory, continue installing the fire, fitting a spigot restrictor where necessary. After fitting the fire carry out a check for spillage as detailed in Chapter 5.

Light the fire, close all doors and windows in the room, switch on any extractor fan and leave the fire on at full-on rate for five minutes. Check along the canopy for spillage with a smoke match. At least some of the smoke should be drawn into the canopy.

If there is any doubt about the results, check again after leaving the fire on for a further ten minutes. If the products are still not clearing, remove the spigot restrictor and repeat the test. If the fire is still spilling, disconnect it and leave it disconnected until a thorough investigation of the flue has been carried out and the problem has been cured.

Pre-cast Flues

Gas fires may be fitted into flues made from pre-cast concrete blocks. These are usually one of the three sizes:

- 380 mm × 63 mm
- 300 mm × 63 mm
- 196 mm × 70 mm.

Because the flue ways are narrow, temperatures greater than 66° C may occur on the surface of the wall unless a cooling device, usually in the form of a 'cooler plate', is fitted to the rear of the closure plate, Fig. 30. Cooler plates may be made up from aluminium sheet and other similar devices are available from some of the appliance manufacturers.

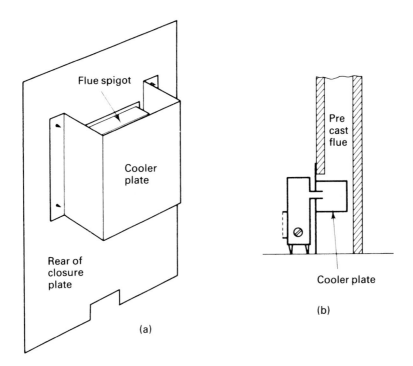

Fig. 30   Cooler plate: (a) plate fitted to back of closure plate;
(b) positioned in pre-cast concrete flue

Hearth and Wall Fixing

Any fire which stands on the floor should be placed on incombustible material, usually a tiled hearth. Where there is no hearth, a sheet of material complying with BS 476 must be provided. It should be of not less than 12 mm thickness and should project at least 300 mm forward from the back plate of the fire and at least 150 mm at either side.

Fires must be levelled and they may be held against the closure plate by means of small rectangles of plastic fixed to the hearth in front of the feet of the fire by means of a suitable contact adhesive.

Most fires can be mounted on a wall, usually by keyhole slots in the back panel of the fire. For wall mounting, the fireplace hearth and surround may be removed and the builder's opening bricked up or filled in by insulation board, leaving an aperture over which the closure plate may be fitted. Space must be left behind the plate, below the level of the spigot opening, in case soot or rubble should fall down the chimney.

Wall-mounted fires should be fitted so that the top of the burner is at least 225 mm above the floor, carpet or any other combustible material.

The screws used to secure the fire must have a satisfactory key in the wall and metal wall plugs should preferably be used. The gas supply should be run to a central position in the opening or in accordance with the manufacturer's instructions.

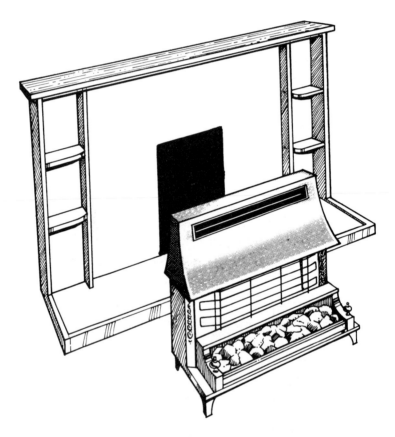

Fig. 31   Fire surround and gas fire

Fire Surrounds

Many gas fires are fitted to proprietary fire surrounds. A few fires, because of their design, are not suitable for fixing to surrounds and it is important for the installer to ensure that the fire and surround are compatible. It is vital for any shelf or overhang above the fire to be at least a minimum distance from the top of the fire and it should not project more than a permitted distance in front of the fire. These distances vary from fire to fire and it is necessary to consult the installation instructions for the fire and the surround to ensure the safety of the installation. In order to create an adequate seal, the floor surface where the hearth is to be fitted should be level. The wall against which the surround is to be fitted should be plumb. To ensure a good seal, the edges of the surround and the base of the hearth have a foam strip applied to them. The surround is usually secured to the wall by mirror clips or fixing brackets which are recessed into the plaster work. Where necessary the hearth should be supported.

Behind the surround there must be an adequate fire opening leading to a suitable flue. The closure plate is usually taped to the front of the fire surround. It can however be taped over the fire opening behind the surround but provision may then have to be made to examine and clean out the flue catchment area when the fire is serviced and it might also be necessary to extend the appliance flue spigot outlet to ensure that it protrudes at least 25 mm into the fireplace opening. Fig. 31 shows a fire surround and gas fire.

Fig. 32   LPG fire for installation in a caravan or non-permanent dwelling (Spinflow)

General Points on Installation

There are many ways of fitting gas fires in rooms with or without existing flues. A balanced flued fire (Fig. 16) may be fitted on a suitable outside wall. A gas fire with a fan-assisted flue (Fig. 18) may be installed. If a customer requires a specific conventional fire, then a false fireplace may be built in the angle of the wall or on the face of the wall and a 125 mm flue pipe used, the flue pipe being hidden behind suitable panels. Fireplaces and openings may be covered with in-fill plates of various designs and materials.

All installations should conform to current BS codes of practice.

Some fires have special requirements and manufacturers' installation instructions should be read carefully before you start work.

Finally when the job is completed, make sure that the customer knows how to operate the appliance.

LPG Fires

Most LPG fires installed in permanent dwellings are 2nd family gas fires which have been modified to burn LPG However, in caravans, mobile homes and non-permanent dwellings, usually only a smaller type of fire is fitted and some manufacturers produce a fire for this purpose. Fig. 32 shows such a fire.

Installation of LPG fires in caravans and non-permanent dwellings

The installation of these fires must be in accordance with the recommendations in the following British Standards.

BS 5482: Code of practice for domestic butane- and propane-gas-burning installation.
Part 2:   Installations in caravans and non-permanent dwellings.
BS 5601: Code of practice for ventilation and heating of caravans.
Part 1:   Ventilation.
BS 5871: Specification for the installation of gas fires, convectors and fire/back boilers and decorative fuel effect gas appliances.

The fire can usually be wall or floor mounted. When floor mounted provision is made to screw the appliance to the floor.

When fitting the flue pipe between the appliance and flue terminal the following points must be observed:

- provision must be made for periodic examination of the flue.
- under no circumstances should there be a fall or horizontal run in the flueing. All runs must incline upwards towards the flue terminal.

- bends should be as large as possible. The minimum radius should be 60 mm.
- the flue terminal should be no less than 1.83 m above the base of the fire and must be firmly fixed to the flue pipe using hose clips.
- support the flue away from any combustible material by a minimum of 50 mm using suitable brackets.
- the cupboard, false wall etc. that the flue runs through should have a vent of at least 1290 mm^2 open area at the top and bottom. Where possible both vents should be to the outside.
- after installation always check for the absence of spillage.

A typical installation is shown in Fig. 33 and a flue support in Fig. 34.

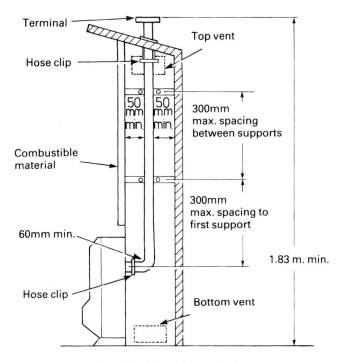

Fig. 33   Typical installation

Regulation 30(2) and (3) of the Gas Safety (Installation and Use) Regulations 1998 (see end of Chapter 5) also applies to the sleeping areas of caravans.

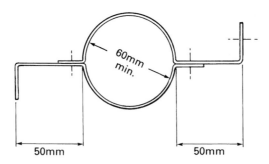

Fig. 34   Flue support

Servicing Gas Fires

Routine servicing should include the following operations:

- check ventilation requirements and flue termination
- clear the work area and lay down a dust sheet
- disconnect and remove the fire
- remove the outer case and the radiants or fuel
- check for soot and rubble behind the closure plate and clean out if necessary (excessively large quantities can block the fuel spigot and cause spillage)
- reseal the closure plate if required, ensuring that the relief opening is clear
- clear dirt and lint from the base of the fire where the cool air enters from the room
- clean dust and lint from the burner and the injectors
- check the heat exchanger for corrosion damage and clean the visible flueway
- grease gas tap if it is stiff to turn
- reconnect the fire and check for gas soundness
- ensure fire is level and secure
- test the ignition, clean or renew any batteries or parts as necessary
- check any electrical components; wires should be properly insulated, connections clean and sound
- test any spark gap and reset if necessary
- check gas rate by burner pressure and adjust governor if required
- check the operation of any flame supervision device, if fitted
- check that thermostat phial is correctly located and that thermostat operates satisfactorily
- replace radiants and case
- check flue pull

- clear up the work area.

Fault Diagnosis and Remedy

Gas fires may develop any of the faults which can occur on burners or control devices and were described in Vol. 1. There are, however, a number of faults which are characteristic of gas fires.

Radiants

'Ghosting' and 'pluming' are terms used to describe flames emerging from the top of a radiant. Some fires ghost all the time but if ghosting develops over a period of use it may indicate that the gas is not getting sufficient air to burn correctly. The burner may be linted or the air guide plate damaged or dirty. Ghosting may occur if the base of the fire is restricted, as with a thick rug.

Uneven radiant appearance may be a fault in the burner or the radiant. Change the radiants over to pin-point the cause. One radiant may be more restrictive than the others. On a duplex burner one injector may be partially blocked or one section of the burner may be restricted.

Lack of Warm Air or Room not Warm Enough

If this is simply a lack of convection air, the fire may need a spigot restrictor.

It may, however, be due to the thermostat (if fitted) shutting down before the room has been properly heated. This can be due to:

- faulty thermostat
- phial or capillary tube displaced by customer when cleaning and touching the hot firebox
- cool room air not circulating properly up through the fire.

This latter fault may be caused by the closure plate not being properly sealed or by the fire being fitted behind a very high curb so that room air is heated by the hot curb and hearth before it reaches the fire. It may be necessary to put the fire on a plinth to improve the circulation of cool air.

Ignition

Faults on ignition systems were dealt with in Vol. 1 and electric spark generators are covered in Chapter 13. Fault-finding algorithms are included in the BGC multi-meter instruction book.

The commonest faults on piezo-electric ignition are associated with the spark gap and with the current tracking to earth from the high tension lead.

On filament ignition, check batteries or transformer and the orientation and position of the igniter head relative to the main burner.

CONVECTOR HEATERS

Flueless Convectors

These are simple heaters consisting essentially of a sheet steel box, forming a combustion chamber and containing the burner, in a pressed steel outer case, Fig. 35.

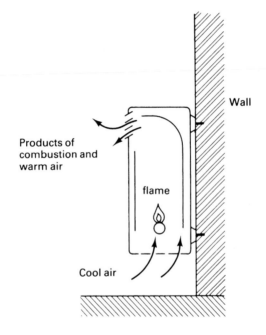

Fig. 35   Fixed flueless convector

The warm air, mixed with the products of combustion, is discharged into the room through louvres in the case. Because it is flueless, the input is limited to 50 W per cubic metre of room space, in a room, or 100 W/m^3 in another location. So heaters are produced with either a 1.5 kW or a 2 kW input and usually fitted in hallways with stairs leading to an upper floor. This location has the volume specified by the ventilation requirements (Chapter 5).

Balanced Flued Convectors

Natural Draught, Natural Convection

This was the most common type in the past and many appliances were imported from Holland and Germany. A typical design is shown in Fig. 36.

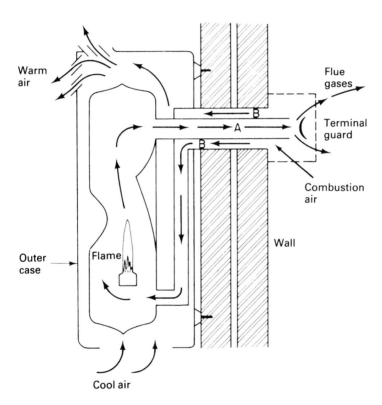

Fig. 36   Balanced flued convector, natural draught

The air duct and flue duct are often concentric. The inner duct, A, is usually aluminium and the outer, B, galvanised steel. Heat exchangers are commonly vitreous enamelled mild steel. Some are made from cast iron or aluminised steel.

The majority of these appliances have thermoelectric flame supervision incorporated in a multifunctional control. This also houses a pressure regulator, thermostat, piezo ignition and a two-stage turndown facility.

Because of its balanced flue there are no ventilation problems so the heater may be sized for the full heat requirement of its location. A large heater in a hall can supply background heating for upstairs bedrooms.

An added advantage is that, because it does not rely on electricity, the heating service is not lost during power cuts.

Fanned Draught, Fanned Convection

These heaters are the product of British research and development and are more compact than the previous category. They are termed 'fan-assisted' appliances and may have one or two fans. With a fan-assisted combustion circuit, the intake of fresh air and the discharge of products is forced, so the resistance of the heat exchanger and the flue ducting may be high. For example, up to about 0.6 mbar. This

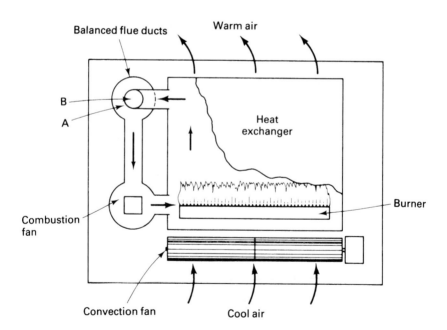

Fig. 37   Balanced flued convector, fanned draught, fanned convection, separate fan motors

allows the use of smaller flues of about 65 to 75 mm diameter, compared with the 125 to 175 mm diameter needed for a natural draught heater.

One type of assisted heater is shown in Fig. 37. This has a combustion fan drawing fresh air from around the terminal through duct A, to the heat exchanger. Products of combustion are then discharged at the top of the heat exchanger, through the flue duct B.

The convection fan is near the base of the heater and warm air is forced out at the top.

A similar model is illustrated in Fig. 38, but the fans for the combustion circuit and for the convection air are both driven by the same electric motor. In this heater the warm air is discharged at the base.

A further type, Fig. 39, has its combustion fan on the flue outlet from the heat exchanger and not on the air inlet.

Gas, supplied at atmospheric pressure through a zero governor, is introduced into fresh air ducted from the air inlet. Mixing occurs in a

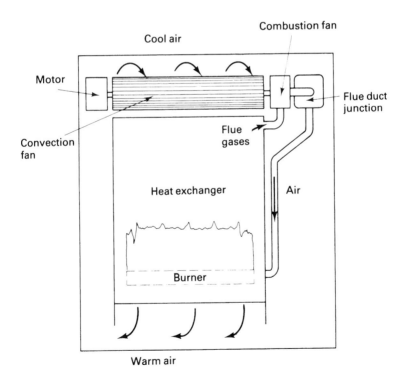

Fig. 38   Balanced flued convector, fanned draught, fanned convection, single fan motor

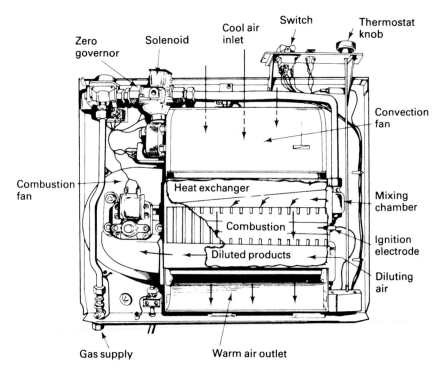

Fig. 39   Balanced flued convector, fanned draught, fanned convection, premix air and gas

special chamber at the inlet to the heat exchanger. The air/gas mixture passes into the heat exchanger, where combustion takes place without the need for a burner.

Products of combustion pass into a chamber below the heat exchanger where they are diluted with some of the fresh air from the inlet duct. The combustion fan expels the waste gases through the flue duct.

The controls used on fan-assisted heaters vary considerably from one model to the other. Ignition may be by spark, permanent pilot or glowcoil. Some heaters have automatic sequence control which ensures that the heat exchanger is purged and the flame established before locking-on the gas supply and switching on the convection fan.

Flame supervision devices can include those based on flame conduction or rectification, mercury vapour pressure and thermoelectric effect. Thermostats and clocks are available with many heaters. However, if it is required to fit a clock or thermostat to a heater to which they are not normally fitted, the maker's advice should be sought. It is generally not possible to fit them into the live supply to the heater.

Protection is usually provided to guard against overheating caused by blockage of the outlet louvres or convection fan failure. This usually takes the form of a bimetal overheat switch.

Room-sealed, fanned draught appliances can usually be fitted with extended flue and air ducts. These may be up to 1.5 to 1.8 m in length. Some appliances are designed for side outlet flues so that the heater may be fitted adjacent to an outside wall.

Fanned Draught, Natural Convection

This type of heater is similar to the previous category. Having fan-assisted combustion enables the flue way to be kept small. Not having a convection fan reduces the overall output. Having only one fan reduces the noise level of the heater.

Open-Flued Convectors

Fanned Draught, Fanned Convection

The fan on this appliance collects the products of combustion as they emerge from the heat exchanger, Fig. 40. Diluent air from the room is also drawn into the fan and the gases are discharged through the flue ducting. The fan maintains a level of pressure in the ducting so that, at wind velocities on the terminal of up to 48 km/h, the products will not be blown back into the room.

Heaters of this type are generally available with horizontal flues which may be up to 3 m in length. The appliances are usually fitted with electronic control systems.

There are a number of large, fanned convectors which are used for commercial applications and which are dealt with in Vol. 3, Chapters 9 and 11.

Installing Convector Heaters

Gas Supply

Similar fittings are used to those required for gas fires. The connection is often $R_c^{1/4}$ and unless an appliance isolating cock is supplied a union cock needs to be fitted to allow the supply to be disconnected for servicing purposes.

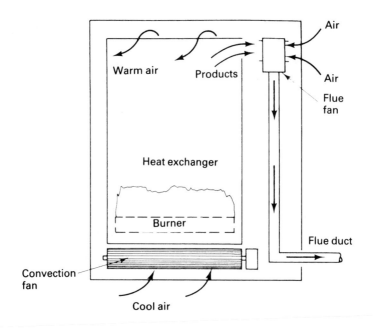

Fig. 40   Open flued convector, fanned draught, fanned convection

Siting the Heater

Read the manufacturer's instructions carefully, most of them give the clearance distances which must be observed.

Clearances are required to allow:

- the case to be removed
- the burner to be removed, often to one side
- the heater and assembled flue to be manoeuvred into position
- the gas supply to be fitted or, later, disconnected
- the heater to circulate warm air successfully
- curtains to clear the heater by 25 mm
- good access for lighting up and servicing.

The recommendations with respect to the flue termination were covered in Chapter 5, together with the precautions to be taken where walls are made of combustible materials.

Fitting the Heater and Flue Duct

Mark out on the wall the position of the flue opening and the holes for the heater securing screws. Manufacturers often supply a template

for this purpose. It is advisable to cut out the hole for the flue duct
first and then check the position of the holes required for the fixing
screws, relative to the flue duct.

If the heater has a cylindrical flue duct of about 65 to 75 mm
diameter, the hole is best made with a power drill and masonry cutter
called a 'core drill', Fig. 41. A pilot hole is drilled first and leads the
cutting drum through the wall. Take care to prevent chipping the
bricks on the outer wall face.

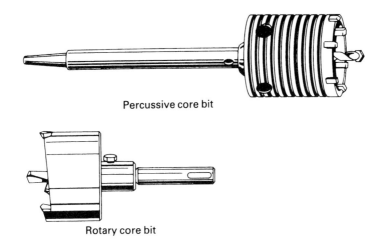

Percussive core bit

Rotary core bit

Fig. 41   Core drill

If the hole cannot be drilled, it may be cut out with a hammer and
chisel. Start by cutting away the mortar around the brick in the centre
of the hole. When the mortar is removed the brick may be cut into
pieces and removed. With this method there is considerably more
making-good to seal the duct in the wall.

Follow the manufacturer's instructions implicitly when making the
ducts to fit the particular thickness of wall. Some ducts may be cut
only at one end and this depends on whether they fit on the heater or
the terminal. Some terminals fit on the ducts before the assembly is
passed through the wall, others are fitted after the duct is positioned.
Flue ducts may be available in standard lengths with a measure of
telescopic adjustment.

For heaters to operate satisfactorily, the ducts must be the right
length and all the joints must be properly sealed. Possibly the most
important joint is that between the outer duct and the heat exchanger.
It is the most difficult one to get at and, if it leaks, the gas may keep
going out.

Heaters must be securely fitted to the wall, using a method recommended by the manufacturer and appropriate to the wall material. Toggle bolts, expanding fixings or white metal wall plugs are suitable.

When the heater is secured, fit the gas and, if necessary, electrical connections. Check the operation of the heater and instruct the customer.

Servicing Convector Heaters

The servicing procedure is similar to that for the radiant convector fires with the following differences.

In practically all cases, flange plates, which are screwed on to the heat exchanger, must be removed to give access to the burner or flue-ways. Ignition devices are often mounted on flanges and must be removed for cleaning and resetting. These flanges must be carefully resealed and this often involves fitting a new gasket. For ease of reassembly, gaskets can often be held in place with a few spots of contact adhesive.

On any heater with electrical components, turn off the electrical supply or remove the plug before taking off the case. There are usually a number of exposed spade terminals and perhaps bimetal switches which are live when the current is switched on. The heater can be operated without the case but it is safer to switch off and get the cleaning done first. Then visually check the electrical installation and test the tightness of the connections before switching on and checking the operation of the heater. Electrical testing is dealt with in Chapter 12.

Fan units require servicing regularly. Once a year is usually adequate. Clean the fan by brushing the blades gently with a soft brush until the dust is removed. Take care not to distort the blades. Do not oil the bearings with ordinary machine oil. Many fans do not require lubricating. Check the manufacturer's instructions to make sure.

Fault Diagnosis and Remedy

Convector heaters are subject to the usual faults which can occur on their igniters, control devices and electrical components. These are all dealt with separately in their appropriate volume and chapter.

Those of them which have balanced flues, which is the majority, may also develop faults in the flue ducts or the heat exchanger. These usually result in the burner and/or the pilot flames being extinguished and may, at first sight, appear to be due to a faulty flame protection device. It is necessary to check the controls first, but if the fault persists it may be due to one of the following causes.

Faulty Flue Termination

The flue terminal is too close to an obstruction. This may be due to a faulty installation, but it may also be caused by something, like an electrical switchbox or alarm, being fitted afterwards. It is likely to result in the pilot being blown out.

Blocked Flue

Anything which restricts the air inlet supply or the flue outlet will starve the burner of combustion air. Inevitably the burner will smother and the flames will go out.

The same effect can be caused by a gas leak on the burner or injector inside the heat exchanger. The leaking gas will ignite and burn in the flue ways, so robbing the burner of air and extinguishing it.

Faulty Sealing

This is, perhaps, the most common fault on balanced flued heaters of any kind. If there is wind pressure on the terminal, or a higher static pressure outside the building, air from the ducts can be blown into the room. The draught is surprisingly strong and can easily extinguish the pilot. In some cases the leakage may interfere with the combustion of the main burner and this, too, can be extinguished.

On some heaters the most vulnerable joint is that at the rear where the ducts are connected to the heat exchanger. This can sometimes be broken if the heater securing screws become loosened. Always check that the heater is securely fastened to the wall when carrying out a service.

As with many gas appliances, the major responsibility for the satisfactory operation of the heater rests firmly on the man who carries out the initial installation. Good fixings, flueing and sealing are essential for trouble-free operation.

Water Heating – Instantaneous Appliances

Chapter 8 is based on an original draft by R. Clowes

Water Supply Byelaws (Second Edition) 1989

When work is carried out on water installations it must conform to the relevant regulations. The basis of the regulations covering water supplied in the United Kingdom are the Water Supply Byelaws (Second Edition) 1989. Their purpose is to prevent the wastage or undue consumption and misuse or contamination of the water supply.

Although the byelaws will be particularly relevant to chapters 8, 9 and 10 of this volume, they should be adhered to when carrying out any work that may affect the water supply.

The byelaws cover the following areas that are likely to be encountered by engineers working on domestic gas appliances:

- unacceptable materials for potable (drinkable) water supplies, e.g. lead, linseed oil etc.
- cistern protection, to ensure hot water of potable quality
- protection against backflow/backsiphonage
- unvented hot water systems
- flushing volumes for W.C.s etc.
- identification of water pipes in non-domestic premises.

Introduction

Gas water heaters began to be developed on a wide scale in the early 1930s, following the country's recovery from the depression. Many houses, particularly in the south of England, had been built without bathrooms. Normally the sole means of heating water was the kettle on the stove for small quantities and the 'copper', usually heated by coal and holding about 45 litres for boiling the washing and providing the bath water. There was, therefore, a ready market for the easily installed, inexpensive, instantaneous gas water heaters. Some of the earliest were the bath heaters or 'geysers' followed by sink heaters and finally multipoints.

In some northern areas where coal was plentiful and cheap, the back-boiler behind the living room fire provided an apparently free hot water service in the winter. In summer, a gas sink heater could supply water for the kitchen, leaving the coal fire to cope with 'bath-nights'.

An 'instantaneous' water heater produces instant hot water. That is, it heats the water to the required temperature as the water passes through it. Cold water goes in and hot water comes out.

The temperatures at which water is normally required are:

- washing and bathing   33 – 43° C
- washing up   60 – 70° C
- household cleaning   30 – 70° C
- clothes washing   30 – 95° C
- tea making   100° C.

To provide water at the temperatures required, and to avoid any danger of the water boiling and producing steam, modern heaters are designed to raise the temperature of the water through not more than 55.5 deg C. In the past some single point instantaneous water heaters were designed to supply boiling water for tea making.

If the average temperature of incoming cold water is 10° C, then the average temperature of the hot water will be 65.5° C.

Since, in most heaters, the gas rate does not vary, the temperature of the water may be altered by adjusting the rate of flow. The faster the water flow, the lower the temperature rise and vice versa.

For example, if a heater with a gas input of 29.3 kW produces a temperature rise of 55.5 deg C in a water flow of 5.6 litres per minute, it will give a temperature rise of 44.4 deg C if the flow is increased to 7 litres per minute.

So: 5.6 litres/min gives 55.5 deg C rise

and 7 litres/min gives 44.4 deg C rise

But, $5.6 \times 55.5 = 310.8$

and, $7 \times 44.4 = 310.8$

so, rate of flow × temperature rise = constant (for a particular heater)

Therefore, the temperature rise at any rate of flow can be calculated:

At 6 litres/min the temperature rise will be:

$$\frac{310.8}{6} = \textbf{51.8 deg C}$$

In symbols this may be expressed as:

$$F_1\, T_1 = F_2\, T_2$$

where F_1 and F_2 are the rates of flow and T_1 and T_2 are the corresponding temperature rises.

The complete calculation of heat output from an instantaneous heater is based on the formula given in Vol. 1, Chapter 5. It becomes:

$$\text{Heat output} = \frac{\text{Rate of flow}}{\text{of water}} \times \frac{\text{Temperature}}{\text{rise}} \times \frac{\text{Specific heat}}{\text{capacity of water}}$$

In units,

kW = litres/second × deg C × (4.186) kJ/kg deg C

For Example,

To find the heat output of a water heater giving a flow of 7 litres/min at a temperature rise of 44.4° C.

$$\text{Heat output} = \frac{7}{60} \text{ litres/second} \times 44.4 \times 4.186$$

$$= \textbf{21.68 kW}$$

Since the heat input of the heater was 29.3 kW, the thermal efficiency will be:

$$\text{Efficiency} = \frac{21.68}{29.3} \times 100$$

$$= \textbf{74\%}$$

Both the rate of flow of water through the heater and the incoming cold water temperature will affect the temperature of the hot water delivered. In summer the cold water may be nearly 20° C and in winter it could be 4° C. So if the heater gives a temperature rise of 50 deg C the outlet temperatures would be:

- summer, 70° C
- winter, 54° C.

To compensate for this, most heaters are fitted with a temperature selector which allows the rate of water flow to be altered to maintain a constant outlet temperature.

Types of Heater

There have been many different sizes and designs of instantaneous water heater and some of the old types may still be found on the district. Since most wet central heating systems also provide domestic hot water, the market has declined with the growth of central heating. There are now three main sizes or types of heater all of somewhat similar design.

These are:

- multipoint, full service water heater
- bathroom heater
- sink water heater.

All instantaneous water heaters are now manufactured to BS 5386: Part 1 1976 (with amendments 1979 and 1987) (EN 26) Gas burning appliances for instantaneous production of hot water for domestic use.

Fig. 1 (exterior)   Multipoint water heater (Chaffoteaux)

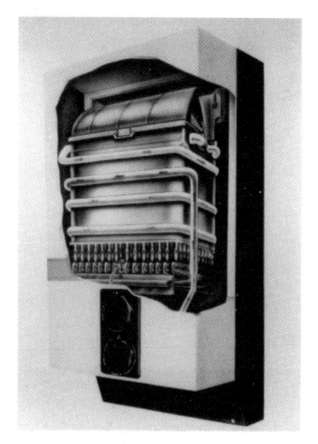

Fig. 1 (interior)   Multipoint water heater (Chaffoteaux)

Multipoint Water Heater

This normally supplies hot water to taps at the bath, wash basin and sink, Fig. 1, and may also be used to supply hot water for showers and to automatic washing machines and dish washers. It has a heat input of about 29.3 kW and an output of about 22.6 kW. The water flow rate is 5.8 litres/min raised 55.5 deg C, or about 9 litres/min raised 36 deg C.

Whilst this rate of flow is perfectly adequate for one tap, it is not sufficient to supply more than one tap at a time. In this respect, instantaneous water heaters are at a disadvantage when compared with storage water heaters which can be made to supply hot water at any required rate of flow.

Instantaneous multipoint heaters may either be connected to a mains water supply or to a suitable cold feed cistern. The water pressures normally required are about:

- 345 to 485 mbar for mains supply models
- 2.5 to 3.5 m head for cistern supply models.

The minimum pressure (head) quoted in the manufacturers' installation and servicing instructions, for instantaneous water heaters, is generally for the correct operation of the appliance only. Allowance for the resistance of pipework, fittings etc. must often be made especially where showers, washing machines or dishwashers are being supplied.

Some continental models may require pressures up to 1 bar. The inlet connections are usually 22 mm diameter for gas and both cold

Fig. 2   Fanned draught balanced flue multipoint water heater (Chaffoteaux)

water inlet and hot water outlet. Some models may have 15 mm cold water inlet connections and may require a higher operating pressure.

Multipoint water heaters are now room-sealed appliances. In addition to the natural draught balanced flued appliances, which have to be sited on an outside wall, fanned draught balanced flued multipoint water heaters, see Fig. 2, are also available. This type of appliance can be sited in other positions by utilising the flue supplied by the appliance manufacturer. The terminal position must comply with the guidelines given in Chapter 5. When fitted in a room containing a bath or shower the I.E.E. wiring regulations must be borne in mind. Figures 3 (a), (b) and (c) show the circuit, illustrated and functional flow wiring diagrams of the fanned draught balanced flue multipoint water heater shown in Fig. 2.

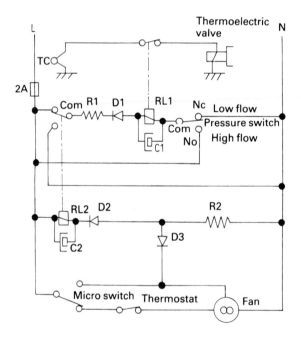

Fig. 3   (a) Circuit wiring diagram

Combi (Combination) Boilers

This type of appliance supplies at any one time either full central heating or domestic hot water. The appliance operates in a similar manner to the multipoint water heater when in the domestic hot water mode. The hot water supplied to the taps is, however, heated indi-

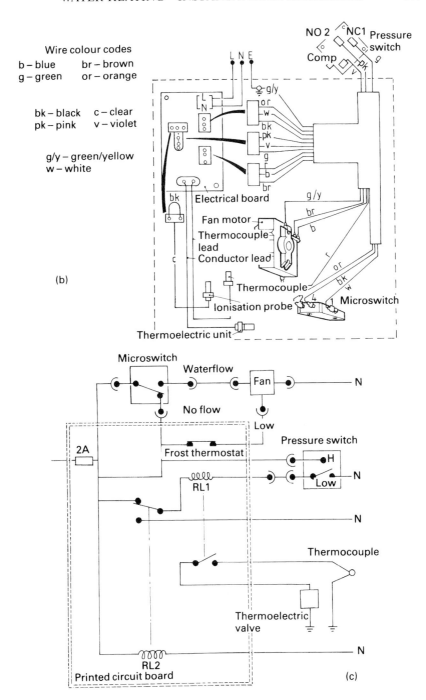

Fig. 3   (b) Illustrated (pictorial) wiring diagram and
              (c) Functional flow wiring diagram

rectly through a calorifier. A full description of the basic operation of this type of appliance is given in Chapter 10.

Bathroom Water Heater

A smaller edition of the multipoint heater, this supplies water to a bath and a wash basin. It may be connected to the bath and basin taps. Older types of this heater were sometimes fitted with a swivel spout controlled by an integral tap, while retaining the multipoint connection facility, Fig. 4.

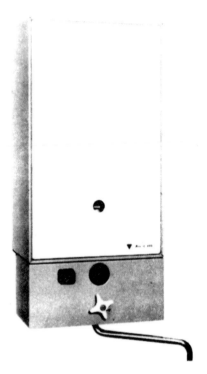

Fig. 4   Bathroom water heater (balanced flue)

The heat input is in the region of 20.5 kW and the output is about 16 kW, or 4.2 litres/min raised 55° C. Heaters may operate on mains or cistern water supplies requiring water pressures of about 415 – 480 mbar on mains or about 3 m head on cisterns.

All bathroom water heaters are now room sealed. Gas and water connections are usually 15 mm.

Sink Water Heaters

These heaters may supply hot water to one or two sinks or wash basins or to one shower. They may be fitted with a swivel spout and integral hot and cold water taps and/or be designed for multipoint connection, Fig. 5. The heat input is in the region of 11 kW and the output about 8 kW. The water flow rate is 2.2 litres/min raised 55 deg C. Because they have an input of less than 12 kW, they may be fitted in an adequately ventilated kitchen without a flue.

An unflued appliance must have a warning label attached to it advising the customer not to use the heater continuously for more than 5 minutes. If however the periods of operation will exceed 10

Fig. 5   Sink water heater (Vaillant)

minutes, then a flue is required and additional ventilation must be provided. It must be remembered that not all models can be fitted with a flue. Since 1st January 1996 the Gas Safety (Installation and Use) Regulations have prohibited the installation of instantaneous water heaters (except for room sealed type) in any room unless they are fitted with a device which turns off the flow of gas to the burner before a dangerous level of fumes can build up (Regulation 30(3)). Such a device would operate using an oxygen depletion system incorporated into the thermo-electric flame failure system. These units are commonly called anti-vitiation devices or oxy-pilots. See the Gas Safety (Installation and Use) Regulations 1998 at the end of Chapter 5.

The connections are usually:

- gas 12 mm
- water 15 mm
- flue 75 to 90 mm, into a draught diverter.

Construction and Operation

Instantaneous water heaters generally consist of the following parts:

- heating body, comprising the combustion chamber and heat exchanger
- water section, incorporating water filter, governor or throttle, automatic valve with or without thermostat, slow ignition device, temperature selector
- gas section, with governor or throttle, main gas and pilot taps
- burner
- ignition device
- flame supervision device
- outer case.

All heaters, except sink heaters, also have balanced flue ducts and terminals.

Heating Body

This is the combustion chamber surmounted by a gas-to-water heat exchanger in one component, Fig. 6. It was described in Vol. 1, Chapter 10.

Heating bodies are usually made of copper, protected by a lead/tin coating. The combustion chamber is cooled by a coil carrying the incoming cold water and the heat exchanger has a single-path water tube passing three or four times through banks of fins.

Because the water control taps are on the outlet, the heating body must withstand mains water pressure and all heating bodies are now of the multipoint type. On earlier model sink and bath heaters the control taps were on the inlet and heating bodies sometimes had annular waterways. The 'open' type of heater in which water poured over a cone or cylinder in the path of the combustion products was also popular.

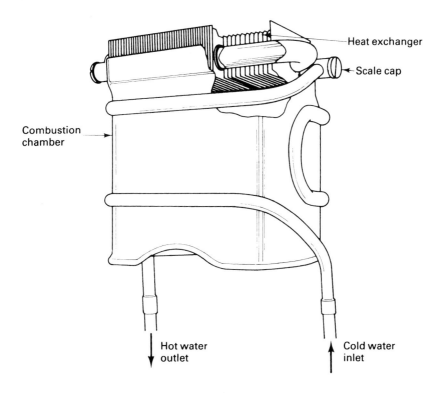

Heat exchanger

Scale cap

Combustion chamber

Hot water outlet

Cold water inlet

Fig. 6   Heating body

Water Control

Automatic Valve

The automatic valve has a dual purpose. It turns on the gas when a tap is turned on and turns the gas off when the water is turned off. So the heater is operated automatically by turning any hot water tap on and off.

The valve also protects the heater by limiting the temperature rise to about 55 deg C, so preventing steam formation and excessive scale

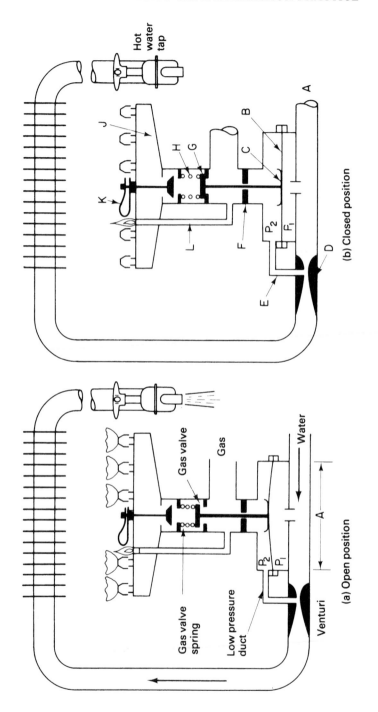

Fig. 7 Automatic valve, differential pressure type: (a) open; (b) closed.
A: Diaphragm housing; B: Rubber diaphragm; C: Bearing plate; D: Venturi; E: Low pressure duct; F: Stuffing box
or gland; G: Gas valve; H: Gas valve spring; J: Burner; K: Flame supervision device (bi-metal type); L: Pilot

deposits. It does this by preventing gas from passing to the burner until the quantity of water flowing through the heater has reached a pre-determined 'minimum flow'.

Most contemporary heaters use automatic valves which are oper- ated by a differential pressure created by the water flow through a venturi. Some employ a thermostat to limit the temperature of the water delivered by controlling the flow of water. A diagrammatic sketch of a differential pressure valve is shown in Fig. 7.

It consists of a cylindrical diaphragm housing A, containing a flat, rubber diaphragm B. The stem of the gas valve G, rests on a bearing plate C, on top of the diaphragm. The stem passes through a packing gland at F.

A venturi is positioned in the inlet water supply at D with its throat connected to the diaphragm housing above the diaphragm by a low-pressure duct, E. When no water is flowing the pressures above and below the diaphragm are equal and the gas valve is held on to its seating by the spring H.

When water is turned on it flows through the venturi and the pressure at the throat is reduced as the velocity increases. This causes a reduction in the pressure above the diaphragm, while the underside is subject to full water pressure. When minimum flow is reached the pressure differential is sufficient to force the diaphragm upwards, so lifting the gas valve off its seating against the tension of the spring. If the pilot is lit and the bimetal cut-off K is open, gas will pass to the main burner, J.

Immediately the rate of water flow falls below the minimum required to produce the necessary pressure differential, the spring will close the gas valve and prevent gas passing to the burner.

There have been many designs of differential pressure valve. Some have the diaphragm positioned vertically. Others use a pouch-shaped diaphragm which is made to inflate by the pressure difference.

An adequate head of water must be available to give the flow rate necessary to produce the pressure differential required.

Thermostatic Valve

Some multipoint water heaters use a vapour-pressure thermostat to modulate the water flow rate through the heater and so control the outlet water temperature. The temperature is usually set at 60° C.

Figure 8 shows the operation of the valve. A small, cylindrical throttle, situated in the inlet waterways, forms a restriction. When water flows around the throttle a pressure loss is produced on the outlet side. The reduced outlet pressure is communicated by a duct to the upper side of the diaphragm providing the differential pressure to operate the automatic valve. So the throttle acts in a similar way to a venturi. When the throttle is closed, initially, the annulus around it allows the minimum flow of water to pass.

The temperature-sensing element of the thermostat is located in the heat exchanger waterways. It is connected by a capillary tube with the bellows which controls the throttle. When water is turned on and becomes heated, the bellows expands and opens the throttle against the tension of a spring. Water can then flow above the throttle as well as around the annulus. So the flow rate is increased and the temperature is reduced.

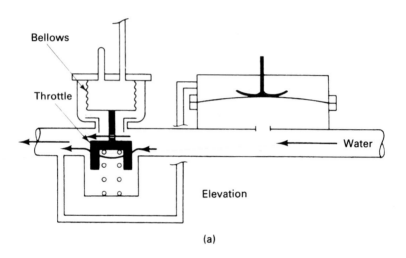

(a)

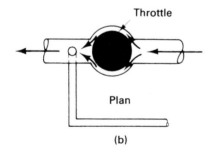

(b)

Fig. 8   Thermostatic valve: (a) elevation; (b) plan

Temperature Selector

On heaters without thermostats the temperature of the hot water delivered can be adjusted by varying the flow rate using the hot water tap. If the tap is partially closed, the water temperature will increase. However, if the tap is turned too far, so that less than the minimum flow required to operate the valve can pass, the gas will be shut off and the water will suddenly run cold. A temperature selector prevents this problem.

A typical selector is shown in Fig. 9. It consists of a manually operated, screw-in throttle, situated in a bypass around the venturi. With the throttle screwed in, all the water passes through the venturi. This operates the automatic valve and gives the highest temperature rise.

As the throttle is screwed out it allows a small quantity of water to bypass the venturi and the hot water temperature is reduced.

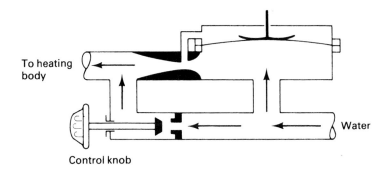

Fig. 9   Temperature selector

Water Governor and Water Throttle

Heaters which are designed to be fitted to a cistern water supply will always have a low and constant inlet water pressure. Adjustment of water flow rate is usually by means of a screw-in throttle, Fig. 10. The throttle may also incorporate the water filter, as shown.

Heaters fitted to mains water supplies may be subject to high or fluctuating water pressures and the water flow rate is normally controlled by a water governor. The governor may be a separate device, fitted on the water inlet to the heater or it may be incorporated into the automatic valve, Fig. 12.

A separate water governor is shown in Fig. 11. This is similar to a compensated, constant pressure gas governor with double valves.

The body is made from brass, chromium plated for strength and corrosion resistance. It houses a reinforced synthetic rubber diaphragm with adjustable spring loading. The valves and seatings are of nearly the same diameter, the lower seating being just large enough to

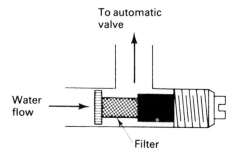

Fig. 10   Adjustable water throttle

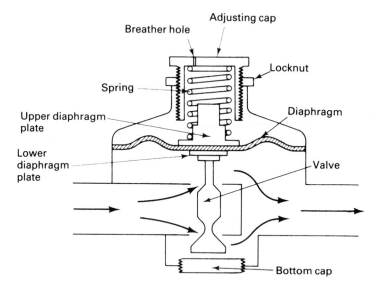

Fig. 11   Water pressure governor

allow the upper valve to pass through. This allows the valve to be withdrawn for cleaning or when renewing the diaphragm.

The governor gives a constant outlet water pressure and, if used in conjunction with a fixed restriction on its outlet, will provide a constant volume or rate of flow.

A water volume governor which is integral with an automatic valve is shown in Fig. 12.

The valve unit screws into the water section in the path of the incoming water, below the diaphragm. A light spring holds the valve stem in contact with the diaphragm so that the valve moves as if it were attached to the diaphragm.

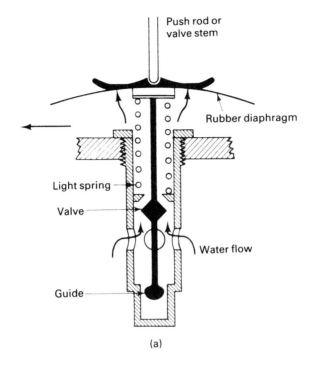

(a)

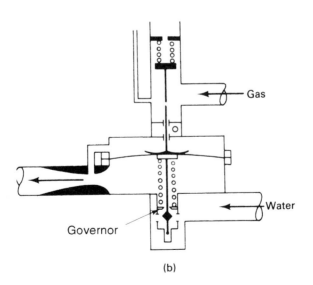

(b)

Fig. 12   Water volume governor: (a) governor; (b) governor in automatic valve

When water flows, the differential pressures cause the diaphragm to rise and open the gas valve. The diaphragm will continue to rise, bringing the governor valve towards its seating until the upward force is balanced by the gas valve spring. Any variation in inlet water pressure will move the diaphragm and the governor valve will be automatically adjusted to restore the desired outlet pressure.

Since water at constant pressure is fed through the fixed orifice of the venturi, a constant flow rate is achieved.

Slow Ignition Device

It is essential that the main gas burner should ignite safely and quietly. To ensure this, most automatic valves incorporate a 'slow-ignition device'. This device is designed to make the gas valve open slowly and close quickly. The slow opening ensures quiet, non-explosive ignition and the quick closing prevents overheating the water remaining in the heater. There are several designs of slow ignition device, some fixed, others adjustable. They are located in either the high-pressure or the low-pressure ducts to the diaphragm.

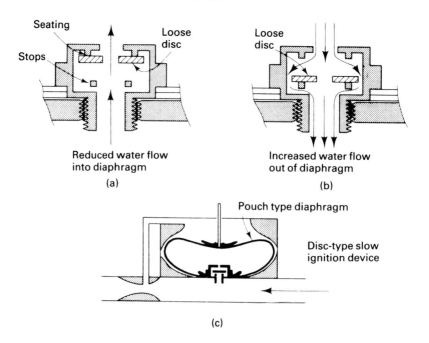

Fig. 13   Slow ignition device, disc type: (a) valve opening; (b) valve closing; (c) device in automatic valve

A fixed type is shown in Fig. 13. It consists of a small, circular disc with a hole in it which can float up and down in a cylindrical, brass housing. The device is located in the high-pressure inlet to the diaphragm. When a hot water tap is turned on, the rate of lift of the diaphragm is controlled by the disc. After allowing an immediate partial lift, the disc is forced against the top seating and water can only pass through the small centre hole. So the valve continues to lift slowly, turning the main gas fully on, after ignition.

When the diaphragm moves downwards, the disc falls to rest on the stops. Water can pass through the centre hole and around the edges of the disc, so the gas valve closes quickly.

An adjustable slow-ignition device is shown in Fig. 14. This has a loose ball instead of a disc and it is usually situated in the low-pressure duct. Screwing the device in or out regulates the amount of water which can bypass the ball and so adjusts the rate of lift of the diaphragm. The ball moves in a similar way to the disc to restrict the waterways when opening and open them when closing.

In association with a slow-ignition device the design of the gas valve can contribute to slow ignition and quick extinction of the flames. Figure 15 shows a valve with a cylindrical section, often called

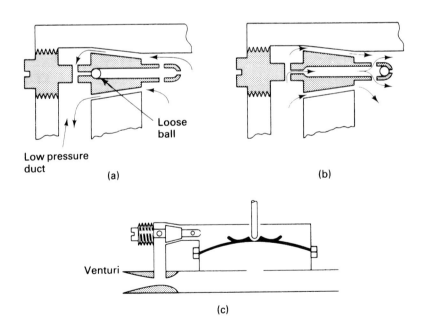

Fig. 14   *Slow ignition device, ball type: (a) valve opening; (b) valve closing; (c) device in automatic valve*

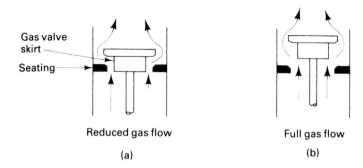

Reduced gas flow   Full gas flow

(a)   (b)

Fig. 15   Gas valve: (a) partially open; (b) fully open

a 'skirt', immediately below the valve. When the valve is partially open the skirt restricts the flow of gas to the burner so that only sufficient gas for quiet, effective ignition is allowed to pass. As the valve continues to lift to its fully open position, the flames increase to their full size.

Gas Control

Gas Taps

Most heaters have a single plug-type tap controlling both main gas and pilot supplies.

Gas Regulator and Throttle

A number of heaters rely on the meter regulator and fixed injectors to provide correct gas rate control. Other heaters make use of a volume governor as an integral part of the appliance, Fig. 16.

The governor shown consists of a lightweight, aluminium disc float which can slide freely over a vertical tube. The tube has a number of gas ports which are closed as the float rises and opened when it falls. The device is contained in a housing and the annular space between the bore of the housing and the outer edge of the float disc forms a fixed orifice. When the required volume of gas is flowing upwards through the annulus, the pressure difference just supports the float.

If the supply pressure is increased, the rate of flow increases and so does the pressure differential across the float. So the float rises to close the gas ports until the flow rate is restored to the required volume.

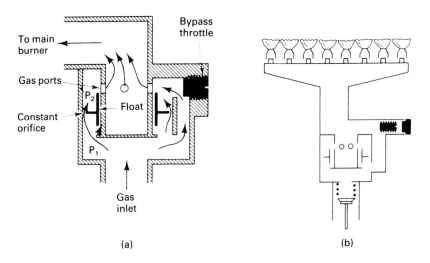

Fig. 16   Gas volume governor: (a) governor; (b) governor in gas supply

If the supply pressure falls, the rate of flow is reduced and the pressure difference can no longer support the float. The float moves downward, opening the gas ports and increasing the rate of flow to the volume required.

Volume governors are frequently provided with a screw-in throttle which can be adjusted to allow a volume of gas to bypass the float and so increase the flow rate to the burner.

Other Components

Flame Supervision Device

Instantaneous heaters are usually fitted with either bimetal or thermo-electric flame supervision devices (Vol. 1, Chapter 11). The flame sensor (bimetal strip, thermocouple etc.) is commonly heated by the single pilot flame which ignites the main burner.

Ignition Device

Some heaters still require manual ignition from an outside source but many make use of piezo-electric igniters. The impact type is commonly used, operated by a separate push-button (Vol. 1, Chapter 11).

Burners

These are normally stainless steel, multiple-bar aerated burners (Vol. 1, Chapter 4). Each bar usually has its own injector. A typical example is shown in Fig. 17.

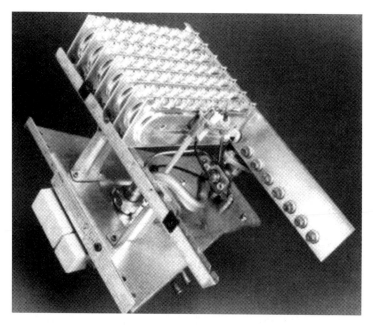

Fig. 17   Instantaneous multipoint water heater burner assembly, showing half burners removed to expose injectors

Outer Case

The casing is frequently made in two parts with the rear section serving as a back-plate for fixing the appliance to a wall. The front casing may be stainless steel sheet or sheet steel finished in white vitreous or stoved enamel.

Location of Components

Figure 18 is a diagrammatic drawing showing the layout of the components of a single-point sink instantaneous water heater, without thermostatic control.

A multipoint heater with a thermostat is shown in Fig. 19.

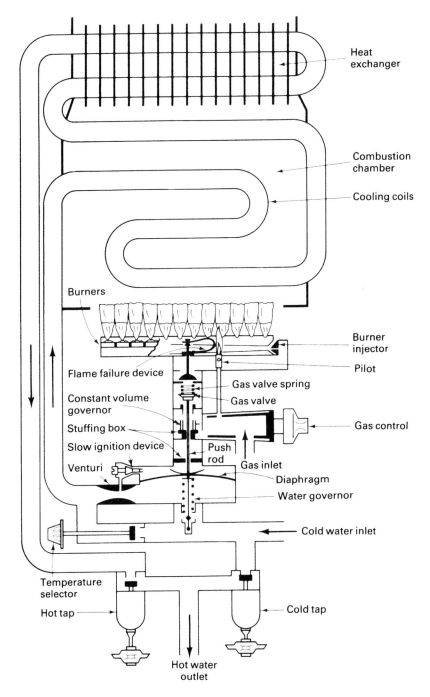

Fig. 18   Components of sink instantaneous water heater

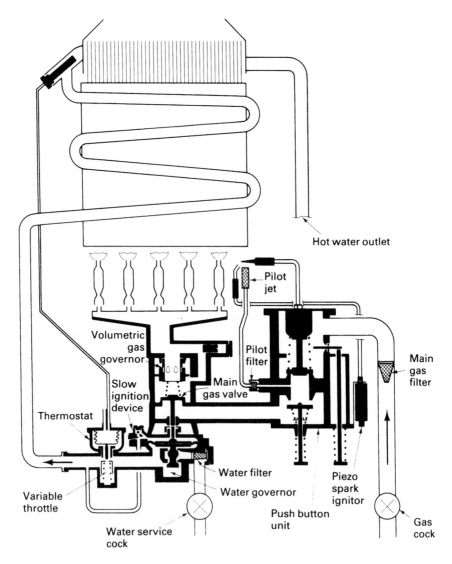

Fig. 19   Multipoint instantaneous water heater with thermostatic control

Installing Multipoint Water Heaters

General

Detailed recommendations for the installation of domestic gas water heaters are given in the following documents:

BS 6891 : 1998 Installation of low pressure gas pipework of up to 28 mm (R1) in domestic premises (2nd family gas)

BS 5482 :   Domestic butane- and propane-gas-burning installations.

   Part 1 : 1994 Specification for installations at permanent dwellings.

BS 5440 :   Installation of flues and ventilation for gas appliances of rated input not exceeding 60 kW (1st , 2nd and 3rd family gases)

   Part 1 : 1990 Specification for installation of flues

   Part 2 : 1989 Specification for installation of ventilation for gas appliances.

The Water Supply Byelaws (2nd edition) 1989.

The Gas Safety (Installation and Use) Regulations 1998.

When installing any gas water heater, the manufacturer's instructions must be followed carefully.

Before connecting any heater directly to the main water supply, the local water authority must be informed.

When multipoint heaters were open-flued appliances, the need to provide a satisfactory flue system was of paramount importance when selecting the location. Now that the heaters are balanced-flued, the heater must generally be located on an outside wall.

Some models may be fitted under draining boards in kitchen units, Fig. 20, but heaters are usually hung on the wall, clear of any working surface. This locates the balanced-flue terminal up out of harm's way and provides the heater with easy access for maintenance or repair. Heaters fitted with spouts must be positioned at a height above the sink or basin to give a type 'A' air gap (see Table 30) between the outlet of the spout and the highest spillover level of the receiving sink or basin. If a spout discharges into a sink it is usually positioned so that it may be used to fill a bucket.

Although the heater may be fitted in a bathroom, it is best located in the kitchen, adjacent to the most frequently used draw-off tap.

The appliance must be fixed to a sound wall and be clear of any combustible material or be suitably insulated in accordance with the manufacturer's instructions.

LPG Appliances

BS 5482 Part 1 states that appliances fired by LPG should not be installed in cellars or basements. It states that only room-sealed appliances should be installed in garages (although the Gas Safety (Installation and Use) Regulations 1998 no longer stipulate this). It

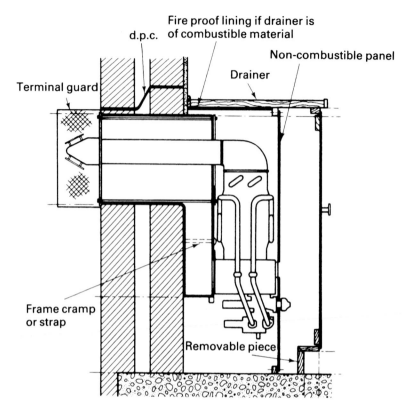

Fig. 20   Balanced flue multipoint water heater

also states that where an instantaneous sink water heater is fitted to supply a shower unit, the appliance shall be flued to atmosphere, and shall be room sealed.

Water Supply

The water pressure in the supply must be adequate to provide the rate of water required to operate the automatic valve and give a satisfactory temperature rise. Manufacturers' instructions always state the minimum head of water required to operate the valve or minimum inlet working pressure.

Pipes, fittings and materials used in the installation of instantaneous water heaters must comply with the water supply byelaws with regard to potable (drinkable) water, e.g. where solder fittings are used they must only use lead-free solder, when connecting to an existing lead pipe care must be taken to prevent galvanic action and a special compression fitting from lead to copper, as described in Chapter 1, should be used. Jointing compounds must not be linseed oil based etc. If the appliance takes its cold supply from a cistern, then the cistern must comply with the water supply byelaws. The positioning and accessibility of all water pipes etc. must also comply with these byelaws.

The inlet water pressure is required to overcome the pressure loss when water flows through the:

- cold water supply to the heater
- heater
- hot water supply to the draw-off.

The pressure loss will depend on:

- rate of flow
- size, length and type of pipes
- number, size and type of pipe fittings
- size and type of taps and valves
- type of heater.

The pressure in mains water supplies is usually adequate and high-pressure heaters rarely require more than 1 bar. If mains pressure is inadequate a low-pressure heater may be used on a cistern supply. Some heaters may be connected to either mains or cisterns.

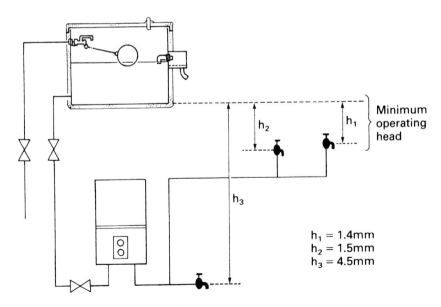

Fig. 21   Head of water to operate automatic valve

The pressure in a cistern supply is measured from the level of the highest draw-off tap to the water level in the cistern, Fig. 21. This is because the head at this tap is lower than that at the other two (Vol. 1, Chapter 6). So if the heater can operate on the flow rate from the highest tap, it should operate on any other tap which is likely to have a higher flow rate.

In practice because the water level in the cistern may fall as water is drawn off, the measurement is usually taken from the bottom of the cistern.

The actual location of the heater, below the cistern, has no effect on the operation of the valve. The heater is similar to any other fitting in the supply and creates the same pressure loss in any position.

The pressure recommended by the manufacturer is usually between 2.5 and 4 m head and this is the total pressure loss through the heater and an average installation. Sometimes the head specified is the pressure loss through the heater alone, usually about 1 to 1.5 metres. An additional 1.5 m must then be added to allow for the pressure loss in the cold water supply to the heater and the hot water supply to the draw-off.

On installations with exceptionally long pipe runs it may be necessary to use pipes of a larger diameter in order to reduce the pressure loss. When connection is made to an existing installation it may be necessary to renew the pipework or raise the height of the cistern. If the cistern supplies other taps in addition to the heater supply, the pipes must be adequately sized to avoid excessive pressure loss when the taps are in use.

Cold water supplies should be flushed out before being connected to the heater.

Water, supplied from the cistern shown in Fig. 21, will have to be of potable (drinkable) quality. In order to comply with Byelaw 30 of the Water Supply Byelaws, the cistern would need to be covered (to prevent dirt etc. dropping into the water) and insulated (against frost). The warning (overflow) pipe and any air inlet must be screened to prevent the ingress of insects etc., see Fig. 22.

Hot Water Draw-offs

The length of the hot water draw-off pipes should be kept to a minimum in order to minimise the time delay before hot water is delivered and to avoid wastage of water. The recommended lengths of hot water draw-offs are shown in Table 28.

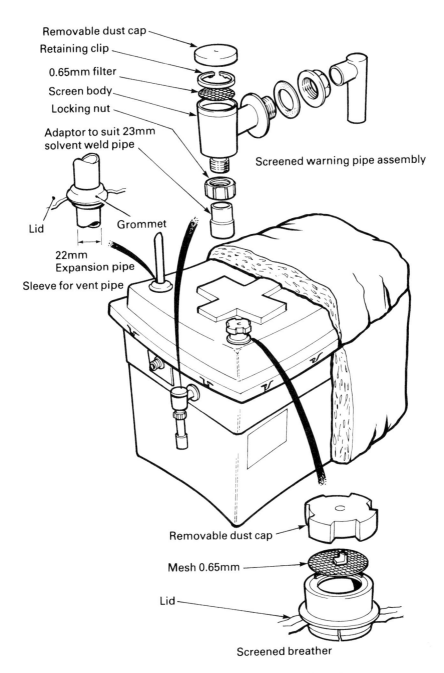

Removable dust cap
Retaining clip
0.65mm filter
Screen body
Locking nut
Adaptor to suit 23mm
solvent weld pipe

Screened warning pipe assembly

Lid
Grommet

22mm
Expansion pipe
Sleeve for vent pipe

Removable dust cap

Mesh 0.65mm

Lid

Screened breather

Fig. 22   Cistern complying with Byelaw 30

TABLE 28 Hot Water Draw-offs, Pipe Sizes and Recommended Maximum Lengths of Pipe Runs

Nominal Pipe Size	Maximum Length
Not above 22 mm	12 m
Above 22 mm but not above 28 mm	7.5 m
Above 28 mm	3 m

Both cold and hot water supplies will normally be run in light gauge copper tubing and should be secured at the intervals given in Table 29.

TABLE 29 Maximum Intervals between Pipe Supports (Lightweight Copper)

Nominal Pipe Size	Intervals	
	Horizontal run	Vertical run
15 mm	1.2 m	1.8 m
22 mm	1.8 m	2.4 m
28 mm	1.8 m	2.4 m

(BS 5546 : 1990)

Gas Supply

The supply pipework should be sized in accordance with the manufacturer's recommendations. The gas meter must be capable of passing the gas rate of the heater in addition to the demands of all other appliances.

Flue

Details of balanced flue installation were given in Chapter 5 and the layout of a typical flue is shown in Fig. 23.

Most manufacturers provide sets of flue ducts which are adjustable to fit different thicknesses of wall. In this way heaters may be fitted on walls from 75 mm up to 530 mm thick.

Room-sealed appliances do not require air vents in the room containing the appliance.

Service Cocks

Control cocks should be fitted on the inlet gas and water supplies, immediately adjacent to the heater. The cocks are known as 'service cocks' because they are used to turn off the gas and water during servicing operations.

Service cocks are often integral with the appliance and separate cocks are not then required.

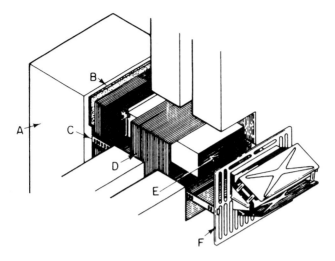

Fig. 23   Balanced flue installation. A: Case; B: Case seal; C: Slotted strap; D: Liner; E: Flue duct; F: Terminal

Layout of Installation

A typical multipoint water heater installation is shown in Fig. 24.

Installing Sink Water Heaters

The documents listed under the general subsection of Instantaneous Multipoint Water Heaters, also apply to Sink Water Heaters.

Location

This depends on whether the heater is a single-point or a multipoint model. The single-point heater, which has integral hot and cold taps and a swivel spout must be sited with the spout over the sink or basin. A type 'A' air gap (see Table 30) must be allowed between the outlet of the spout and the highest spillover level of the receiving sink, bath or basin. The heater should be fitted to one side of the sink with the spout ending at a convenient position over and above the top edge of the sink, Fig. 25. The standard spout is about 230 mm long but most manufacturers can supply alternative spouts up to about 600 mm in length.

When it is necessary to locate the heater at a greater distance from the sink, the multipoint model should be used. This can be connected to the hot tap, Fig. 27.

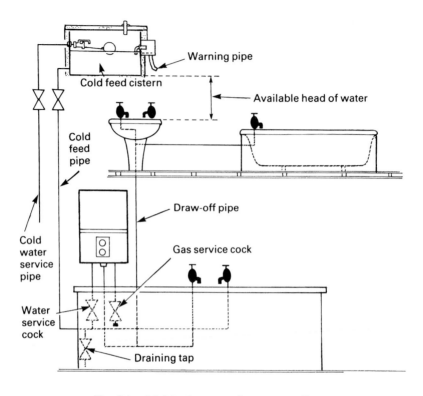

Fig. 24   Multipoint water heater installation

If a multipoint heater is used to supply a sink and a wash basin it should be conveniently sited for both draw-offs and preferably nearest to the one most frequently used. This is normally the sink, which also requires the hottest water.

Water Supply

The points made on the water supply pipes, fittings and materials, for multipoint water heaters, also apply to sink water heaters.

Sink heaters are invariably connected to the mains water supply and require inlet pressures of about 550 mbar. The connection may be made as shown in Figs. 25 and 27. Alternatively a 'heater tee' may be used. This screws into the boss behind the cold water tap.

Hot Water Draw-off

The length of draw-off should be as short as possible. Manufacturers' installation instructions often specify a maximum length, which may be about 4.5 m.

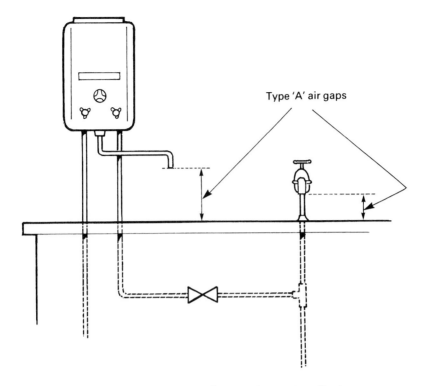

Fig. 25   Single-point sink water heater installation

In order to comply with the water supply byelaws steps must be taken to prevent backsiphonage of water from open, possibly contaminated vessels into the mains water supply. Any draw-off tap must have a type 'A' air gap, see Fig. 26, between the outlet and the spillover level of the receiving vessel (sink, bath, basin etc.). The air gap must be at least the vertical difference given in Table 30.

TABLE 30 Type 'A' and Type 'B' Air Gaps

Bore of pipe or outlet	Vertical distance between point of outlet and spillover level
1. Not exceeding 14 mm	20 mm
2. Exceeding14 mm but not exceeding 21 mm	25 mm
3. Exceeding 21 mm but not exceeding 41 mm	70 mm
4. Exceeding 41 mm	Twice the bore of the outlet

Type 'A' air gap

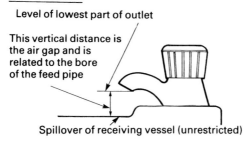

Level of lowest part of outlet

This vertical distance is
the air gap and is
related to the bore
of the feed pipe

Spillover of receiving vessel (unrestricted)

Type 'B' air gap

Lowest part of outlet   Cover

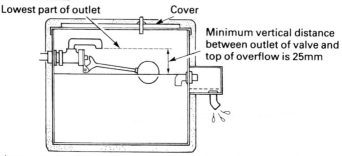

Minimum vertical distance
between outlet of valve and
top of overflow is 25mm

Highest water level is when the warning pipe is passing the maximum
rate of inflow to the cistern.

Note that if the warning pipe is adequately sized the water level will
be no higher than the top of the warning pipe.

Fig. 26   Type 'A' and Type 'B' air gaps

Gas Supply

The gas meter and the supply should be capable of passing about
1.2 m^3/h in addition to other requirements. The supply pipe to the
heater is usually 15 mm.

Flue

Heaters which have an input rate of 12 kW or less may be installed,
without a flue, in a room with an openable window and with a room
volume of more than 6 m^3.

For rooms with volumes of 6 to 11 m^3, a vent of at least 3,500 mm^2
area must be fitted. If there is another flueless appliance in the room,
the vent area should be 12,000 mm^2.

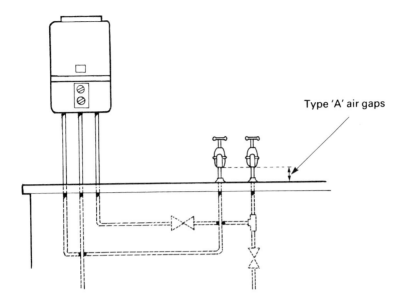

Type 'A' air gaps

Fig. 27   Multipoint sink water heater installation

For normal, intermittent use the heater does not require a flue. However, if the period of operation will exceed 10 minutes, the flue cap should be replaced by a draught diverter and a 75 or 90 mm flue should be fitted, see Chapter 5.

Warning labels must be attached to advise the customer not to exceed periods of use of more than 5 minutes, Fig. 28.

Fig. 28   Warning label

Installing Showers

Most instantaneous water heaters may be used to operate showers. When sink heaters are used the following points should be noted:

- the heater must be flued
- open flued heaters must not be installed in the same room as the shower
- it should be connected to the mains water supply, with the permission of the water authority
- the heater should supply the one shower only
- if necessary the cold supply to the mixer valve should be governed.

When installing any multipoint heater, the following points apply:

- both hot and cold water supplies to the shower should come from the same source, that is, both cistern or both mains
- cistern supplies should give a head of about 5 to 6 m at the shower rose

Fig. 29   Hose constraining ring (Mira showers)

- if the head from the cistern is too low, an electric pump may be used
- on mains supplies, a water governor should be fitted, on the cold feed to the shower, pressures required are about 1.4 to 1.7 bar
- follow the manufacturer's recommendations on the type of shower and mixing valves which may be fitted
- some large multipoints may be capable of supplying more than one shower rose.

To prevent the possibility of backsiphonage of water, when the shower head is supplied through a flexible hose, it may be necessary to fit either a hose constraining ring, Figs. 29 and 30, or a check/double check valve in an appropriate part of the system. The hose constraining ring prevents the shower head from discharging water at any point less than the distance given in Table 30 above the spillover level of the shower tray, bath or basin being supplied. Where the shower head is not prevented from reaching the shower tray etc. then check/double check valves must be fitted.

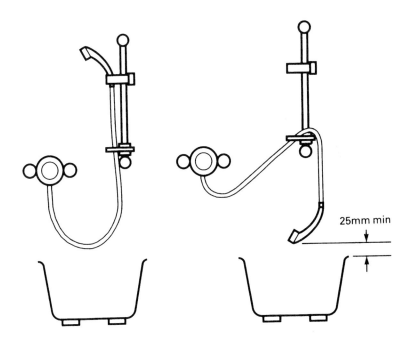

Fig. 30   Operation of hose constraining ring

Figures 31 and 32 show the pipework layouts for mains and cistern supplies to a shower.

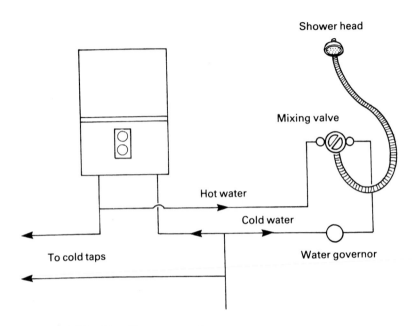

Fig. 31   Shower installation, mains water supply

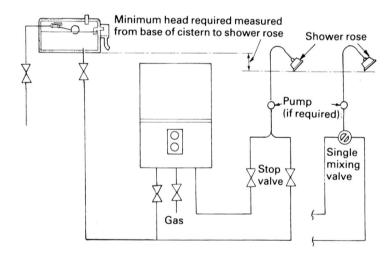

Fig. 32   Shower installation, cistern water supply

Installing Washing Machines and Dishwashers

Large multipoint water heaters may be connected to suitable automatic washing machines or dishwashers. Sink water heaters are generally not suitable.

The washing machine or dishwasher should:

- contain a heating element to boost the water temperature for certain programmes requiring water above 60° C
- not operate on a 'timed-fill cycle'
- be controlled by a pressure level switch which cuts off when the drum has the correct level of water
- not use a mixing thermostat where the hot and cold solenoids are rapidly switched on and off to mix the water
- have the flow restrictor washers removed from the hot water inlet
- be installed to comply with water supply byelaws 22 and 23.

The heater and the machine should both be connected to the same mains water supply, Fig. 33.

The pressures required are about 1.2 to 1.5 bar.

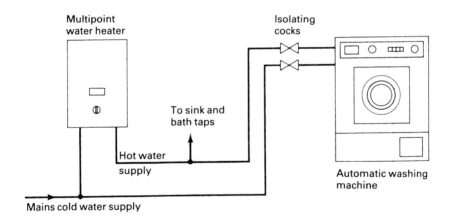

Fig. 33   Connection to washing machine

Commissioning Instantaneous Heaters

When the installation has been completed test the inlet gas and water supplies for soundness and purge. Then follow the procedure below:

- check that heater gas control is off
- turn on all hot taps
- turn on water service cock
- when water flows freely, turn hot taps off

- check heater and draw-offs for water leaks
- turn on gas and light pilot, check ignition device
- check length and position of pilot flame, operation of flame supervision device and anti-vitiation device
- turn gas control to full on position
- check for gas soundness on heater joints
- check for correct ventilation
- open hot tap, heater should light up
- check burner pressure and adjust, if necessary
- check the flame picture and, if necessary, check the gas rate by the meter test dial
- check the water flow rate and temperature rise, adjust water throttle if necessary, check thermostat, if fitted
- on a sink heater with an open flue, check the draught diverter for spillage, on room-sealed appliances, check joints and seals
- turn hot tap on and off to check slow ignition, adjust if necessary
- turn off hot tap
- check that heater operates on all taps
- turn off pilot and check operation of flame supervision device
- turn on and relight pilot
- instruct the customer on the operation of the heater and on the need for regular servicing.

Servicing Instantaneous Heaters

Heaters should be regularly serviced to maintain efficiency and ensure safe operation. Servicing should be carried out annually and should include the following operations:

- check the operation of the heater and discuss its performance with the customer
- turn off gas and water service cocks
- remove outer case
- open hot taps and remove drain plug
- remove heating body and wash out the flueways
- remove and clean main burner
- clean pilot burner and examine electrodes or thermocouples
- clean gas valve and grease stem if necessary
- check gas and water filters
- ease and grease control cocks if required
- if removing the water section on a multipoint fit temporary continuity bond across water supplies
- renew all jointing washers when reassembling

- when reassembled, check gas and water for soundness
- test the ignition device and adjust if necessary
- check pilot flame, test flame supervision device and anti-vitiation device
- check burner pressure and test control devices
  - governor
  - thermostat
- check operation of the automatic valve and adjust slow ignition device if required
- check lighting doors or windows
- check flue seals on room-sealed heaters or draught diverter for spillage on open-flued
- check that ventilation grilles are correctly sized and unobstructed
- check cartridge in scale reducer if fitted, see Chapter 9
- clean and replace outer case
- check appropriate warning label is attached
- clear up the work area and leave the appliance in working order.

Fault Diagnosis and Remedy

Faults on instantaneous heaters may be associated with the:

- gas supply or gas section
- water supply or water section.

Faults on the pilot or pilot supply are obviously in the first category but the causes of many other faults may be found in both categories. It is therefore always advisable to check that the pilot is alight and the flame supervision device operative before checking the water section.

Common faults and their remedies are as follows:

Symptom	Action
No pilot light	*Check that gas is on and present at the appliance tap*
	Check that pilot supply, filter and burner ports are clear
	Check pilot adjustment

Pilot ignites but does not stay alight	*Check size of flame and its position on flame supervision device* *Check thermocouple connections and output voltage* *Check for excessive draughts or faulty seals on balanced flue*
Pilot correct but no gas to main burner	*Check flame supervision device* *Check gas valve stem clean, free and undamaged* *Check that water is on and present at the appliance* *Check for adequate water flow:* • *satisfactory head of water* • *pipes adequate and clear* • *no air locks* • *jumper washer undamaged on hot tap* • *heating body clear of scale* *Check automatic valve:* • *filter clear* • *venturi and low-pressure duct clear* • *slow-ignition device clear and correctly adjusted* • *bearing plate and push rod free and undamaged* • *diaphragm not perished, distorted or punctured*
Noisy or rapid ignition	*Check length and position of pilot flame* *Check operation and adjustment of slow-ignition device*
Water not hot enough	*Check gas valve fully open* *Check flame supervision device* *Check gas rate to burner* *Check thermostat setting* *Check rate of water flow, operation of water governor, position and operation of temperature selector*
Gas valve closes slowly or incompletely	*Check gas valve* • *seating clean and undamaged* • *stem clean and lubricated, push rod undamaged* • *spring satisfactory*

Check slow ignition device clean and
correctly adjusted
Check diaphragm not distorted and water
governor valve spindle clear and free
Check venturi and low-pressure duct clear

Water Supply Faults

If, when the hot tap has been turned off, the residual water in the heat exchanger becomes overheated, the water will try to expand. The pressure which results will normally be dissipated by the hot water expanding back into the inlet cold supply.

If, however, the water service cock or stop cock has a loose jumper, the cock can act as a non-return valve and prevent free expansion of the water. The increased pressure will be contained within the heater and its installation and water joints may be burst.

This is likely only to occur where a stop cock with a loose jumper is associated with a slow or incompletely closing gas valve or with a heavily scaled heating body. It is advisable always to pin the jumper on any cock fitted at the inlet to a multipoint instantaneous water heater.

Similar problems to those described above, can also occur through dead legs or air entrainment. Dead legs are lengths of water pipe that have been cut off but have not had all of the air evacuated from them.

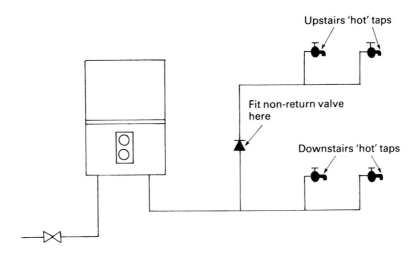

Fig. 34   Method of preventing air entrainment

Air entrainment usually occurs when upstairs and downstairs hot taps are opened simultaneously resulting in no water flow at the tap with the highest resistance to flow, usually the upstairs tap. The result is an ingress of air into the system. If the upstairs hot tap is then turned off first, the air is trapped in the system and forms a temporary dead leg. With air entrapped in a system, through a dead leg, when the appliance is turned off, by closing the hot tap, water will continue to flow through the system and the appliance until the air is fully compressed. In certain cases this flow keeps the gas valve in the open position even though the hot taps are closed.

Dead legs should be cut off as short as possible in order to overcome the problem.

Air entrainment can be eliminated by fitting a check (non-return) valve in the pipe feeding the upstairs hot taps, see Fig. 34.

CHAPTER 9

Water Heating – Storage Appliances

Chapter 9 is based on an original draft by B. J. Whitehead

Introduction

Over the years, many different types of storage water heaters, circulators and boilers have been developed. Some were designed to give a full hot water service, others supplied water only to a sink or a bath.

Since small bore central heating began to take up the domestic water heating load, there has been a change in the number and types of water heaters being produced. Currently the models available include:

- medium rated circulators
- back circulator units
- multipoint storage water heaters.

All these appliances give a full hot water service and offer an alternative to instantaneous water heaters with the following advantages.

- water is delivered at a constant temperature, usually 60° C, at all times of the year
- higher temperatures, up to about 82° C, can be provided if required
- the rate of water delivery depends only on pipe and tap size and can be much higher than that from instantaneous heaters
- systems can be designed to meet any customer's demands

WATER SUPPLY SYSTEMS

Cold Water Supply

In some of the water authority areas in this country, it is normal practice to fit a cold water storage cistern in domestic premises. This

403

prevents the mains being overloaded at times of peak demand. The cistern usually has a nominal capacity of not less than 114 litres which may be increased if the water requirements of the household call for larger quantities to be available.

A typical system is shown in Fig. 1. The cold taps at the bath and the wash basin, together with the lavatory cisterns are supplied from the cold water storage cistern. The tap at the sink is connected directly to the mains supply in order to provide fresh water for drinking and cooking. This system ensures that adequate water is always available to meet peak demands.

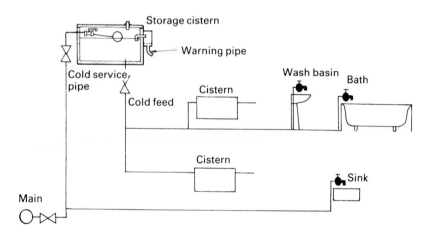

Fig. 1   Cold water supply system (cistern fed)

BS 6700 : 1997 gives recommendations for the design, installation, testing and maintenance of services supplying water for domestic use. The water supply byelaws are designed to prevent waste, undue consumption, misuse and contamination of water. These byelaws were made and are enforced by the water supply authorities, companies and councils throughout Great Britain. The byelaws are not retrospective but any work on new or existing installations must conform to the relevant regulations.

Cisterns

The water level in the storage cistern shown in Fig. 1, is controlled by a ball valve which discharges into the air above the highest water level in the cistern, Fig. 2, ensuring a type 'B' air gap in line with Table 30 in Chapter 8. This ensures that any water in the cistern, which could become dirty or contaminated, cannot be drawn (siphoned) back into the main supply.

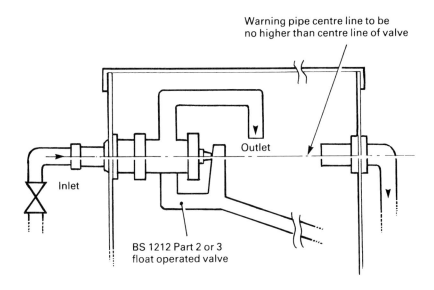

Fig. 2   Installation of float operated valve in a cistern

Where there is no such air gap, between the point of discharge and the water level, a check (non-return) valve must be fitted in the cold water service pipe to the ball valve. Similarly the tap supplying drinking water to the sink must be fitted so that its outlet spout is a minimum vertical distance, in line with the type 'A' air gap given in Table 30 in Chapter 8, above the overspill level of the sink. Where this minimum distance cannot be attained then the cold supply must be protected in the same way as the supply to the ball valve described earlier.

The lid on the cistern and the screens in the warning pipe connection and air inlet, are all designed to prevent the ingress of dirt, insects etc., Fig. 3.

The insulation around the cistern is to protect against extremes of temperature, frost etc. which can lead to subsequent wastage. The warning pipe tells the customer when the ball valve is in need of maintenance thus preventing damage to the building around and below the cistern as well as the wastage of water.

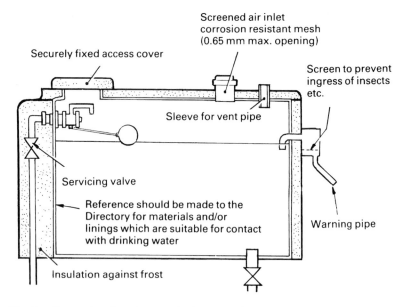

Fig. 3   Example of a cistern which meets the requirements of byelaw 30

Hot Water Direct Systems

Hot water supply systems were introduced in Vol. 1 when hydrostatic pressure and circulating pressure were discussed. A typical direct system is shown in Fig. 4. Although there may be many direct systems on the district, modern gas systems are generally indirect to avoid problems of scaling up.

For a normal domestic dwelling the hot water cylinder should have a capacity of at least 136 litres or 45 litres per person. The cold feed cistern should have a capacity at least equal to that of the cylinder so that all the hot water can be drawn off even if the mains water supply fails. If the cistern also supplies the cold water system its capacity should be not less than 228 litres.

When storing water for domestic purposes the cistern must comply with byelaws 24, 30 and 31.

The connection of the cold feed pipe should be about 25 mm above the base of the cistern and, in the case of a combined storage and feed cistern, about 12 mm above the cold water supply connection. This reduces the risk of debris entering the cold feed pipe. The cold water feed to the storage cylinder should be fitted with a 'spreader' end so that incoming water will not disturb the stratification. It should dip before entering the cylinder to prevent hot water circulating back up the cold feed pipe by convection currents set up inside the pipe.

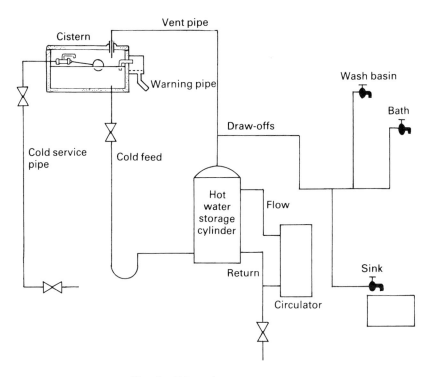

Fig. 4   Direct hot water system

All draw-offs should be taken from the vent pipe, never from flow or return pipes. The vent pipe should be 22 mm diameter. It should rise steadily from the cylinder before bending over the cold feed cistern and terminating through a grummet in the cover on the cold feed cistern. No valves should be fitted on the vent pipe or between the circulator or boiler and the vent pipe connection. Various combination storage units are available which reduce the cost of the installation by combining hot and cold water storage in a prefabricated unit. An example is shown in Fig. 5.

Draw-offs are usually 22 mm to bath taps and 15 mm to wash basins or sinks. The maximum length of a draw-off from the storage cylinder, should not exceed the lengths given in Table 28, Chapter 8. Where this is not possible, a 'secondary circulation' should be provided, Fig. 6. By having secondary flow and return pipes from and to

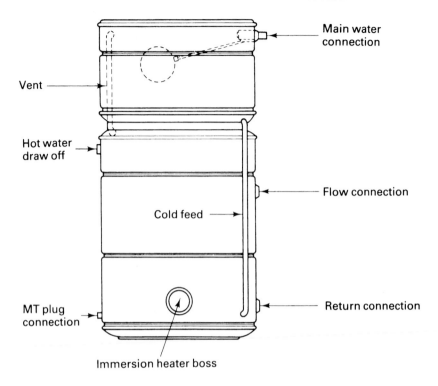

Fig. 5   Combined hot and cold water storage system

the hot water storage cylinder, the dead leg of pipe in which water would cool down is substantially reduced, so preventing wastage of water.

The secondary return should be connected into the hot water cylinder at a point not more than one quarter of the height of the cylinder, from its top and, on a direct cylinder, level with the primary flow.

Primary flow and return pipes from a circulator or boiler should be sized and connected in accordance with manufacturer's instructions. Generally the minimum circulating head is about 300 mm and the maximum about 6 m. With circulating heads of 1.5 m or more it will be necessary to restrict the circulation in order to reduce the rate of water passing through the heat exchanger and so obtain a flow temperature of about 65° C.

The circulation may be restricted by means of:

● using 28 mm flow and return pipes for low heads and 22 mm pipes for heads in excess of 2 m

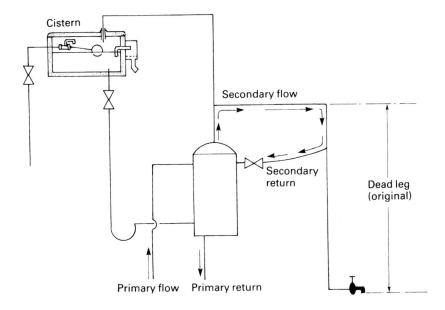

Cistern

Secondary flow

Secondary return

Dead leg (original)

Primary flow   Primary return

Fig. 6   Secondary circulation

- discs with small drillings inserted into the union connection on a heater return
- adjustable circulation restrictor valves.

No valves, other than a circulation restrictor or a drain cock should be fitted on flow or return pipes.

More than one circulator may be fitted to the same cylinder on a direct system. This is sometimes done on a large commercial installation where it is more convenient to employ two heaters which together can supply the full load and where one may be adequate at off-peak times. The heaters may be maintained one at a time, so ensuring an uninterrupted supply of hot water.

In domestic premises a number of combined systems have been used. One of the most common was the fitting of a gas circulator to supply summer water heating to a solid fuel system of water heating and central heating. In both cases the gas circulator must be connected to the cylinder by separate flow and return pipes, Figs. 7(a) and (b). In the case of the central heating system, the circulator is connected to the cylinder and not to the calorifier or heating circuit. A valve is fitted on the heating circuit to prevent heat being lost, in the summer, by a natural convection circulation from the cylinder.

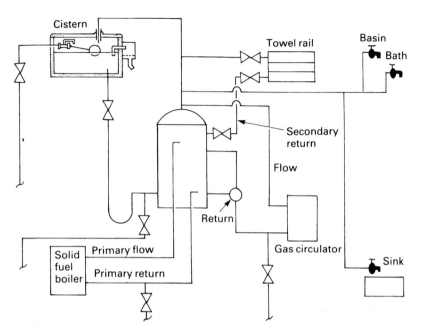

Fig. 7(a)   Combined systems: direct cylinder, domestic water heating boiler

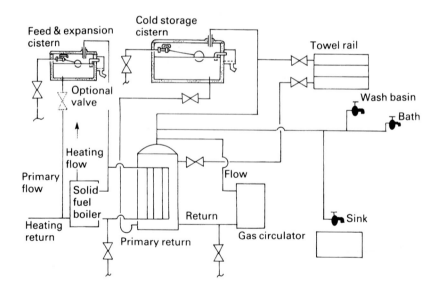

Fig. 7(b)   Combined systems: indirect cylinder, central heating boiler

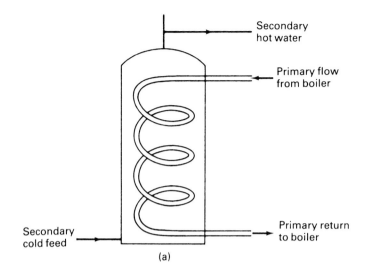

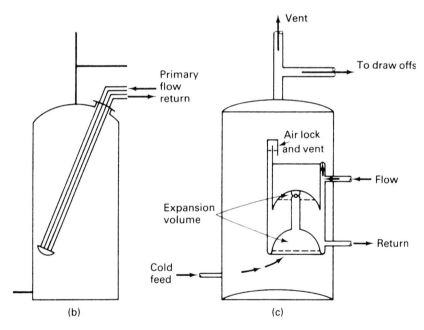

*Fig. 8   Indirect cylinders and calorifiers: (a) double feed; (b) immersion
calorifier; (c) single feed*

Indirect Hot Water Systems

In an indirect system the water in the hot water cylinder is heated by means of a calorifier (Vol. 1, Chapter 10), so there is no direct contact between the water in the circulator and the domestic hot water. Examples are shown in Fig. 8.

Because the circulating water stays in the circulation and is not continually replaced by fresh water, scale is only deposited during the first circulation.

This system is used where domestic water heating is combined with central heating (see Chapter 10). It also has advantages in hard water areas where it can minimise the scale deposit in circulators or boilers and in flow and return pipes. It is becoming general practice to use indirect cylinders, usually of the single feed type, for all storage systems.

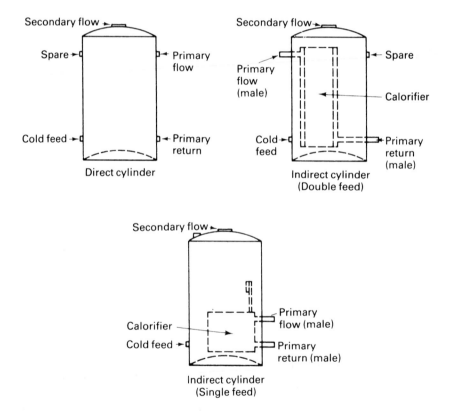

Fig. 9   Connections to direct and indirect cylinders

Indirect cylinders are described in more detail in Chapter 10. Basically there are two types:

- double feed
- single feed.

The double feed cylinder requires two cisterns. One is the cold water feed cistern for the domestic hot water and the other is a feed and expansion cistern for the water circulating in the calorifier and the circulator.

The single feed type has one cistern and the circulating water is separated from the domestic water in the cylinder by air bubbles trapped in the calorifier. Water can be taken into the circulating system but cannot come out again. Expansion is contained within the calorifier and a small internal vent. The size of cylinder is determined by the water capacity of the circulator or boiler and the primary circulation. If too small a calorifier was used it would not be able to accommodate the expansion of the water in the primary system. So the air locks would be displaced and mixing would occur between the primary and the domestic hot water.

Immersion calorifiers are only used on pumped primary circulations and are dealt with in Chapter 10.

A comparison of the connections to direct and indirect cylinders is shown in Fig. 9.

Multipoint Storage Heater Systems

There are a number of multipoint storage water heaters which combine a heat exchanger with a hot water storage capacity of about 90 litres. They require to be connected to a cold feed cistern, vent pipe and draw-offs, Fig. 10.

Some combined units can provide central heating as well as water heating and incorporate boiler, hot water storage, cold feed cistern and feed and expansion cistern. they require connecting to mains water supply, draw-offs and heating circuit.

Unvented Hot Water Systems

Changes, introduced in the 1985 Building Regulations (England and Wales) and the 1986 Model Water Supply Byelaws, have seen the introduction of gas heated unvented hot water storage systems. Previous legislation limited the connection of systems containing no more than 15 litres of water from being directly connected to the

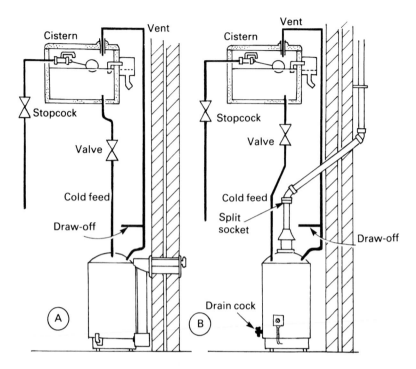

*Fig. 10   Multipoint storage water heater systems: (a) balanced flued;
(b) open flued*

mains water supply. This allowed instantaneous multipoint water heaters to be connected directly to the mains but excluded the connection of storage systems.

Requirement G3 of Schedule 1, Part G (1992) of the Building Regulations 1991 deals with the safety aspects of unvented hot water systems while byelaws 90 to 95 of the Water Supply Byelaws cover the matters concerning wastage and contamination of water supplies. G3 states that only a competent person shall install an unvented hot water system and there shall be precautions:

(a) to prevent temperature of stored water at any time exceeding 100° C; and
(b) to ensure that the hot water discharged from the safety devices is safely conveyed to where it is visible but will not cause danger to persons in or about the building.

Fig. 11 illustrates the differences between the traditional vented system and the unvented storage system.

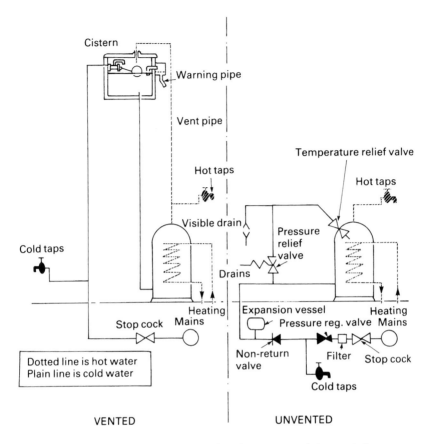

Fig. 11   Comparison between a vented and an unvented domestic hot water storage system

These differences require changes in the equipment and controls used in the unvented system. The system can be heated directly using appliances similar to those in Fig. 10 or heated indirectly, using a circulator, or boiler that may also be supplying a heating system. In any unvented system the storage vessel has to be constructed to withstand much higher pressures than vessels used in traditional vented storage systems.

In addition to the changes in the storage vessel the following controls and equipment are needed:

• line strainer (water filter)
• check (non-return) valve
• expansion vessel

- expansion valve*
- temperature relief valve*
- energy cut-out
- operating thermostat (cylinder thermostat).

Unvented systems have the following advantages over vented systems:

- higher water pressure at hot taps
- equal pressures in hot and cold supplies – easier temperature control of showers
- no cistern or pipework in roof space – reducing possibility of frost damage, no contamination or cistern noise
- possible reduction in pipe sizes
- location of storage vessel more flexible.

Unvented systems also have the following disadvantages:

- not suitable in areas with limited mains water pressure and flowrate
- no storage back-up
- installations require discharge pipe to a visible drain
- extra controls and pipework
- some maintenance necessary.

*The expansion and temperature relief valves are connected through tundishes (air break devices) to discharge pipes. Any water discharged from these valves must be directed in such a manner that it can easily be seen without creating a danger to anyone in or about the building.

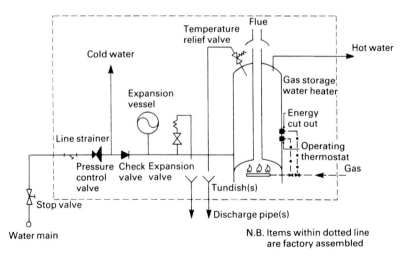

Fig. 12 (a)   Directly heated gas-fired unvented hot water scheme 'Unit'

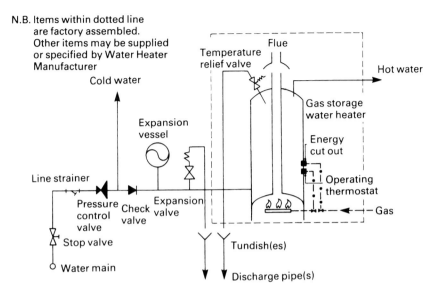

N.B. Items within dotted line
are factory assembled.
Other items may be supplied
or specified by Water Heater
Manufacturer

Cold water

Flue
Temperature
relief valve

Hot water

Gas storage
water heater

Energy
cut out

Operating
thermostat

Gas

Line strainer

Expansion
vessel

Pressure
control
valve

Check
valve

Expansion
valve

Stop valve

Water main

Tundish(es)

Discharge pipe(s)

Fig. 12 (b)   Directly heated gas-fired unvented hot water scheme 'Package'

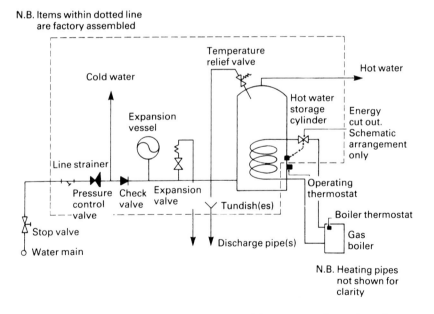

N.B. Items within dotted line
are factory assembled

Temperature
relief valve

Cold water

Hot water

Hot water
storage
cylinder

Energy
cut out.
Schematic
arrangement
only

Expansion
vessel

Line strainer

Operating
thermostat

Pressure
control
valve

Check
valve

Expansion
valve

Tundish(es)

Boiler thermostat

Stop valve

Water main

Discharge pipe(s)

Gas
boiler

N.B. Heating pipes
not shown for
clarity

Fig. 12 (c)   Indirectly heated unvented hot water scheme 'Unit'

An unvented hot water storage system is available either as a unit –
equipment and controls assembled at the factory – or as a package –

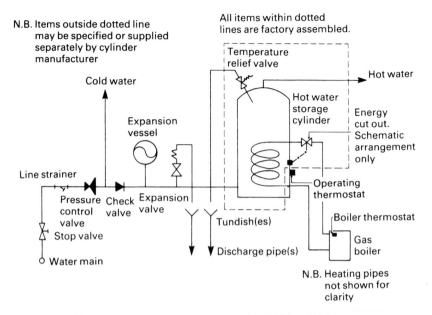

Fig. 12 (d)   Indirectly heated unvented hot water scheme 'Package'

some of the functional controls installed on site. Fig. 12 (a), (b), (c) and (d) illustrates the differences between units and packages.

Figs. 13 and 14 illustrate schematic layouts of copper and steel unvented indirect hot water cylinders with their associated controls.

Details of the controls and equipment, illustrated in the figures on unvented hot water storage systems, are as follows:

Line Strainer – filter to prevent foreign bodies etc. from contaminating the system, also ensuring efficient operation of other controls, see Fig. 15 (a) and (b).

Pressure Control Valve – inlet water pressure reducing or limiting valve. Both reduce the mains water pressure to the working pressure required in the system – usually around 2 bar for copper cylinders and 3 bar for steel, Figs. 16 (a) and (b).

Expansion Valve – any excessive over pressurisation of the system – e.g. malfunction of pressure control valve etc. – will be relieved through the expansion valve, Fig. 17.

Check Valve – non return valve which prevents any backflow of water from the system into the cold water pipes or mains if there should be a pressure reduction in the cold water mains, Fig. 18.

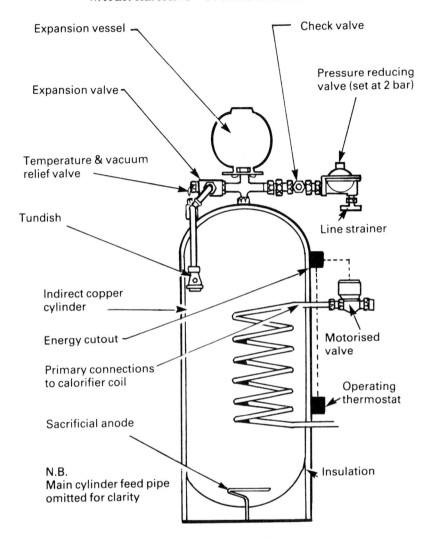

Fig. 13 Schematic layout of unvented indirect hot water storage copper cylinder

Expansion Vessel – prevents wastage of water by accommodating the increase in the volume of expanded water under the normal heating conditions. The charge pressure of the vessel should be set at the nominal operating pressure recommended by the cylinder manufacturer. This vessel prevents a build-up of excessive pressure when the water is heated and expands in an unvented system just as the vent pipe does in an open vented system, Figs 19 (a) and (b).

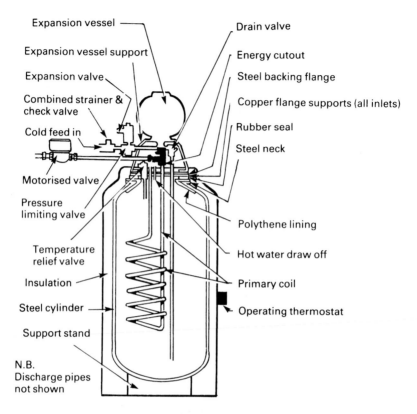

Expansion vessel

Expansion vessel support

Expansion valve

Combined strainer & check valve

Cold feed in

Motorised valve

Pressure limiting valve

Temperature relief valve

Insulation

Steel cylinder

Support stand

N.B.
Discharge pipes
not shown

Drain valve

Energy cutout

Steel backing flange

Copper flange supports (all inlets)

Rubber seal

Steel neck

Polythene lining

Hot water draw off

Primary coil

Operating thermostat

Fig. 14   Schematic layout of unvented indirect hot water storage steel cylinder

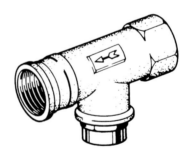

Fig. 15 (a)   'In line' pattern line strainer

Temperature Relief Valve – protects the system if the control thermostat and energy cut-out fail to operate. The valve operates below 100° C, normally it is fully open at 95° C. Some temperature

Fig. 15 (b)   'Angle' pattern line strainer

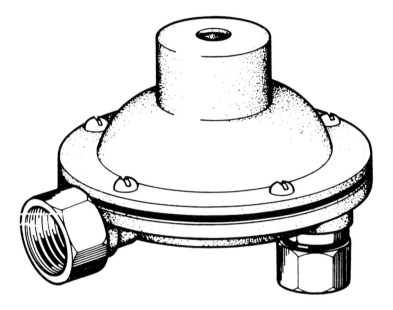

Fig. 16 (a)   Pressure reducing valve

relief valves also incorporate a pressure relief function but this does not obviate the need for a separate expansion valve, Fig. 20.

Tundish – air break device fitted into the discharge pipes from both the expansion and temperature relief valves. A blocked discharge from either of these valves could negate their effective operation, Figs 21 (a) and (b).

In addition to these controls there will be an operating thermostat, an energy cut-out device and in some cases a vacuum relief valve.

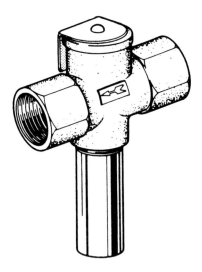

Fig. 16 (b)   Pressure limiting valve

Fig. 17   Expansion valve

The operating thermostat is often a strap-on type used to control the temperature of the stored water. The recommended temperature setting of this device is 60° C to 65° C.

The energy cut-out device is a back-up thermostat that operates at a temperature of 85° C to 90° C. It is normally arranged for this device to cut off the supply of primary heat to the cylinder, i.e.

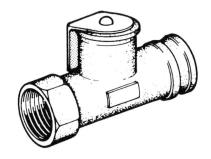

Fig. 18   Check valve

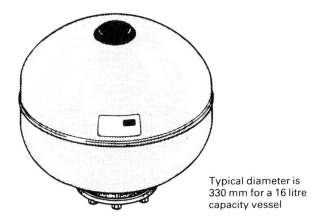

Typical diameter is
330 mm for a 16 litre
capacity vessel

Fig. 19 (a)   Expansion vessel

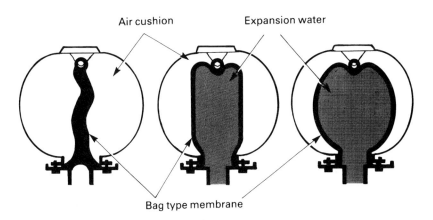

Fig. 19 (b)   Principle of operation of an expansion vessel

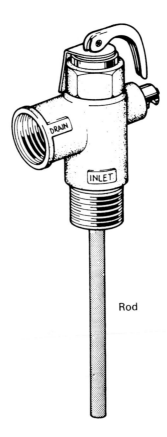

Fig. 20   Temperature relief valve

through a 2 port valve fitted in the primary circuit. The control must have a tripping device that requires to be reset manually after it has operated.

The vacuum relief valve is generally incorporated in the temperature relief valve. Its function is to equalise the pressure in a storage vessel if it is subjected to a vacuum condition. Situations that can create a partial vacuum in the storage vessel are:

- draining the cylinder down without venting the system, e.g. through a draw-off tap
- drawing off water faster than the inlet can replenish it
- the system cooling down after the water supply isolating valve has been closed.

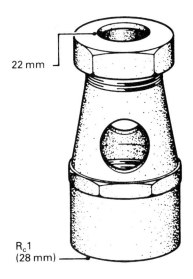

Fig. 21 (a)   Tundish, brass bodied

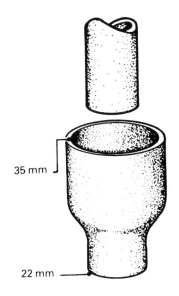

Fig. 21 (b)   Tundish, using an inverted copper reducing coupler

Thermal Storage or Water Jacketed Tube Heaters

Another method of supplying hot water at mains pressure is through a tube heater. Fig. 22 schematically illustrates the operation of a storage tube heater. The system which usually operates in conjunction with a heating system, works in the reverse of the traditional indirect hot water cylinder. The domestic (secondary) hot water passes through a tightly coiled, finned tube, heat exchanger that is contained within a small cylinder through which the primary water passes. After passing through the finned tube heat exchanger the secondary water flows through a thermostatic mixing valve before flowing to the hot water outlets. The system is basically an instantaneous water heater with a heat bank reserve in the primary cylinder.

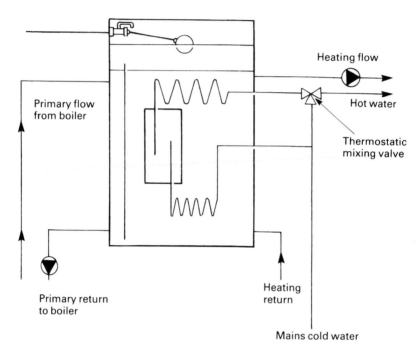

Fig. 22   Thermal storage tube heater

APPLIANCES

Circulators

Gas circulators may be open or balanced flued. The open flue models generally take a 75 mm diameter flue pipe and usually incorporate a draught diverter. Balanced flue ducts are available to suit a variety of construction and thickness of wall. Typical circulators are shown in Fig. 23 (a) and (b).

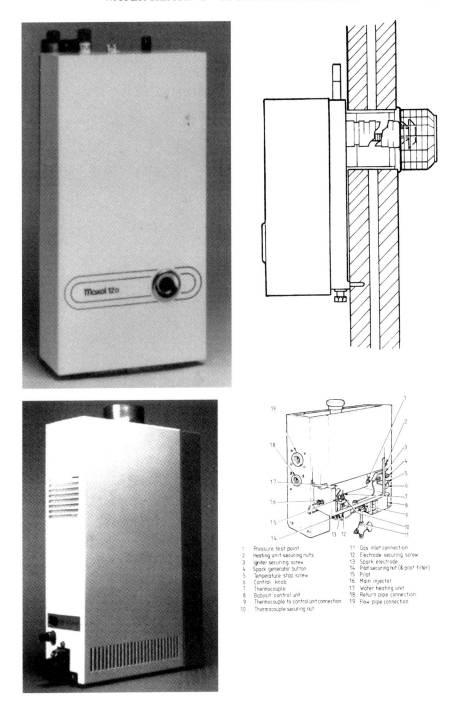

1	Pressure test point	11	Gas inlet connection
2	Heating unit securing nuts	12	Electrode securing screw
3	Igniter securing screw	13	Spark electrode
4	Spark generator button	14	Pilot securing nut (& pilot filter)
5	Temperature stop screw	15	Pilot
6	Control knob	16	Main injector
7	Thermocouple	17	Water heating unit
8	Babysit control unit	18	Return pipe connection
9	Thermocouple to control unit connection	19	Flow pipe connection
10	Thermocouple securing nut		

Fig. 23   Circulators: (a) balanced flue (Maxol); (b) open flued (Main)

The construction of the heat exchanger varies from the conventional copper tubes and fins, Fig. 24, to the annular heater body with stainless steel fins and baffle, Fig. 25.

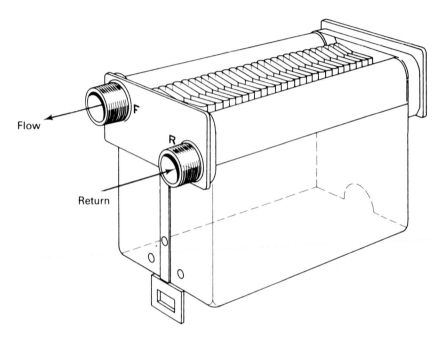

Fig. 24   Heat exchanger, finned tubes

Some models have enamelled, pressed-steel cases for fitting in a kitchen or an airing cupboard. Others are incorporated into warm air heating units and share a common flue.

Heat input rates are about 4.4 kW and the outputs are about 58 litres/hour or 3.3 kW.

The gas supply is 8 to 12 mm diameter. Domestic circulators are suitable for connecting to direct or indirect cylinders from 90 to 270 litres capacity. Maximum operating heads may be 15 to 18 m. The control devices fitted to circulators include the following.

Thermostats

These may be bimetal snap action, fitted in the top of the hot storage cylinder, or vapour pressure types. On circulators fitted to warm air units the controls are usually 24 V electric and an electric immersion thermostat is used.

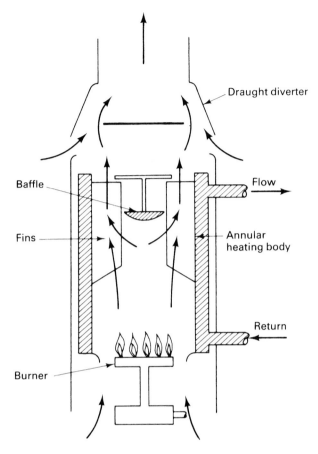

Fig. 25   Cross-section through heat exchanger

Flame Supervision Devices

Usually either thermoelectric or vapour pressure devices are fitted. They may be incorporated into a type of multi-functional control.

Igniters

These are usually piezo-electric.

Regulators

Some of the older types of appliances had a spring-loaded regulator controlling the appliance gas rate. Most modern domestic circulators rely on the meter regulator to give the correct working pressure at the burner injector, to ensure that the appliance gas rate is correct.

Gas Control

The user's control cuts off the main burner but can leave the pilot alight to save re-igniting. Circulators on warm air heaters may have a 24 V solenoid valve controlled by the thermostat and the user's switch control in series. Some solenoids may be wired to a programmer (see Chapter 12).

Back Circulators

A back circulator unit is designed to fit into a fireplace opening, usually behind a gas fire. It may replace a solid fuel back-boiler and if the existing flow and return pipes are adequately sized, in good condition and suitably installed, it may be connected to them.

The fireback must be removed and any rubble or debris cleaned out. This can be done without removing the tiled surround. The brickwork in the openings should be made good and the throat of the chimney should be 'gathered in' evenly. That is, it should slope up

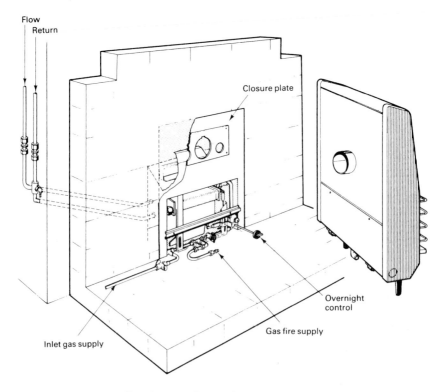

Fig. 26   Back circulator installation

from the fireplace opening to the brick flue. The unit almost fills the standard 400 mm fireplace opening at the front and is about 280 mm wide at the back and 220 mm deep, Fig. 26.

A special closure plate is fitted flush on the face of the tiled surround and a gas connection is provided so that an independently controlled gas fire may be fitted in front.

It is essential that the brick chimney should be checked before installation. Ensure that:

- no downdraught problems exist
- the chimney has been swept
- updraught is satisfactory
- the gather is satisfactory and there are no signs of defective brickwork
- the height of the chimney conforms to the manufacturer's installation requirements.

The flue and ventilation must be adequate for the maximum combined rating of the back circulator and the gas fire. The likelihood of condensation should be checked assuming that each appliance in turn is the only one in use (see Chapter 5).

Multipoint Storage Heaters

There are a variety of heaters on the market, several produced by North American manufacturers. They consist essentially of a storage cylinder or chamber with a gas burner and heat exchanger below. The flue on some models may pass up through the centre of the storage vessel. A cross-section of a typical model is shown in Fig. 27(b).

Storage heaters may have open or balanced flues and some models have fanned draught balanced flues. Some heaters may be installed under a draining board or fitted with a case, as part of a kitchen unit layout.

The storage cylinder may be copper or vitreous enamelled or glass-coated steel. The heat exchangers are commonly aluminium, stainless or aluminised steel.

Typical heat inputs are 6.5 kW or 8 kW. Capacities may be 72 litres to 90 litres. Hot water outputs are about 112 to 125 l/h raised to 60° C.

The connections are generally:

- gas supply, 15 mm ($R_c^{1/2}$)
- cold feed, 22 mm
- draw-offs and vent 22 mm
- flue pipe, if fitted, 100 mm.

Fig. 27(a)   Multipoint storage water heater (A•S Andrews)

A gas-fired multipoint storage water heater (GSWH) is simple in its concept, in its design and in its application.

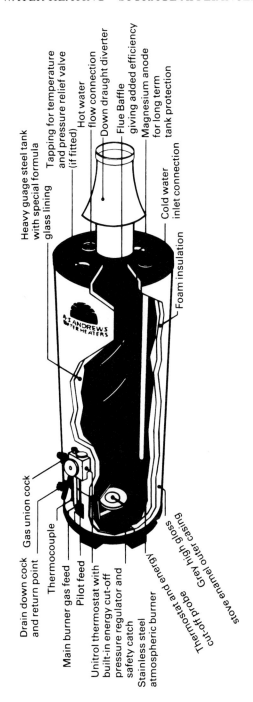

Fig. 27(b)   Multipoint storage water heater (A•S Andrews)

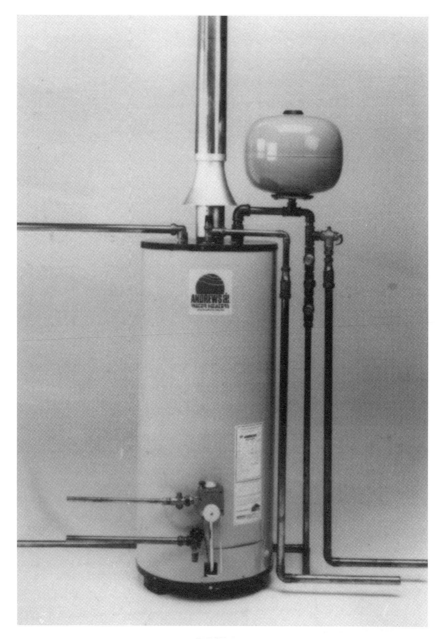

Fig. 28   Unvented multipoint storage water heater

A typical water heater comprises a glass-lined water cylinder beneath which is an atmospheric gas burner venting via a central flue, with a flue gas retarder (baffle) for maximum heat transfer, to atmosphere. This cylinder is well lagged with a 'sandwiched' foam jacket. Four water connections are provided, e.g. R1 feed and supply at the top (the feed taking hot water internally to the base of the cylinder and the draw-off at high level); $R_p\frac{3}{4}$ return tapping and a low level $R\frac{3}{4}$ drain down cock. There is one other aperture into the cylinder, that being for the magnesium anode which, through electrolysis in the water ensures that the glass-lined cylinder has long life protection.

The gas supply is via a $R_c\frac{1}{2}$ gas cock to a multifunctional gas control. This control is normally non-electric and provides in one unit: thermostatic operation with settings from 40 to 65° C, flame supervision, overheat protection and energy cut-off. The burner can be removed from the low level access at the front of the unit for easy maintenance. The casing is finished in baked enamel. High temperature units are suitable for temperatures up to 82° C. They are similar in construction and operation to the unit already described but having a hand access to the cylinder for inspection and cleaning purposes, larger 'bar' burners and a higher recovery rate – usually in excess of 550 litres/h. These larger capacity appliances will be covered in Vol. 3.

Fig. 28 illustrates a multipoint storage water heater for use on an unvented system.

ANCILLARY VALVES

Ball Valves

Ball valves control the flow of water into cisterns. They consist of a float or ball attached to a lever which is pivoted at the opposite end. The float rises as the cistern fills up and the pivoted end moves a piston which closes the water inlet. Fig. 29 (a) shows the 'Portsmouth' pattern which is manufactured to BS 1212 Part 1. In the past this type of valve was the most common but with the introduction in 1986 of the Water Supply Byelaws the $R_L\frac{1}{2}$ size did not meet the requirements of Water Supply Byelaw 42. It failed because (i) it will not withstand a backsiphonage test when the water level is as high as the centre line of the valve and (ii) the valve float should be fitted with an easily adjustable device for setting the water level, bending the float arm is not an acceptable method. To renew the washer on the piston type of valve, remove the end cover and pull out the split pin which will release the lever (float arm). The piston may then be extracted and unscrewing the cap will allow the washer to be removed.

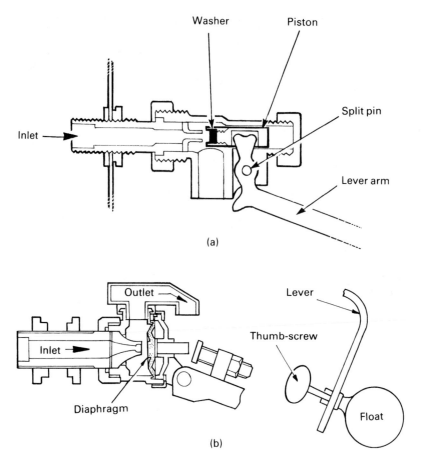

Fig. 29   Ball valves: (a) Portsmouth (piston) type; (b) diaphragm type

Fig. 29 (b) shows a diaphragm pattern valve which is manufactured to BS 1212 Part 2. As can be seen with this type of valve, when the water level is in line with the centre of the valve there is a type 'B' air gap between the outlet and water level and an adjustable device (thumb-screw) is used to set the water level. This valve meets the requirements of byelaw 42. Fig. 30 shows four ball valves, three piston types and one diaphragm type.

Drain Cocks

A drain cock or 'M.T. plug' is a small valve with a simple, screw-in plug, Fig. 31. The outlet is ridged to take a hose connection.

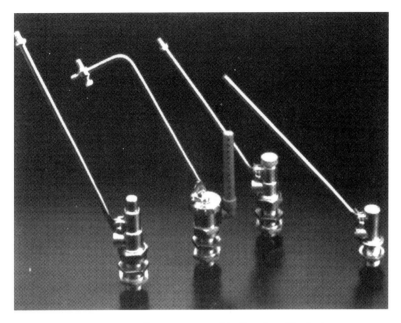

Fig. 30   Different types of ball valve (Pegler)

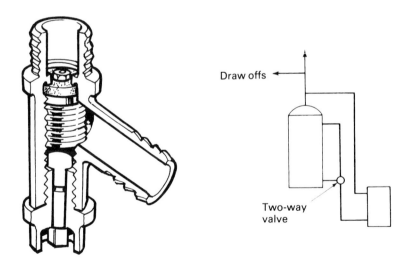

Draw offs

Two-way
valve

Fig. 31   Drain cock   *Fig. 32   Economy valve*

Drain cocks are fitted at low points in the system to allow all the water to be removed if necessary.

Draining taps should conform to BS 2879.

Economy Valves

Up to the early 1980s economy valves were often fitted to storage water heating systems. They are still encounted on the district but they are not usually installed on modern systems as improvement in the methods of insulating cylinders has reduced the need for them.

An 'economy valve' is a two-way valve fitted on the return pipe of a circulator so that a small quantity of water may be kept hot when the whole of the storage capacity is not required. During a normal day the demand for domestic hot water is limited to quantities of 10 to 20 litres at a time. Only when baths are needed is the whole storage necessary. Some economy can be effected by reducing heat losses from the cylinder by cutting down the volume of stored hot water.

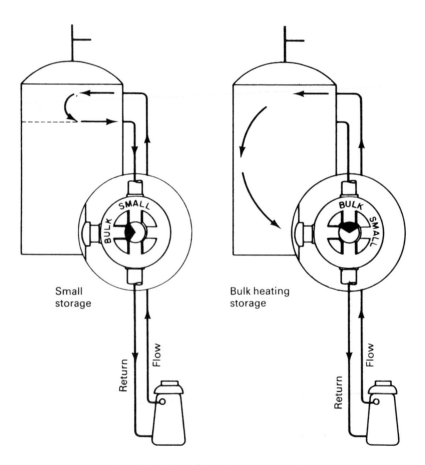

Fig. 33   Flow through economy valve

This is done by using two alternative return pipes, one from the base of the cylinder and another from about a quarter of the way down, Fig. 32. The two returns are connected to the two inlets of the economy valve and the outlet is connected to the common return to the circulator. The valve plug may be turned so that one of the alternative returns is open and the other closed, Fig. 33.

Secondary Circulation Valve

In order to avoid heat losses from a secondary circulation when the system is shut down overnight a valve may be installed. This is fitted on the return, close to the cylinder. It is sometimes referred to as a 'night valve', Fig. 34.

The valve may be operated manually or by a motorised valve controlled by a clock.

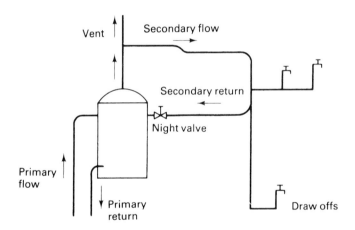

Fig. 34   Night valve

Vent Pipe Air Valve

Where the head of water at the feed cistern is very low, air may be drawn in at the draw-off branch on the vent pipe. This will give a reduced, intermittent flow at the draw-off tap which can be overcome by fitting an automatic air release valve, Fig. 35.

The valve is fitted vertically in the vent pipe and consists of the phosphor-bronze ball resting on a seating. The ball lifts to allow the water to expand and any air to escape, but prevents air being drawn

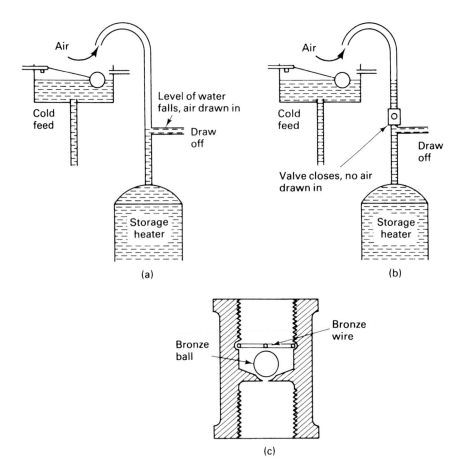

Fig. 35   *Vent pipe air valve: (a) air drawn into water supply; (b) valve installed in vent; (c) automatic air valve*

in. The highest draw-off should be opened to permit the entry of air when draining down the system. This method is very rarely encountered.

Stop Taps and Draw-Off Taps

Most stop taps and bib or pillar taps have a washer fitted to a jumper located in a screw-down spindle, Fig. 36. The jumper is free to rotate but is commonly clipped in its socket so that it is lifted off the seating when the tap is turned on.

To exchange the washer:

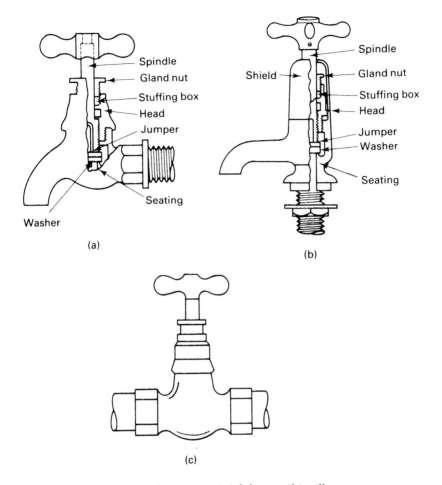

Fig. 36 *Draw-off taps and stop tap: (a) bib tap; (b) pillar tap; (c) stop tap*

- turn off the water supply and drain down
- open the tap fully
- remove any shield or cover
- unscrew the head by its hexagon flats
- remove the washer from the jumper and renew; the washer is often secured to the base of the jumper by a nut.

When reassembling, tighten the shield by hand pressure only.
Taps should conform to BS 1010.

Installing Storage Water Heating

The manufacturer's installation instructions for each particular heater should be followed implicitly. They will normally specify the suitable locations and minimum dimensions of any compartment or airing cupboard into which the appliance may be fitted. For some circulators special fixing kits and prefabricated flow and return pipes are available.

In airing cupboards, surfaces adjacent to an appliance should be insulated with sheets of fireproof material and the cupboard shelf may require to be modified or raised. Purpose-made cylinder stands may be required to adjust the height of the cylinder. Grilles should be provided to admit combustion air at low level and ventilate the compartment at high level (see Chapter 5).

Pipework should be run so as to avoid air locks and normally not buried in walls or floors. Pipes may preferably be concealed in ducts or chases with removable covers. They should be located to avoid damage by frost. Where they pass through walls or floors, pipes should be sleeved. Sweep fittings should be used to aid the flow of water.

For efficient operation and to conserve energy the hot water storage cylinder and the piping of the primary and any secondary circulation should be lagged. Most cylinders are lagged with a sectional jacket 50 mm in thickness. Glass fibre or foamed plastics are commonly used in jackets and prefabricated pipe sections. Cylinder jackets should have fireproof panels immediately adjacent to a circulator in the airing cupboard.

Commissioning

The procedure for commissioning is as follows:

- flush out the system thoroughly
- check all water joints and make good any leaks
- re-check after heating up
- ensure that all air is released
- test the gas installation for soundness and purge
- check for correct ventilation
- check operation of combustion fan if fitted
- light up the appliance in accordance with the manufacturer's instructions
- check ignition device
- check that pilot flame is correct and stable
- check the burner pressure and flame picture; check gas rate by meter test dial if necessary

- check for spillage at the draught diverter, or check flue seals on room-sealed appliances
- check the controls
    - flame supervision device
    - thermostat
- set the thermostat to the temperature required
- instruct the customer on the method of lighting and operating.

Servicing

Servicing is normally carried out annually and should include the following operations:

- question the customer and examine the appliance for faults
- if the appliance is on, check the water temperature at a draw-off
- isolate the appliance electrically
- clean the main burner and pilot
- renew the pilot filter
- clean the heat exchanger
- check gas tap and grease if necessary
- check gas soundness
- check for correct ventilation
- test the ignition device and light the appliance
- check combustion fan, if fitted
- check that pilot flame is correct and stable
- check the burner pressure and flame picture, adjust if necessary
- test the controls
    - flame supervision device
    - thermostat
    - solenoid or relay valve
- check flue
    - open flue for spillage
    - room-sealed for flue seals
- check that the appliance is securely mounted and stable
- check pipework and valves for water leaks
- clear up the work area and leave the appliance working correctly.

Fault Diagnosis and Remedy

Faults on the burner and controls of gas storage water heating appliances have been covered in previous chapters.

The other faults which can occur are generally due to:

- corrosion
- scale deposits

● faulty installation.

Corrosion

This was discussed in Vol. 1, Chapter 12. It occurs when metals of different electrochemical potential are brought together in an electrolyte, for example, copper and steel submerged in slightly acidic water.

Usually, in water heating, the same metal is used throughout the appliance and the entire system to avoid electrolytic corrosion. However, some multipoint storage heaters which have steel storage vessels are fitted with sacrificial anodes of magnesium which corrode instead of the steel and may be renewed when necessary.

Scale

Scale is formed when water is heated and is deposited in the heat exchangers of appliances and in pipes and storage vessels. It has the effect of slowing down the circulation of water in flow and return pipes. This increases the temperature rise in the water and results in an even greater deposition of scale until the appliance overheats and the thermostat shuts down.

Scale deposits may be minimised by keeping the hot water at a temperature of not more than 60° C.

Indirect systems are not subject to heavy deposits of scale. The initial heating removes the chemical responsible from the circulating water, which is not replenished, and no further scale is formed.

Faulty Installation

Obviously there can be many different kinds of faults which can affect the circulation or the temperature of the water or the delivery from the draw-off.

Some typical faults are as follows:

If the circulating head is too low the water in the circulator will overheat. Circulation may take place for a time in the reverse direction and then stop altogether. If the thermostat is in the return the circulator will shut off without heating up the cylinder. If the water boils there will be loud rumbling and banging noises. The circulator must be refitted with an adequate circulating head.

If the circulating head is too high the circulation will be too fast and the water will have to circulate through the whole system several times before it becomes heated to the required temperature. So it will not be possible to draw off hot water within a short time of lighting up, instead the whole cylinder will be full of lukewarm water.

The circulation should be restricted or shortened.

If the open vent terminates over the cold storage cistern at too low a level it may be possible for the expanding water to reach and cover the end of the vent. If this occurs, a circulation may be set up with hot water rising up the vent and cold water passing down the cold feed. In this way the water in the cold storage cistern may be heated whilst that in the cylinder remains cold. Cold taps supplied from the cistern will supply hotter water than the hot taps!

The obvious solution is to raise the vent pipe termination.

If the branch for the draw-offs is taken from too high a position on the vent, air may be drawn into the pipes. When a draw-off tap is opened the water level in the vent pipe falls immediately before the water begins to flow up, out of the cylinder. If the draw-off is too high, air can be drawn in down the vent and will mix with the hot water giving a spluttering and intermittent supply.

The solution is to alter the position of the branch, or, if the head is low, fit a vent pipe air valve.

Characteristics of Water

Constituents

Water consists of hydrogen and oxygen,

$$2H_2 + O_2 = 2H_2O$$

Because it is an extremely good solvent, it is never found naturally in a completely pure state but always contains some dissolved mineral substances.

Rainwater is natural distilled water which may contain only the gases it has dissolved from the air. However, near to the sea it may contain salt and in industrial areas it collects soot and other pollutants.

River and lake water contains different quantities and types of dissolved matter, depending on the nature of the strata over which it flows. In limestone districts it contains quantities of calcium bicarbonate, $Ca(HCO_3)_2$, whilst in granite areas it contains only a small amount of dissolved minerals. Rivers may be contaminated by waste matter.

Spring and deep well water is usually drinkable, the impurities and bacteria being filtered out by the passage of rainwater down through the earth. At the same time the water dissolves chalk and other minerals.

Treatment

The only completely pure water is 'distilled water' which has been boiled off and condensed again.

For most practical purposes distillation is not necessary.

Most water supplies are subjected to sedimentation and filtration to remove the suspended solids and most bacteria. The water is then sterilised by treating it with chlorine or chlorine and ammonia. Dissolved solids may be removed by water softening.

Water is said to be 'hard' when ordinary soap does not produce an immediate lather.

The hardness is due to:

- calcium (or magnesium) sulphate — permanent hardness
- calcium (or magnesium) bicarbonate — temporary hardness

Temporary hardness is so called because it can be removed by heating the water. This changes the soluble calcium bicarbonate into the insoluble calcium carbonate $CaCO_3$, which is deposited in the form of 'scale'.

Permanent hardness cannot be removed by boiling.

Water acquires its temporary hardness when rainwater flows over limestone or chalk. Because the rainwater contains dissolved carbon dioxide from the atmosphere, it is able to convert the chalk or calcium carbonate into the soluble bicarbonate.

$$CaCO_3 + CO_2 + H_2O = Ca(HCO_3)_2$$

Water may be softened by a number of processes.

1. Adding Soda and Lime

This process is used commercially. The soda removes the permanent hardness and the lime removes the temporary hardness.

2. Base Exchange Method

This is used commercially and in domestic water softeners. Both permanent and temporarily hard water may be softened by using zeolites. A natural form of zeolite is sodium aluminium silicate and similar substances are produced artificially. For example, 'Permutit' which is a proprietary brand of sodium zeolite.

When hard water flows through a zeolite, the sodium ions are replaced by calcium ions so a calcium zeolite is produced.

$$\frac{\text{Calcium}}{\text{bicarbonate}} + \frac{\text{Sodium}}{\text{zeolite}} = \frac{\text{Sodium}}{\text{bicarbonate}} + \frac{\text{Calcium}}{\text{zeolite}}$$

The reaction is reversible by adding brine (sodium chloride) to the calcium zeolite. This produces calcium chloride, which is run off to waste, and sodium zeolite, which can be used again.

3. Adding Sodium Hexametaphosphate

This mineral, $(NaPO_3)_6$, is available commercially under the trade names of 'Calgon' and 'Micromet'. It can dissolve scale and, when added to water before heating, holds the carbonate in suspension in the water so that it is carried through by the flow.

Sodium hexametaphosphate has been used in 'scale reducers' which have been fitted to gas water heater installations. The scale reducer contains the substance in a renewable, wire mesh cartridge and is fitted in the cold water supply to the appliance or system, Fig. 37.

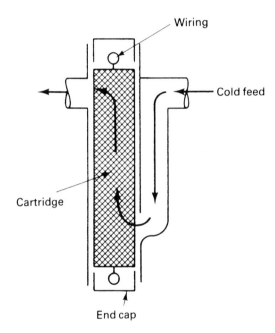

Fig. 37   Scale reducer

Degree of Hardness

Hardness of water is estimated by 'titrating' the water with a standard solution of soap dissolved in dilute alcohol. That is, by adding small quantities of the solution until a lather is formed which lasts for 3 minutes.

The degree of hardness may be measured in degrees Clark. One degree is equivalent to 1 g of calcium carbonate in 100 litres of water. That is, 1 part per 100,000 or 10 parts per million.

Waters are generally considered to be:

- soft   : up to 6°
- medium   : 6 to 10°
- hard   : above 10°

Soft waters, which are collected from moorland and peaty soil, are slightly acidic. Hard water from chalk and limestone districts is alkaline.

Acidic waters may dissolve small quantities of copper and lead. If small particles of dissolved copper come in contact with galvanised steel, they destroy the zinc and the exposed steel will rust. Lead is a cumulative poison which does not pass out of the body, so dissolved lead, even in small quantities, can eventually become dangerous.

Alkaline waters coat the pipes with a film of lead carbonate or sulphate which prevents any further reaction.

Descaling

In hard water areas, calcium salts deposit themselves in the form of scale in domestic hot water appliances. The rate at which the scale is formed increases with the rise in water temperature. At a temperature of 45° C, scale forms quite quickly and as the temperature rises it increases more rapidly.

With storage heaters the problem is not so great, as in hard water districts, the stored water is usually heated indirectly. The main problem is with instantaneous water heaters. In areas where the water is particularly hard descaling may have to be done every two years. Several proprietary makes of chemical descalant are available, all are highly acidic and must not be allowed to come into contact with galvanised pipes, tanks or cylinders which could be damaged by corrosion. The solution could also affect thermostat probes and automatic valves, so it is generally only passed through the heat exchanger or heating body of the water heater.

The equipment used consists of a 20 litre plastic tank, fitted with an integral chemical (electric) pump. Two transparent plastic tubes are connected permanently to the tank and two coupling tubes are supplied to connect the transparent tubes to the heat exchanger connections. The operator must be suitably protected before handling any of the decalcifying fluids, he should wear protective clothing, eye shields and rubber gloves.

After filling the tank with the decalcifying acid (suitably diluted to the manufacturer's instructions), the operator now connects the tubes to the heat exchanger. The pump is switched on and circulates the fluid once or twice a minute through the heat exchanger. After about one minute the fluid becomes white and effervescent, this is caused by gases from the dissolved lime (scale). The decalcifying period can be 5 to 15 minutes depending upon the amount of scale and size of appliance. Decalcifying is complete when the effervescence stops and the liquid is clear in the transparent hose from the heat exchanger outlet. After disconnection, the heat exchanger pipes must be thoroughly flushed with clean water to remove all traces of the decalcifying fluid.

CHAPTER 10

Central Heating by Hot Water

Chapter 10 is based on an original draft by A. Jones

Introduction

Central heating is the heating of a number of rooms by means of a single heat source.

It can be effected by heating a quantity of air, water or steam, circulating it into the rooms where it gives up its heat and then returning it to the heating unit for reheating and recirculating.

Central heating may be divided into four categories:

- full central heating
- partial central heating
- selective heating
- background heating.

Full Central Heating

This is the heating of all rooms simultaneously, to full comfort temperatures. The heating unit must be capable of achieving this with an outside temperature of −1° C.

The temperature and air changes required for full central heating standard in the various rooms are given in Table 31. These temperatures have increased as living standards have been raised and the temperatures and air changes shown are those recommended by BS 5449 and the Institution of Heating and Ventilating Engineers Guide. These are 'environmental temperatures' instead of air temperatures. Environmental temperature takes into account the temperature of the floor, walls and ceiling as well as inside air.

If required, it can be obtained approximately from the formula:

$$t_e = \frac{2}{3} t_r + \frac{1}{3} t_a$$

450

where t_e = environmental temperature

t_r = mean radiant temperature of all surfaces

t_a = air temperature

Partial Central Heating

This is the heating of some of the rooms, or the hall, simultaneously, to a full heating standard.

Selective Central Heating

In selective heating it is possible to convey heat to all the rooms, but the heating unit can provide full heating to only one group of rooms at any one time. The customer might direct the full output of the unit to the living area during the day and to the bedrooms before retiring. This form of heating is more commonly applied to warm air systems which are dealt with in Chapter 11.

Background Central Heating

This is the simultaneous heating of all or some rooms to a temperature below that shown in Table 31. Maintaining this background

TABLE 31 Temperatures and air changes on which heat loss calculations should be based

Room	Room temperature*	Air changes †
	° C	
Living room	21	1
Dining room	21	2
Bedsitting room	21	1
Bedroom≈	18	$\frac{1}{2}$
Hall	16	$1\frac{1}{2}$
Bathroom§	22	2
Kitchen§	18	2
Toilet§	18	$1\frac{1}{2}$

* These temperatures apply only to whole house central heating and for heated rooms with part house central heating.

† Local building regulations may require a specific rate of air change for particular rooms. Although the values given for air changes may not agree with the values given in BS 5925 : 1980 Code of practice for design of buildings: ventilation principles and designing for natural ventilation. They are the values that are to be used when calculating heat requirements for ventilation.

≈ When used part time as bedsitting rooms or for study purposes additional means should be provided for maintaining a higher room temperature.

§ Where continuous mechanical ventilation is provided due allowance for greater air change should be made.

temperature enables individual space heaters to raise the temperature to full heating standard very quickly.

Hot Water Circulating Systems

The hot water in a 'wet' central heating system can be circulated in two ways:

- gravity circulation
- pumped circulation.

Gravity Systems

The circulation of hot water by gravity has been used in domestic water heating systems for many years and was employed in the early central heating systems.

Gravity circulation is caused by circulating pressure arising from the difference in density between hot and cold water and height of the columns of circulating water. A method of calculating circulating pressure was given in Vol. 1, Chapter 6.

A more simplified formula which gives a reasonable value is:

$$P = h\, 0.0981\, (d_2 - d_1)$$

where P = circulating pressure in millibars
h = circulating height in metres between the centres of the heat source and the heat emitter
d_1 = density of water in the flow in kg/m^3
d_2 = density of water in the return in kg/m^3

Circulation tables are available and Table 32 gives an extract which shows how small the pressures are in a normal domestic system.

For example, with a flow temperature of 80° C and a return temperature of 60° C, the temperature difference is 20° C and the pressure is only 1.1 mbar/metre height.

TABLE 32 Circulating Pressures

Flow Temperature (°C)	Temperature difference, flow – return (° C)					
	5	10	15	20	25	30
	Circulating pressure, mbar per metre of circuit height					
70	0.2	0.5				
75	0.3	0.5	0.8			
80	0.3	0.6	0.8	1.1		
85	0.3	0.6	0.9	1.1	1.4	
90	0.4	0.7	1.0	1.3	1.3	1.8

This very low circulating pressure requires the use of large diameter pipes of about 28 to 54 mm diameter in order to circulate the quantity of hot water required.

The disadvantages of the gravity system are:

- large diameter pipes, which are
  – unsightly
  – difficult to install
  – expensive
- need to run pipes with a minimum slope of 1 in 120 to avoid air locks
- not suitable for long, horizontal runs
- boiler normally never higher than the lowest radiator and usually below or at the same level
- high temperature difference between flow and return water
- large quantity of water in the system means that there is a slow response to the controls.

The main advantage of the gravity system was that it could be totally independent of electricity.

Gravity systems of central heating are no longer installed but, as they may still exist, the circuits could be of interest and are included for information. There are five basic systems:

- one pipe drop
- two pipe drop
- one pipe ring main
- two pipe up feed
- horizontal drop.

Of these, the first two were used in domestic premises, while the remainder were commonly used in larger buildings, churches and schools.

One Pipe Drop (Fig. 1)

The flow pipe rises from the boiler to the roof space and dropping pipes feed the radiators. The drops join the return pipe at ground floor or basement level. Both the flow and return connections of each radiator are connected to the same pipe. So the second radiators are supplied with water at a lower temperature than the first. The radiators must, therefore, be sized in accordance with the temperature drop of the water.

When all the radiators are shut off, water can still circulate through the drops, so wasting heat.

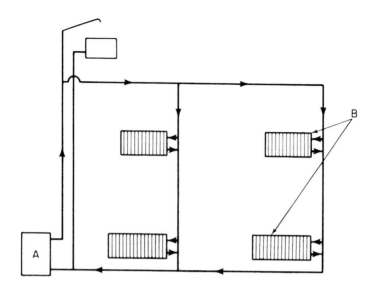

Fig. 1   Gravity system, one pipe drop. A: Boiler B: Radiators

Two Pipe Drop (Fig. 2)

The overall layout is similar to the one pipe drop but the radiators are connected to separate flow and return pipes. Returning, cooler water

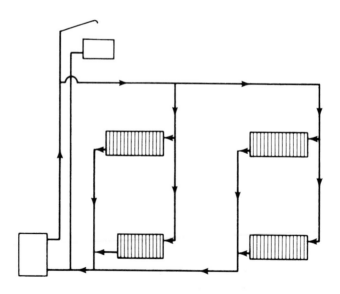

Fig. 2   Gravity system, two pipe drop

does not mix with the flow of hot water, so all radiators are supplied with water at approximately the same temperature. When all the radiators in a drop are shut off, no water circulation is possible in that circuit. So no heat is lost from the pipework.

Pumped Systems

With the introduction of the small, gland-less central heating pump, which was developed through the Coal Utilisation Council, wet central heating systems became easy to install and much less expensive. Prior to this, pumps or 'accelerators' were used generally only on the larger systems. With a large accelerator the pump is separate from the electric motor, which is often mounted vertically above it. With the new pumps the electric rotor is inside the pump and connected directly to the impeller.

The new pumps brought in the use of 'small bore' pipework and copper pipes of 15 and 22 mm diameter became common with a main feed of 28 mm. This was later followed by 'micro bore' or 'minibore' systems in which the radiators were supplied through 6 to 12 mm tubing.

In order to avoid noise or vibration, the velocity of the water in pumped circuits should not generally exceed:

- small bore = 1 m/s
- micro bore = 1.4 m/s.

Small Bore Systems

In the early small bore systems it was usual for the pump to be fitted in the return pipe near to or on the boiler. There are, however, some advantages to be gained by fitting the pump on the flow pipe. This has become more general practice, particularly when the system combines central heating and domestic hot water supply.

Alternative positions for the pump are shown in Figs. 3 and 4. The implications of the two positions are dealt with later in the chapter.

There are basically two forms of small bore heating circuits:

- single pipe
- two pipe.

Single pipe systems are now very rarely installed.

Single Pipe Systems (Fig. 3)

This is a simple, ring circuit, similar to the one-pipe gravity circuit but with the latitude to position the boiler at any height, relative to the radiators. Although the main circulation is pumped, the water circulates through the radiators mainly by gravity. The size of the radiators is calculated from the temperature drop at each successive radiator. Balancing the flow of water to each radiator is a simple process.

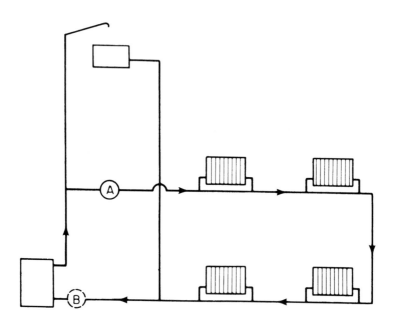

Fig. 3   Small bore system, single pipe. A: Pump position on flow;
B: Alternative pump position on return

Two Pipe Systems (Fig. 4)

In this system the water flows to and from each radiator through separate branches of the main flow and return pipes. So, in effect, each radiator has its own circuit connected to the boiler. Because the lengths of these circuits vary, depending on the position of the radiator, the system is more difficult to balance.

The pump pressure, or 'head', must be adequate to give the required flow of water through the circuit which offers the greater resistance. That is, the longest circuit, called the 'Index Circuit'.

The separate branches distribute water to each radiator at a temperature near to the boiler flow temperature, so radiator sizing is simplified. The two pipe system should be used whenever practicable.

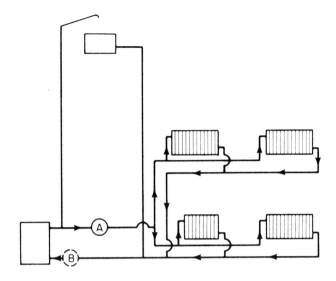

Fig. 4   Small bore system, two pipe. A: Pump; B: Alternative pump position

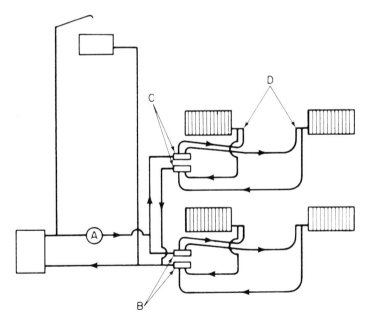

*Fig. 5   Micro bore system. A: Pump; B: Lower floor manifolds;
          C: Upper floor manifolds; D: Twin-entry valves*

Micro Bore Systems

A micro bore system is essentially a two pipe system with the radiators connected to the main flow and return pipes by loops of micro bore tubing, Fig. 5. The main pipework is usually 22 or 28 mm and is kept to the minimum length necessary for it to reach a central position.

The flow and return pipes are fitted with 'manifolds'. These are adaptors with a number of outlets from which the micro bore loops of 6, 8 or 10 mm tubing are run to the radiators, Fig. 6.

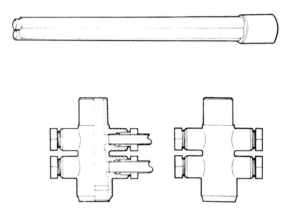

Fig. 6   Micro bore manifolds (Yorkshire Imperial Fittings)

Manifolds are fitted in pairs with the flow and return manifolds beside each other or even made in one unit, Fig. 7.

Fig. 7   Combined flow and return manifold (Wednesbury Tube Company)

In small dwellings all the radiators may be taken from one pair of manifolds, which can accommodate up to nine radiators. It is usual, however, to fit a separate pair of manifolds on each floor in a house and larger properties may have two pairs on each floor. The loops which serve the largest radiators should not be longer than 9 m.

The advantages of the micro bore system are:

- it contains only a small quantity of water and so is quickly heated
- the micro bore tubing is taken from fully annealed coils, usually of copper and is readily bent by hand and easily concealed

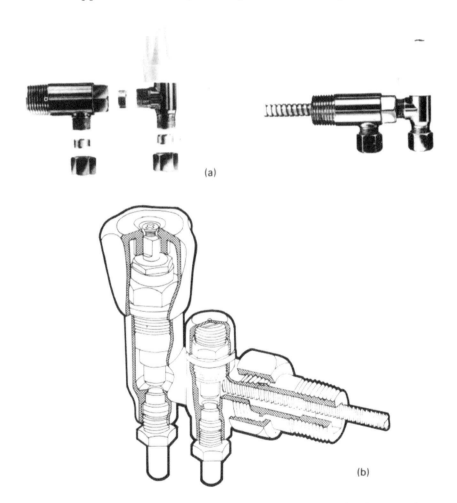

Fig. 8   Twin entry valves: (a) single valve (Wednesbury Tube Company); (b) valve with balancing valve (Yorkshire Imperial Fittings)

- a 'twin-entry' valve may be used to connect both flow and return pipes to the radiator, the flow pipe being situated coaxially in the return connection, Fig. 8
- probable economies in installation and material costs.

Combined Heating and Water Heating Systems

In domestic systems it is usual for the boiler to provide the means of heating both the house and the domestic water. The water is heated by an indirect system as described in Chapter 9. An indirect system can be used in conjunction with gravity, small bore or micro bore heating circuits.

The flow pipe, carrying the heated, primary water from the boiler, divides to feed water to the radiators and also to the calorifier which heats the secondary water in the hot water storage cylinder.

The primary water circulation from the boiler may be either:

- gravity
- pumped.

Gravity Primary

This method of circulation was used originally with gravity systems of heating and subsequently with the earlier small bore systems. The flow and return pipes are usually 28 mm and may be taken directly from the alternative tappings on the boiler. A typical system with small bore heating and a gravity primary is shown in Fig. 9.

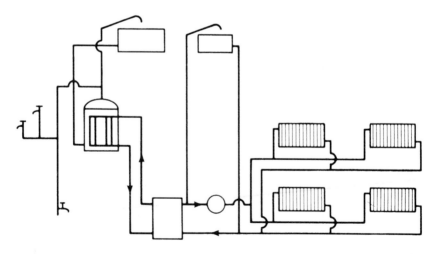

Fig. 9   Gravity primary with pumped heating circuit

With this system the domestic hot water is normally always being heated when the boiler is on. If required, a thermostatically control-led valve can be used to shut off the circulation when the domestic water has reached the temperature required. The heating system can be shut off by switching off the pump, usually by means of the room thermostat. The gravity circuit serves to dissipate the residual heat left in the boiler when the heating circuit has shut down. Gas boilers generally have only a low water capacity and there is no residual heat in the fuel, so gravity primaries are no longer essential, as they are with solid fuel.

Pumped Primary

With this system, flow and return sizes may be reduced to 22 or 15 mm depending on the pump and the size of the cylinder. The cylinder may be fitted in any position relative to the boiler. Separate thermo-stats should be provided to control the domestic hot water and the heating circuit. A balancing valve should be fitted in the primary return. A typical system is shown in Fig. 10. The two circuits can be controlled by 'diverter' or 'zoning valves' (Vol. 1, Chapter 11). A three-port 'diverter' valve can be fitted at the point where the primary flow divides to feed the calorifier (domestic hot water cylinder) and

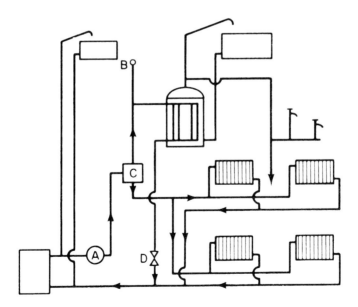

Fig. 10   Pumped primary and heating circuits. A: Pump; B: Air release; C: Diverter valve; D: Balancing valve

the heating circuit (radiators). The valve has one inlet port and two outlet ports. Some models have a valve which closes one or other of the two outlet ports. Other models have a valve which closes one or other of the outlet ports or takes up a position midway between the outlet ports. The first model described allows heated water from the boiler to either the calorifier or the heating circuit, depending upon the demand from the controls (clock, cylinder thermostat and room thermostat). The second model operates in a similar manner except that when all the controls are calling for heat the valve takes up a mid position and the heated water is then shared between the two outlet circuits.

Two-port valves may also be used. These valves have an inlet and an outlet port and the motorised plug either opens or closes the valve. One valve is fitted in the primary flow to the cylinder and controlled by the cylinder thermostat. The other valve is fitted on the central heating circuit and is controlled by the room thermostat.

The control systems and the valves used are dealt with in Chapter 12.

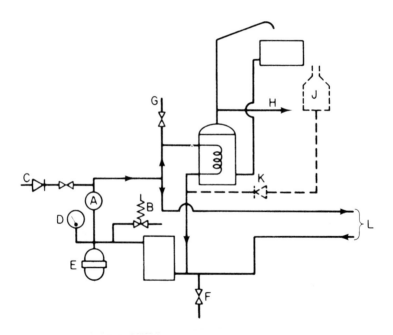

Fig. 11   Sealed system, pump on flow pipe. A: Pump; B: Safety valve; C: Filling point; D: Pressure gauge; E: Expansion vessel; F: Drain cock; G: Air release; H: Drawoffs; J: Top-up bottle; K: Non-return valve; L: Flow and return for radiators

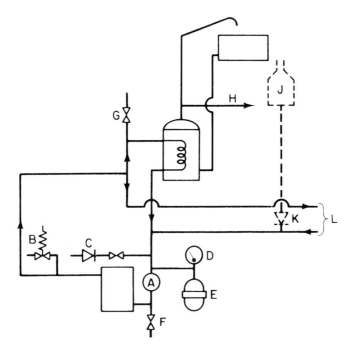

Fig. 12   Sealed system, pump on return pipe

Sealed Systems

The sealed system is basically similar to the open system but an
'expansion vessel' is used to replace the feed and expansion cistern
and its associated cold feed pipe and open vent. Make-up water is
provided either by pre-pressurising the system (see later in this
chapter) or by a top-up bottle that feeds through a non-return valve
and is connected into the system on the return of either the cylinder
or all the radiators. Figures 11 and 12 show typical layouts.

The expansion vessel consists of a small cylinder divided into two
compartments by a rubber diaphragm, Fig. 13. One side of the vessel
is connected to the water circulation and the other contains a charge
of nitrogen or air. The pressure of the charge must be not less than
the static head of water in the system and vessels are available as
follows:

● air/nitrogen charge 0.5 bar suitable for static head 5 m
● air/nitrogen charge 1.0 bar suitable for static head 10 m.

When the water in the system is cold, the air/nitrogen occupies the
whole of the vessel and the diaphragm is pressed firmly against the

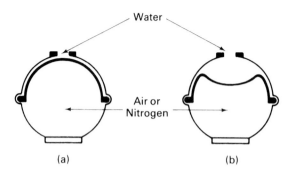

Fig. 13   Expansion vessel: (a) cold; (b) heated

side of the vessel, Fig. 13(a). As the water heats up, it expands and compresses the air/nitrogen, Fig. 13(b). The vessel should allow about two thirds of its volume to be taken up by water and the normal final pressure is up approximately 2 bar.

On cooling, the volume of the water decreases and the air/nitrogen forces it back into the system.

The expansion vessel must be large enough to accommodate the expanded water. If it is too small, the extra expanding water will be forced out of the safety valve or 'pressure relief valve'. This water will be lost and will be replaced from the top-up bottle, which will need to be continually refilled.

The size of the expansion vessel will depend on:

- volume of water in the system
- air/nitrogen charge pressure
- degree of pre-pressurisation, that is, when the water system is filled under pressure
- boiler flow temperature.

The vessel size may be obtained from tables or calculated from the following formula:

$$\frac{\text{Volume of expansion}}{\text{vessel, litres}} = \frac{\text{Volume of heating}}{\text{system, litres}} \times \frac{\text{Pressure}}{\text{factor } P} \times \frac{\text{Temperature}}{\text{factor } T}$$

Where P is:

P	Nitrogen charge pressure, bar	Pre-pressurisation, bar
0.0875	0.5	0
0.15	0.5	1.0
0.11	1.0	0
0.2	1.0	1.5

where T is:

T	Maximum flow temperature, ° C
1.0	93 or above
0.9	less than 93 but not less than 88
0.8	less than 88

Approximate volume of system components

Boiler	– conventional	– 10 litres
	– low water capacity	– 3.5 litres

Pipework	– small bore	– 1.03 litres per kW of system output
	– micro bore	– 7.0 litres total

Radiators	– steel panel	– 7.9 litres per kW of system output
	– low water capacity	– 1.7 litres per kW of system output

Hot water cylinder   – 2 litres

Example,

Average small bore system, volume	= 135 litres
Air/nitrogen charge pressure	= 0.5 bar
Pre-pressurisation	= 1.0 bar
Maximum flow temperature	= 90° C
Volume of expansion vessel	= 135 × 0.15 × 0.9 litres

$$= \textbf{18.2 litres}$$

When using a table, the expansion vessel should be sized in accordance with the following procedure:

(a) The volume of the expansion vessel in litres fitted to a sealed system should not be less than that given in Table 33.
(b) For the purpose of the above calculation, the volume of the system should be determined as accurately as possible using manufacturers' data as appropriate. Alternatively the volumes given previously may be used to give a conservative estimate of the system volume.

If a system is extended an expansion vessel of increased volume may be required unless previous provision has been made for the extension.

TABLE 33   Sizing of Expansion Vessels

Safety valve setting (bar)	3.0								
Vessel charge pressure (bar)	0.5				1.0			1.5	
Initial system pressure (bar)	0.5	1.0	1.5	2.0	1.0	1.5	2.0	1.5	2.0
Total water content of system	Expansion vessel volume (litres)								
litres									
25	2.1	3.5	6.5	13.7	2.7	4.7	10.3	3.9	8.3
50	4.2	7.0	12.9	27.5	5.4	9.5	20.6	7.8	16.5
75	6.3	10.5	19.4	41.3	8.2	14.2	30.9	11.7	24.8
100	8.3	14.0	25.9	55.1	10.9	19.0	41.2	15.6	33.1
125	10.4	17.5	32.4	68.9	13.6	23.7	51.5	19.5	41.3
150	12.5	21.0	38.8	82.6	16.3	28.5	61.8	23.4	49.6
175	14.6	24.5	45.3	96.4	19.1	33.2	72.1	27.3	57.9
200	16.7	28.0	51.8	110.2	21.8	38.0	82.4	31.2	66.2
250	20.8	35.0	64.7	137.7	27.2	47.5	103.0	39.0	82.7
300	25.0	42.0	77.7	165.3	32.7	57.0	123.6	46.8	99.3
350	29.1	49.0	90.6	192.8	38.1	66.5	144.2	54.6	115.8
400	33.3	56.0	103.6	220.4	43.6	76.0	164.8	62.4	132.4
450	37.5	63.0	116.5	247.9	49.0	85.5	185.4	70.2	148.9
500	41.6	70.0	125.9	275.5	54.5	95.0	206.0	78.0	165.5
For systems volumes other than those given above, multiply the system volume by the factor across	0.0833	0.140	0.259	0.551	0.109	0.190	0.412	0.156	0.331

The volume found from the above table shall be multiplied by the appropriate factor from the table below:

Maximum boiler flow temperature	Factor
93° C or greater	1.0
Less than 93° C but not less than 88° C	0.9
Less than 88° C	0.8

Where a vessel of the calculated size is not obtained then the next available size should be used.

When filling or 'pre-pressurising' a sealed system, the mains water supply may be utilised, providing the connection made complies with the Water Supply Byelaws. Byelaw 14 of the Water Supply Byelaws states:

(i)   No closed circuit shall be connected to a supply pipe.
(ii)   Paragraph (i) shall not apply to a temporary connection provided –
        (a)   the connection is made through a double check valve assembly or some other no less effective device which is permanently connected to that circuit; and
        (b)   the temporary connection is removed after use.

Fig. 14 illustrates a typical device used for filling or 'pre-pressurising' a sealed system and also shows the device temporarily connected between the mains water supply and the system.

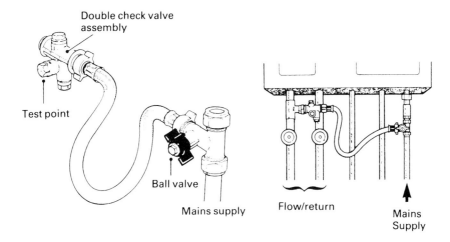

Fig. 14   Method of filling or pre-pressurising a sealed system

Make-up water may be supplied by a top-up bottle located above the highest point of the system. If any water has been lost, by leakage, then, when the system cools down, the non-return valve will open and allow water to be drawn in from the bottle.

An alternative method of providing make-up water is to force extra in after the system is full. The excess water enters the expansion vessel, partially compressing the air/nitrogen and is forced out of the vessel when any water is lost from the system by leakage. This is known as 'pre-pressurisation'.

Because raising the pressure also raises the temperature at which water boils, the sealed system can operate at temperatures above the normal boiling point. Maximum recommended temperatures are 99° C and boilers are fitted with high temperature limit cut-offs which operate at a maximum temperature of 110° C.

At temperatures over 93° C, pipework and radiators are too hot to touch. So pipework must be concealed and natural or fanned convectors must be used in place of panel radiators. Skirting heaters are often used on systems with elevated temperatures.

Where these higher temperatures are not used, the normal operating temperature is about 82° C and the usual small bore or micro bore systems of pipework with panel radiators are employed.

The domestic hot water cylinder must have either a coil or an immersion calorifier.

The advantages of sealed systems are:

- no problems of freezing with cold feed or expansion cisterns
- no problems of pumping water over the vent
- air cannot be drawn into the system, so corrosion is minimised
- high operating temperatures allow smaller heat emitters to be used
- the system can be fitted into bungalows or flats, where head room is limited and open systems could not be installed
- any water leaks are restricted to the amount of water in the system.

Boilers

A boiler should have an output which is at least equal to the total heat emission from the heating circuit, including all the pipework. If the system also provides hot water at the same time, the boiler output rating should be increased by at least 2 kW. Where the system controls give priority to one of the services this increase is unnecessary. Because domestic systems are used intermittently it is usual to add an allowance of about 10%.

When selecting a type of boiler, the choice is often governed by the location in which it can be installed. There are basically five types of domestic gas fired boilers:

- floor standing
- wall mounted
- back boilers
- combined heating and hot water units
- combination (combi) boilers.

Floor Standing Boilers

Free standing boilers are commonly fitted in kitchens and are available with pressed steel casings designed to fit into kitchen unit layouts. They are also supplied without cases for fitting into cup-

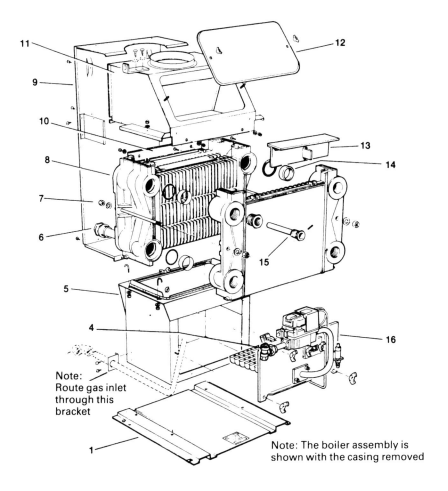

Note:
Route gas inlet
through this
bracket

Note: The boiler assembly is
shown with the casing removed

1. Boiler baseplate	11. Collector hood
4. Gas service cock	12. Cleanout cover
5. Combustion chamber	13. Flue baffle
6. Distributor tube	14. Section alignment
7. Tie rod	rings and 'O' rings
8. Heat exchanger	15. Thermostat pocket
9. Draught diverter	16. Burner and controls
back panel assembly	assembly
10. Rear infill	

Fig. 15   Floor standing boiler with open flue (Ideal)

boards or special compartments. Domestic boilers have outputs from
about 9 to 30 kW and may have open or room-sealed flues. Boiler
design is moving away from open flues and cast iron heat exchangers

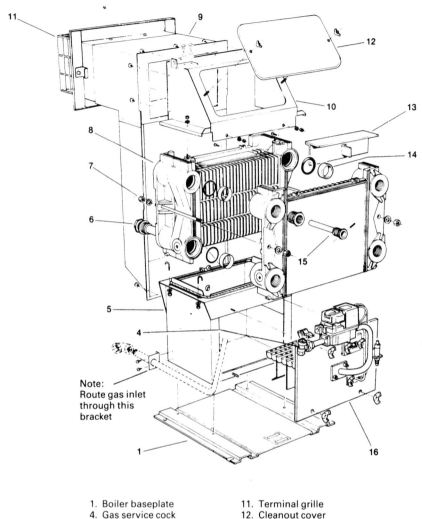

Note:
Route gas inlet
through this
bracket

1. Boiler baseplate	11. Terminal grille
4. Gas service cock	12. Cleanout cover
5. Combustion chamber	13. Flue baffle
6. Distributor tube	14. Section alignment
7. Tie rod	rings and 'O' rings
8. Heat exchanger	15. Thermostat pocket
9. Air and flue duct	16. Burner and controls
assembly	assembly
10. Collector hood	

Fig. 16   Floor standing room sealed boiler (Ideal)

and the trend is towards low water content boilers with integral
pumps and control systems and room-sealed or fanned draught flues.

A typical open flued floor standing boiler (with the casing
removed) is shown in Fig. 15. It consists of a rectangular fluted cast

iron heat exchanger fitted with baffles. A collector hood, fitted over the waterways, carries the products of combustion to the draught diverter or to the flue duct on a room-sealed appliance Fig. 16.

The gas supply is usually $R_c\frac{1}{2}$ into a union or union cock. Gas then passes into a multifunctional control and so to the burner.

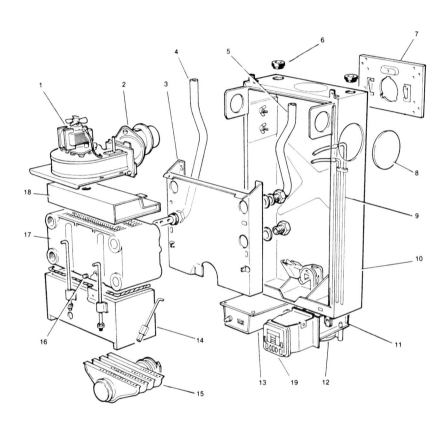

1. Fan assembly
2. Flue outlet elbow
3. Inter-panel
4. Pumped return pipe
5. Pumped flow pipe
6. Rubber sealing grommets
7. Wall mounting plate
8. Sealing plates, 2 off
9. Pressure sensing pipes
10. Back panel
11. Programmer mounting bracket
12. Pressure switch
13. Control box (mounted on the gas control valve)
14. Combustion chamber
15. Main burner
16. Boiler drain port
17. Heat exchanger
18. Collector hood assembly
19. Programmer (optional)

Fig. 17   Wall mounted boiler with fanned balanced flue (Ideal)

Wall Mounted Boilers

These boilers have light, compact heat exchangers often made from finned tubes and containing only about one litre of water. They are known as 'low water content boilers' or 'low thermal capacity boilers'. Because of the substantial reduction in weight they can be fitted on a wall. Some floor standing boilers are of similar design but by mounting the boiler on a wall the choice of location is widened and there can be a saving in kitchen space.

A wall mounted, fanned balanced flue gas boiler is shown in Fig. 17. The controls are mounted at the base of the boiler and above the burner is a single cast iron heat exchanger. Mounted on the heat exchanger is a collector hood, above which is the flue fan assembly.

If the water circulation through a low water capacity boiler is slowed down, the water can overheat and boil. This produces a noise like the 'singing' of a kettle which is called 'kettling'. To avoid kettling, a bypass can be fitted which will ensure that the flow of water is always above the minimum required to prevent overheating.

Bypasses are usually 15 mm pipe and have a regulating or balancing valve incorporated. Manufacturers' instructions regarding the size and location of the bypass must be followed. The bypass should normally be fitted between the main flow and return pipes as shown in Fig. 18. Alternatively, some manufacturers specify a remote location.

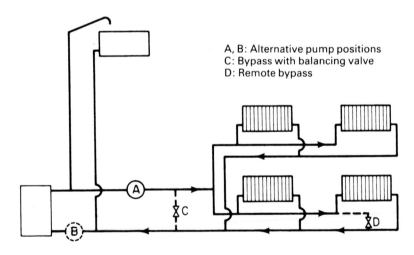

A, B: Alternative pump positions
C: Bypass with balancing valve
D: Remote bypass

Fig. 18   Bypass for low thermal capacity boiler

Overheating can also occur when the boiler has shut down and the residual heat in the heat exchanger causes the small amount of water contained to boil. This can be prevented by the use of a 'pump delay thermostat' or pump overrun device which operates in conjunction with a remote bypass. The thermostat allows the pump to continue running after the boiler has shut down so that water is circulated through the heat exchanger until the residual heat is dissipated through the bypass circuit.

Where boilers are not fitted with pump delay thermostats, overheating can be prevented by setting the pump to run continuously.

Back Boilers

These boilers are fitted into the opening behind a fireplace surround and are usually hidden behind a gas fire. The fire may be integral or separate and the two appliances use the existing flue, suitably lined.

The back boiler saves space in the kitchen, but its size is limited and it usually has an output of about 16 kW. Figure 19 gives an exploded view of the appliance and its installation.

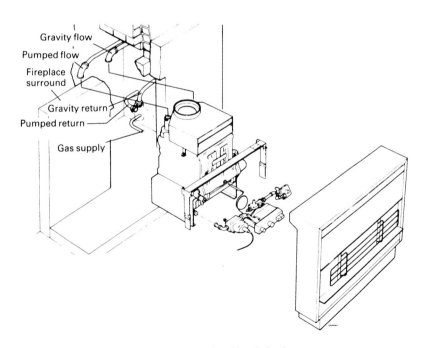

Fig. 19   Gas-fired back boiler

Combined Heating and Hot Water Units

These units usually incorporate the boiler and the hot water storage-cylinder. Some may also include the cold feed and expansion cistern and the domestic cold feed cistern. The controls and connections are built into the unit. A combined unit is illustrated in Fig. 20.

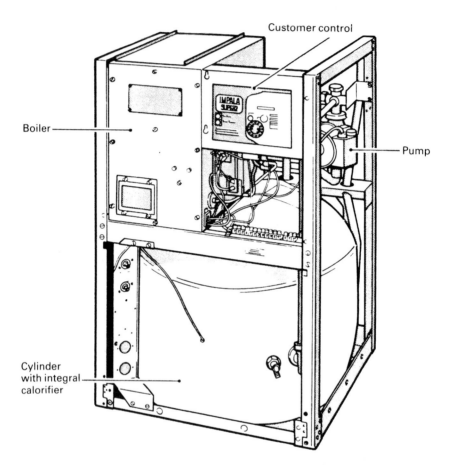

Fig. 20   Combined heating and hot water unit

Combination (Combi) Boilers

This type of appliance, Fig. 21, supplies full or partial central heating through a radiator system and domestic hot water on an instantaneous multipoint water heater basis. The central heating is usually through a sealed system although some appliances will also work perfectly well when connected to an open system. Calls for domestic hot water take priority over central heating demands. Fig. 21 (a) shows a fanned flue room sealed boiler and Fig. 21 (b) a compact balanced flue boiler.

One of the major advantages of this type of appliance is the space saving that can be achieved because there is no domestic hot water cylinder. When a sealed system is used problems associated with minimum head, in bungalows, self-contained flats etc., can be eliminated, plus the fact that there is no need for a feed and expansion cistern, gives a further saving on the space requirements. The major disadvantages are the limitations on the hot water delivery rate when compared to a storage system and the appliance failing to deliver hot water to the heating system when there is a call for domestic hot water.

The combi is basically a central heating boiler that becomes an instantaneous multipoint water heater when there is a demand for

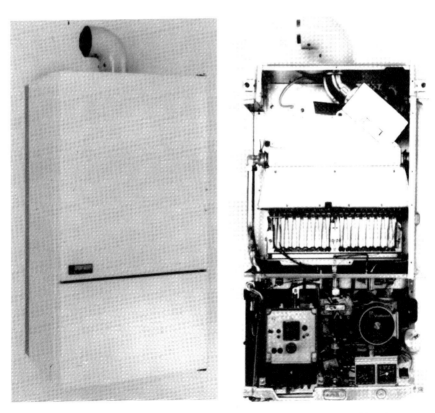

Fig. 21 (a)   Fanned flue room sealed combi boiler (Vaillant)

Fig. 21 (b)   Compact balanced flue combi boiler (Vaillant)

domestic hot water. In fact there are several different types of combi boiler – many originating on the continent – but their basic operations are very similar.

When set for hot water only, or summer, the appliance will only supply domestic hot water. When set on hot water plus central heating the domestic hot water and central heating are supplied with, as previously stated, domestic hot water taking priority, the central heating system being supplied when there are no calls for domestic hot water and when the external controls – time and temperature – are calling for heat. Fig. 22 illustrates a schematic lay out of the water ways and components in an Ideal Sprint room sealed 75 wall mounted combination boiler. Figs. 23 (a) and (b) show functional flow and illustrated wiring diagrams of this appliance.

When there is a call for domestic hot water, i.e. a hot water tap is opened, mains water pressure acting against the flexible diaphragm in the diverter valve assembly, operates the device. This diverts the primary water away from the central heating system and through the domestic hot water (D.H.W.) calorifier. The secondary water, flowing

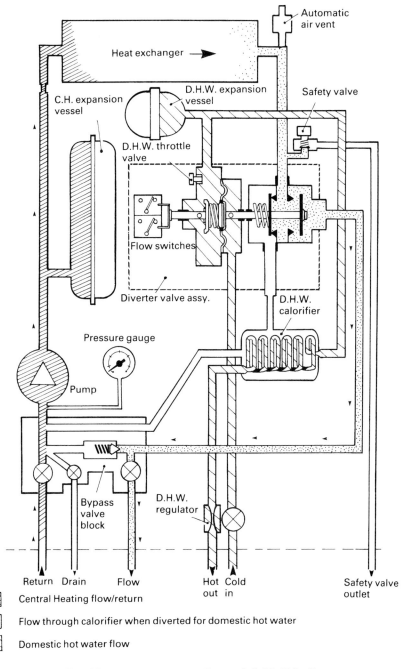

Automatic
air vent

Heat exchanger →

C.H. expansion
vessel

D.H.W. expansion
vessel

Safety valve

D.H.W. throttle
valve

Flow switches

Diverter valve assy.

D.H.W.
calorifier

Pressure gauge

Pump

Bypass
valve
block

D.H.W.
regulator

Return   Drain   Flow   Hot Cold   Safety valve
                                       out in   outlet

▨ Central Heating flow/return

☐ Flow through calorifier when diverted for domestic hot water

▨ Domestic hot water flow

Fig. 22   Waterways in a Sprint RS 75 (Ideal)

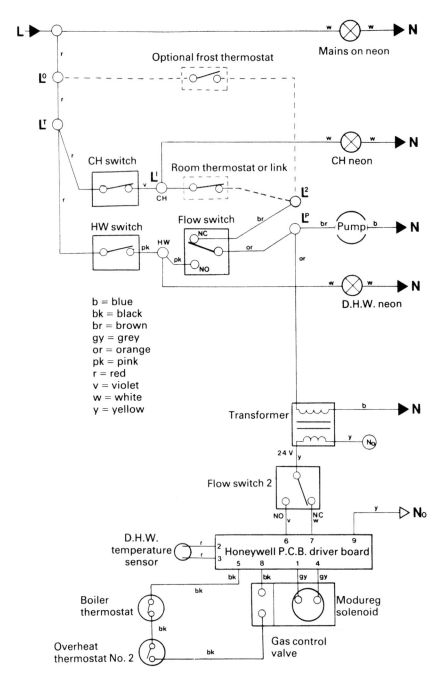

Fig. 23 (a)   Functional flow wiring diagram

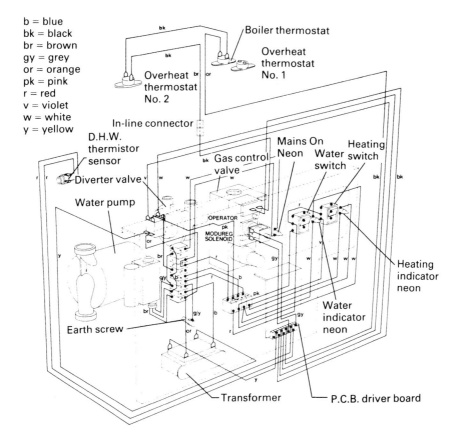

b = blue
bk = black
br = brown
gy = grey
or = orange
pk = pink
r = red
v = violet
w = white
y = yellow

Fig. 23 (b)   Illustrated wiring diagram

from the mains supply because the hot tap is open, passes through the coil in the D.H.W. calorifier where it becomes heated, then through the hot water supply pipes to the hot tap. The movement of the flexible diaphragm and attached spindle in the diverter valve, also makes electrical circuits through the two enclosed flow switches. Flow switch No. 1 (240 V a.c.) causes the pump to run and feeds power to the printed circuit board (P.C.B.) via the boiler thermostat, transformer and flow switch No. 2. The feed from flow switch No. 2 energises the D.H.W. thermistor, the main gas solenoid (24 V a.c.) and initially the modulating (modureg) gas valve at a full 28 V d.c. The D.H.W. thermistor then monitors the temperature of the water and adjusts the voltage to the modureg valve. This valve in turn adjusts the appliance gas rate to ensure that the hot water temperature does not exceed 70° C.

When the hot tap(s) is turned off the appliance returns to the central heating mode. Demands for heat from the external controls – clock and room thermostat – provide a 240 V a.c. electrical supply through the normally closed (N.C.) position of flow switch No. 1. This feeds 240 V a.c. to the pump, plus – through a transformer, the P.C.B. and flow switch No. 2 – 24 V a.c. to the main solenoid and approximately 10 V d.c. to the modureg solenoid. When operating in the central heating mode the appliance does not modulate but switches on and off through the 'Klixon' type boiler thermostat which is set to shut off around 82° C. Overheat protection is provided by one, or on more recent models two, thermostats that interrupt the thermocouple supply when the water temperature reaches approximately 97° C. Once an overheat thermostat has operated, the boiler is shut down until it has cooled and a manual relighting of the boiler has taken place. Causes of the overheating should be investigated and remedied before relighting the appliance.

The type of diverter used in the Sprint boiler described is also used on other combi boilers and is illustrated in Fig. 24.

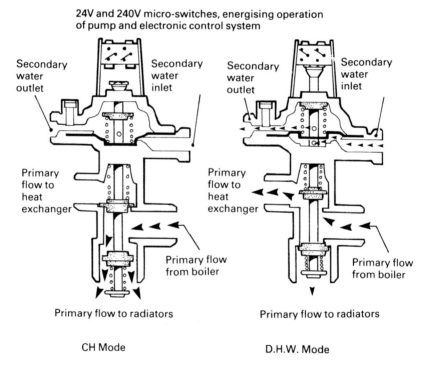

Fig. 24   Diaphragm pass type diverter valve

Other types of diverter valves are used on different combi boilers. Three of the more popular types are shown in Figs. 25, 26 and 27 (a), (b) and (c).

The diverter valve shown in Fig. 25 uses the venturi principle, similar to the instantaneous water heater, to create a pressure difference across a flexible diaphragm. When the water flow through the venturi – caused by opening a hot tap – is great enough to create a pressure difference that overcomes the tension of the spring above the diaphragm, the diverter valve operates and heated water flows through the D.H.W. calorifier instead of the central heating system. When the hot tap is turned off the pressures across the diaphragm equalise and the spring operates valves which close off the supply to the D.H.W. calorifier and open the supply to the central heating

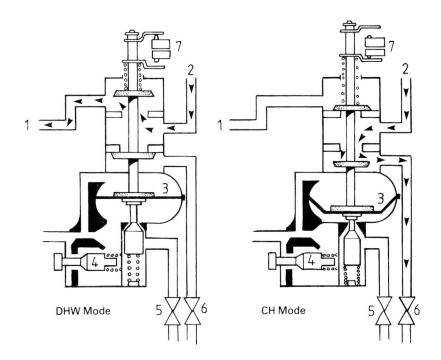

1. Flow to calorifier
2. Flow from boiler heat exchanger
3. Diaphram
4. Hot water regulator
5. Domestic cold water valve
6. CH flow valve
7. Micro switches

Fig. 25   Venturi type diverter valve

system. If there is a demand for heating from the external controls heated water will then flow through the radiators.

The wax capsule type of diverter valve, shown in Fig. 26, is found on some of the Vaillant combi boilers. The valve in this instance is operated by temperature. When heated water is flowing to the central heating system it is in contact with the top 20% of the wax capsule and the wax will expand to lift the diverter valve into the central heating position. When a hot tap is opened the flow of cold water from the mains supply comes into contact with the lower 80% of the wax capsule, causing the wax to cool and contract. The diverter valves then drop and close the hot water supply to the central heating system and open the supply of hot water to the D.H.W. calorifier where the water flowing to the hot taps is indirectly heated.

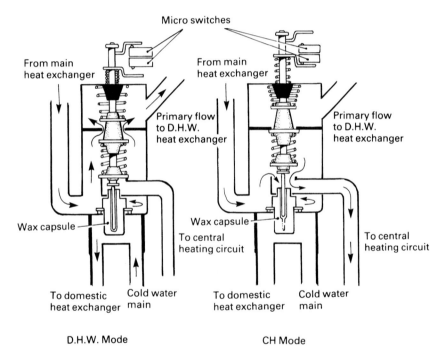

Fig. 26   Wax capsule type diverter valve

Another method of operating a combi boiler in either the central heating or domestic hot water mode, is by means of the hydraulically operated diverter valve illustrated in Figs. 27 (a), (b) and (c). The valve is shown in its three positions:

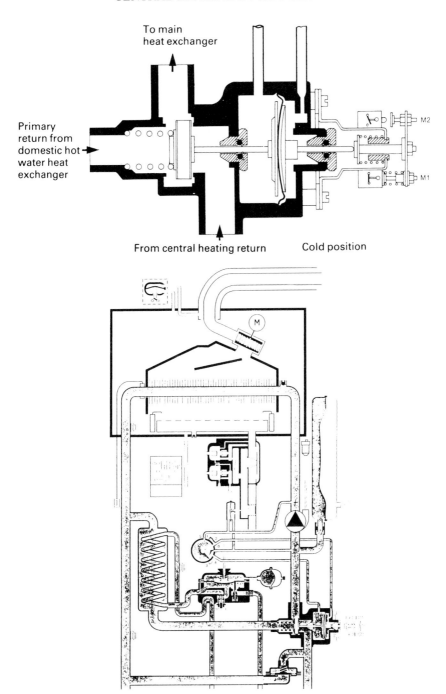

Fig. 27(a) Hydraulic valve: cold position

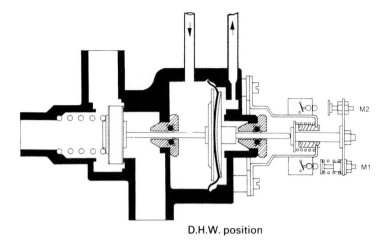

D.H.W. position

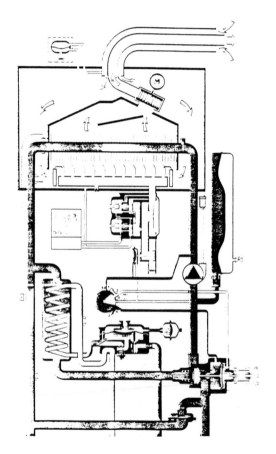

Fig. 27(b)   Hydraulic valve: D.H.W. position

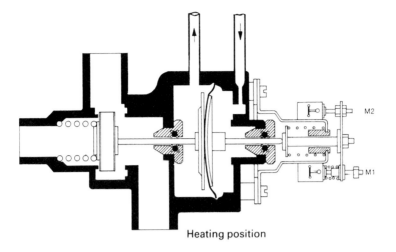

Heating position

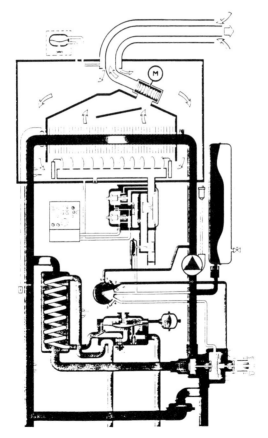

Fig. 27(c)   Hydraulic valve: heating position

(a) cold – no domestic hot water or central heating being called
     for
(b) domestic hot water position
(c) central heating position.

The diverter valve is controlled through a servo control valve which
transmits positive pressure from the outlet of the circulating pump
and negative pressure from the pump inlet.

In the cold position the circulating pump is not running, so the
pressures on either side of the diaphragm in the diverter valve are
equal. The position of the diverter valve and the spindles are controlled by the spring acting against the valve. When domestic hot water is
called for – i.e. a domestic hot water tap is opened allowing sufficient
water to flow and operate the automatic valve – a pressure differential
is created across the diaphragm which lifts a spindle that in turn
positions the servo control valve, operates the circulating pump and
brings on the burner. The positive pressure from the outlet of the
pump is transmitted, via the servo control valve, to the right hand
side of the diaphragm and the negative pressure, from the pump inlet,
to the left hand side of the diaphragm. This causes the diaphragm
and the attached spindles to move to the right. The movement closes
the return from the central heating circuit, opens the return from the
domestic hot water calorifier and operates the appropriate electrical
switches to bring the D.H.W. controls into operation. Heated water is
then pumped through the calorifier where it indirectly heats the
domestic hot water supply.

When there is no call for domestic hot water but a call for central
heating – i.e. time and temperature controls calling for heat – the
circulating pump and the burner operate. The position of the servo
control valve causes the positive pressure, from the outlet of the
pump, to be transmitted to the left hand side of the diaphragm in the
diverter valve and the negative pressure, from the inlet side of the
pump, to be transmitted to the right hand side of the diaphragm. This
causes the diaphragm and the attached spindle to move to the left,
closes the return waterway from the D.H.W. calorifier, opens the
return waterway from the central heating and operates the appropriate electrical switches to bring the central heating controls into
operation. Heated water is then pumped through the heating circuit
until the system is satisfied or there is a further call for domestic hot
water. The hydraulic valve is used on later models of Vaillant combi
boilers.

A further development of the Sprint RS 75, Fig. 22, is the Sprint
Rapide RS 75. This appliance uses a three-port motorised valve as a

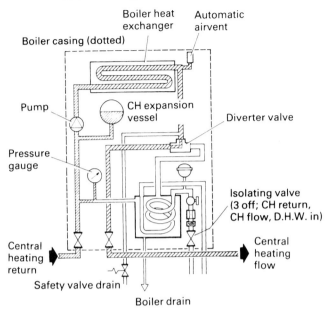

Central Heating (CH) Water Circuit

Boiler heat exchanger
Automatic airvent
Boiler casing (dotted)
Pump
CH expansion vessel
Diverter valve
Pressure gauge
Isolating valve (3 off; CH return, CH flow, D.H.W. in)
Central heating return
Central heating flow
Safety valve drain
Boiler drain

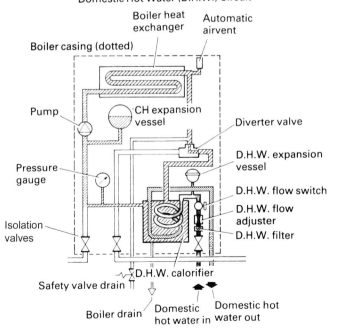

Domestic Hot Water (D.H.W.) Circuit

Boiler heat exchanger
Automatic airvent
Boiler casing (dotted)
Pump
CH expansion vessel
Diverter valve
D.H.W. expansion vessel
Pressure gauge
D.H.W. flow switch
D.H.W. flow adjuster
Isolation valves
D.H.W. filter
Safety valve drain
D.H.W. calorifier
Boiler drain
Domestic hot water in
Domestic hot water out

Fig. 28   Waterways in a Sprint Rapide RS 75 (Ideal)

diverter valve and diagrams of its C.H. and D.H.W. modes are shown
in Fig. 28. The three-port motorised valve is described in Chapter 12.

Yet another method of supplying both central heating and domestic hot water from a combi boiler uses two pumps, one pump
supplying central heating and the other domestic hot water. Fig. 29
(a) and (b) illustrates such an appliance.

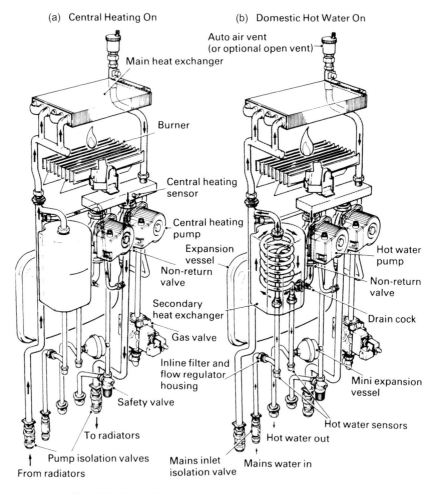

(a) Central Heating On   (b) Domestic Hot Water On

Fig. 29   Combi boiler fitted with two pumps (Myson)

When a hot tap is opened water flows through a coil in the D.H.W.
calorifier and operates a flow sensor. This sensor sends a signal that
operates the domestic hot water pump, switches off the central
heating pump if it is running and operates the burner. Heated water is

then pumped through the D.H.W. calorifier coil and indirectly heats the water in the coil before it passes to the hot water taps. The temperature of the domestic hot water is controlled by a thermistor in the hot water outlet. A system of electronic modulation responds to signals from the thermistor and adjusts the appliance gas rate to suit the water flow and selected water temperature. When the hot water tap(s) is closed the flow sensor sends a signal which shuts off the domestic hot water pump and burner, unless there is a demand for

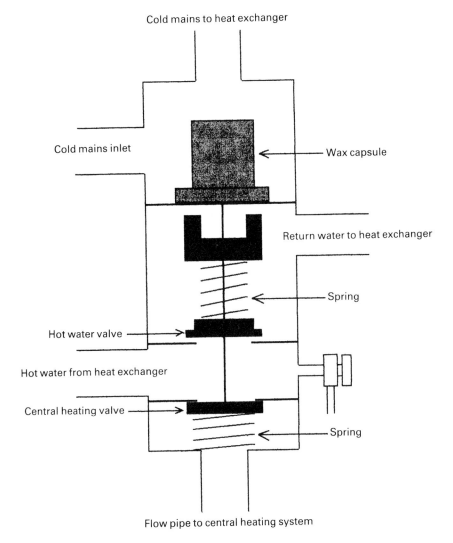

Fig. 30. Potterton/Myson Puma diverter valve

central heating when the central heating pump will be switched on and heated water pumped through the central heating system.

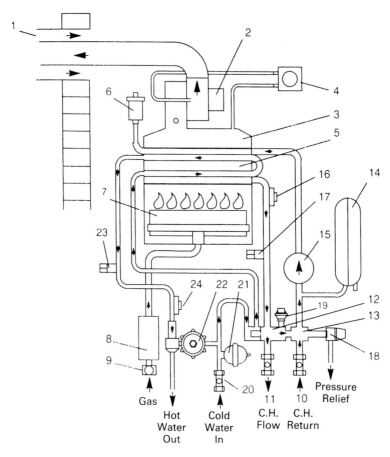

1. Air Duct
2. Fan
3. Fluehood
4. Air Pressure Switch
5. Heat Exchanger
6. Automatic Air Vent
7. Burner
8. Gas Valve
9. Gas Cock
10. C.H. Return Isolating Valve
11. C.H. Flow Isolating Valve
12. 3 Way Diverter Valve
13. Pump Manifold
14. C.H. Expansion Vessel
15. Pump
16. C.H. Overheat Thermostat
17. C.H. Temperature Sensor
18. Pressure Relief Valve
19. Water Pressure Switch
20. Cold Water Supply Cock inc. Automatic Flow Regulator
21. D.H.W. Expansion Vessel
22. D.H.W. Flow Switch
23. D.H.W. Temperature Sensor
24. D.H.W. Limit Thermostat

Fig. 31   Potterton/Myson Puma waterways

New methods of controlling combi boilers are being introduced all the time as manufacturers strive to increase efficiency and reliability. A good example of this is the Potterton/Myson Puma range of boilers, these incorporate two wax capsule operators. One is the motive force for the diverter valve (see figs 30 and 31) and the other holds back the water to the draw off taps until it is hot, this increases efficiency and reduces the time taken for the hot water to be available (see fig 32).

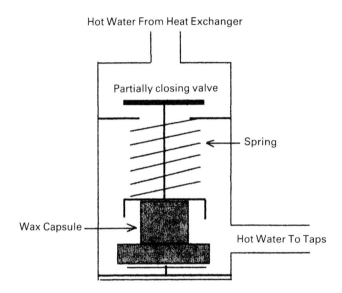

Hot Water From Heat Exchanger

Partially closing valve

Spring

Wax Capsule

Hot Water To Taps

Fig. 32   Potterton/Myson Puma D.H.W. flow limiter

The diverter valve serves two purposes, in conjunction with the domestic hot water flow switch it changes over the boiler from the central heating to the hot water function when a hot tap is opened. Its second function is to hold back the flow into the central heating circuits until the temperature of the water in the heat exchanger reaches 38° C.

The wax capsule is immersed in the mains cold water supply to the heat exchanger, at rest the CH valve closes off the CH flow and the hot water, or by-pass valve, is open giving a short circuit back to the heat exchanger. On a call for CH the boiler and pump start and circulate heated water around the boiler and diverter valve. As the temperature rises the wax capsule is indirectly heated and when its temperature reaches about 38° C it expands, opening the CH valve and closing the hot water valve supplying heated water to the CH circuits. This changeover takes approximately 80 seconds.

When there is a demand for hot water the cold water from the main flows over the wax capsule and cools it, this results in the capsule contracting which has the effect of closing off the CH valve and opening the hot water or by-pass valve. The burner and pump stay on, now controlled via the D.H.W. flow switch which is operated by the flow of water when a draw-off tap is opened. This heats the water within the boiler circuits and heat exchanger transfering it to the mains water which passes to the draw off tap via the hot water flow limiting valve, this valve only allows full flow to the tap when the water is hot.

This wax capsule type flow limiter on the hot water outlet restricts the flow of water to the taps to a trickle until the temperature has reached 38° C. At this point the wax capsule expands and moves a partially closing valve further away from its seating allowing full flow of the now hot water to the taps (see Fig 32).

A disadvantage with some combi boilers is the considerable delay between opening the hot water tap and heated water being delivered. This is a particular problem where the design of the appliance allows the D.H.W. calorifier to cool down after each call for domestic hot water. Some manufacturers have reduced this delay by keeping the water in the calorifier heated when the central heating is running. This can, however, in hard water areas, lead to problems of scale in the hot water coil due to the high temperature, about 82° C, of the water supplied to the D.H.W. calorifier which in turn indirectly heats the static water within the coil to a similar temperature.

A number of different methods are used to modulate the gas rate in combi boilers. Some of these systems are described in Chapter 12.

Fault Finding on Combi Boilers

The complexity of combi boilers and their sophisticated control systems can give the engineer problems when a fault situation occurs. Problems associated with printed circuit boards (PCBs) and other electronic components can be difficult to diagnose. In practice the majority of faults that occur on combis are caused by the central heating system itself creating the problem, sludge for example, or the breakdown of a component on the boiler. Fault finding is further complicated by the wide variety of manufacturers and the different types of operating systems. Before attempting to trace any fault on a combi boiler it is essential that the sequence of operation of the boiler is understood, and any special features that the combi may have are known.

A combi boiler operates in two distinct modes, central heating or hot water, this in itself can give a clue in which general area the fault lies. For example, a complaint of no hot water although the central heating is operating indicates certain parts of the boiler are in good working order, the power supply to the boiler is obviously satisfactory, the gas valve is in working order, the boiler has not overheated, and so on.

Below are a series of procedures that if followed can assist the engineer in tracing and repairing faults on combi boilers and sealed systems (additional information on fault finding can be found in Chapters 12 and 13).

Procedure 1
Symptom – no central heating/no domestic hot water

1. Carry out preliminary electrical safety checks. (see Chapter 13)
2. Switch main electrical supply off for 20 seconds and then switch back on. This will clear any lock out or central heating anti-cycling devices. (some boilers may have a lock out re-set button)
3. Select central heating mode.
4. Check external controls (clock, room thermostat, frost thermostat) are calling and that there is power to the boiler.
5. Check the internal frost thermostat if fitted.
6. Check the gas supply to boiler.
7. Check the water in the system. (see Procedure 4)
8. Check the overheat thermostat. (should be manual re-set)
9. Check that the pump is running.
10. Check the ignition system:

Fully Automatic	**Pilot Type**
Check the operation of the fan, and air flow sensor.	Check that the pilot is established.
Check the ignition.	*Remember that the overheat thermostat could be a thermo-couple interrupter type.*

11. Check the gas valves.
12. Check the boiler thermostat.
13. Check the thermistor(s). **(see Thermistor guide)**
14. Check the printed circuit board, fuses and connectors.

Procedure 2
Symptom – no central heating/domestic hot water operating

1. Carry out preliminary electrical safety checks.
2. Switch main electrical supply off for 20 seconds and then switch back on. This will clear any lock out or central heating anti cycling devices. (some boilers may have a lock out re-set button)
3. Check the water in system. (see Procedure 4)
4. Select the central heating mode.
5. Check that the external controls are calling and that there is power to the boiler.
6. Check the pump. (on twin pump models check the C/H pump)
7. Check the boiler thermostat or thermistor. **(see thermistor guide)**
8. Check the diversion medium is in the correct position and all switches have operated.
9. Check the domestic hot water supply for leaks or dripping taps.
10. Check that any domestic hot water pre-heat system is not operating and keeping the boiler in the hot water mode.
11. Check the printed circuit board, fuses and connectors.

Procedure 3
Symptom – No Domestic Hot Water/Central Heating Operating

1. Carry out preliminary electrical safety checks.
2. Switch the main electrical supply off for 20 seconds and then switch back on. This will clear any lock out or central heating anti cycling devices. (some boilers may have a lock out re-set button)
3. Check the mains water supply.
4. Check the diversion medium is in the correct position and all switches have operated.
5. Check the pump. (on twin pump models check the DHW pump)
6. Check the hot water over heat thermostat if fitted.
7. Check the hot water thermostat or thermistor. **(see thermistor guide)**
8. Check the domestic hot water pre-heat system if fitted.
9. Check the printed circuit board, fuses and connectors.

Procedure 4
Symptom – Sealed System Underpressurised or Overpressurised

Systems are usually pressurised to approximately 1 bar.
Accurate figures can be obtained from
the manufacturers' instructions.

Low Water pressure in System

- **Check** for water leaks.

- **Check** for micro water leaks. Question the customer as to when the system was last topped up.

- **Check** that the customer is not venting the radiators un-necessarily.

- **Check** for discharge at safety valve outlet. If the safety valve itself is in order then any discharge would indicate that the expansion vessel is under pressurised or that the boiler is overheating. One other extreme possibility is that the diaphragm in the expansion vessel is punctured. This can be verified by signs of water at the schraeder valve when it is released.

- **Check** the air pressure in the expansion vessel. This should be between 0.5 bar and 1 bar.
  Note: When making this check the boiler must be drained to release pressure on the water side of the diaphragm.

- **Check** that the pressure gauge is working. This can be achieved by adding water, or draining some off via the safety valve, then checking the reading on the gauge.

High Water pressure in System

- **Check** for filling loop left connected and passing water.

- **Check** the calorifier for break down, mains water passing into the primary water. This can be checked by firstly ensuring that the filling loop is disconnected. Then switch off the power supply to the boiler. Close the C/H flow and return valves, then operate the safety valve to reduce the pressure in the boiler to zero. Any rise in pressure, which will register on the gauge, indicates that the calorifier is leaking internally.

- **Check** that the pressure gauge is working. This can be achieved by adding water, or draining some off via the safety valve, then checking the reading on the gauge.

The fault finding chart below can be used when a customer complains that although the boiler is supplying hot water the temperature is not satisfactory. This a common complaint when combis are installed. Often the customers expectations are greater than the boiler can provide. Working through the chart will confirm whether the boiler is operating satisfactorily or if a fault situation exists.

Combi Fault Finding Flow Chart

Symptom – Boiler apparently working/ Domestic hot water not hot enough

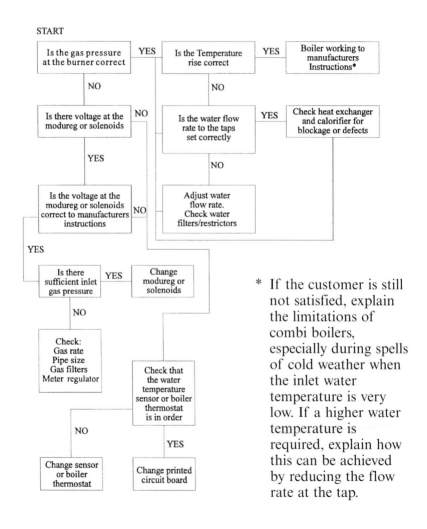

* If the customer is still not satisfied, explain the limitations of combi boilers, especially during spells of cold weather when the inlet water temperature is very low. If a higher water temperature is required, explain how this can be achieved by reducing the flow rate at the tap.

Thermistor Testing Guide

Various types of thermistor sensor are used on combi boilers and different testing methods must be used to check them out. These checks fall into three patterns and are listed below as A, B and C. To test out a suspect thermistor look on the following boiler charts to see which check is required (A, B or C) and then carry out the appropriate check on the thermistor.

TEST A
 – Isolate the boiler from the electrical supply
 – Pull the leads off the thermistor
 – Reconnect the electrical supply
 – Switch on the boiler, in the faulty mode C/H or D/H/W
 – The burner should fire
 – If the burner fires, the thermistor is faulty, replace the thermistor
 – If the burner does not fire then the fault lies elsewhere

TEST B
 – Isolate the boiler from the electrical supply
 – Switch on the boiler in the faulty mode C/H or D/H/W
 – Connect the thermistor leads together
 – Reconnect the electrical supply
 – The burner should fire
 – If the burner fires the thermistor is faulty, replace the thermistor
 – If the burner does not fire then the fault lies elsewhere

TEST C
 – Isolate the boiler from the electrical mains
 – Pull the leads off the thermistor
 – Read the resistance across the thermistor terminals and check the results against the following boiler charts. If the thermistor is outside these limits then replace it
 or
 – If the boiler has thermistors for C/H and D/H/W disconnect the electrical supply, cross connect the suspect thermistor leads to the other thermistor, reconnect the electrical supply then switch the boiler on in the faulty mode, if the burner fires then the thermistor is faulty, replace the thermistor. If the burner does not fire then the fault lies elsewhere

The following list indicates which boilers use thermistors, the type of thermistor used (A, B or C) and their location. To test a thermistor find the boiler on the list, check which test is required (A, B or C), locate the thermistor on the boiler then carry out the appropriate test.

Thermistor Testing Charts

Boiler	Test Method	Thermistor Resistance	Location of Thermistor	GC Part Number
Potterton Lynx 1 Potterton Lynx 2	A	N/A	D/H/W – Hot water outlet pipe C/H–pipe to diverter	337 322 337 322
Myson Gemini	A	N/A	D/H/W – Hot water outlet pipe – C/H – None	392 927
Myson Midas B	C	Between 1k and 30k*	D/H/W – Flow switch outlet C/H – Manifold below burner	392 927
Myson Midas SI Myson Midas SFI	C	Between 3k and 64k*	D/H/W (2 sensors) – Mains water inlet and hot water outlet C/H – Manifold below burner	332 772 332 773 332 774
Worcester 9.24 BF and RSF (old models)	A	N/A	D/H/W – Above calorifier on outlet pipe	381 835
Worcester 9.24 Electronic	C	Between 3k and 13k*	D/H/W – Hot water outlet – C/H – Above safety valve	381 835
Worcester 9.24 RSF E	C	Between 3k and 13k*	D/H/W – Hot water outlet – C/H – Near auto air vent	381 835
Worcester 9.24 Elec BF	C	Between 3k and 13k*	D/H/W – Hot water outlet – C/H – R/hand of gas valve	381 835
Worcester 9.24 Elec RSF	C	Between 3k and 13k*	D/H/W – Hot water outlet – C/H – Near auto air vent	381 835
Worcester 350	C	Between 1.5 and 25k*	D/H/W – Hot water outlet – C/H – Below auto air vent	386 241
Worcester 280 RSF	C	Between 1.5 and 25k*	D/H/W – Hot water outlet – C/H – Below auto air vent	299 199
Worcester 240	C	Between 1.5 and 25k*	D/H/W – Hot water outlet C/H–Flow pipe	378 030

*   The lower resistance is at maximum water temperature. The higher resistance is at cold.

\#   The lower resistance is at cold. The higher resistance is at maximum water temperature.

N/A Not applicable, the resistance is irrelevant.

Thermistor Testing Charts

Boiler	Test Method	Thermistor Resistance	Location of Thermistor	GC Part Number
Ideal Sprint 75 BF and 80 FF	A	N/A	D/H/W – Hot water outlet pipe C/H – Flow pipe	392 927
Ideal Sprint Rapide 90NF	A	N/A	D/H/W – Hot water outlet C/H – Flow pipe	374 067
Ferroli 77 FF and Popular	C	Between 0.5 and 1.5k#	D/H/W – Hot water outlet pipe C/H – Flow pipe	386 818
Vaillant (All Models)	A	N/A	One sensor for both modes D/H/W and CH Flow pipe	263 931
Saunier Duval 620FF/223/225/623	B	N/A	One sensor for both modes Top R/H side of heat exchanger (orange wires)	373 508
Vokera Excell 80E and 96E	C	Between 1k and 30k*	One sensor for both modes low position on flow pipe (red and white wires)	301 227
Glow-Worm Express and Swiftflow 80	C	Between 2.5k and 10k*	D/H/W – Hot water outlet – C/H – LH side of flow pipe	376 987
Glow-Worm Swiftflow 100	C	Between 2.5k and 10k*	D/H/W – Behind pump – C/H – LH side of flow pipe	313 280 376 987
Potterton Myson Puma (All Models)	C	Between 7k and 15k*	D/H/W – Hot water outlet – C/H – Flow pipe	289 528

* The lower resistance is at maximum water temperature. The higher resistance is at cold.

\# The lower resistance is at cold. The higher resistance is at maximum water temperature.

N/A Not applicable, the resistance is irrelevant.

Example in the use of fault finding charts

The fault finding charts are mainly for the use of engineers who are not too experienced in the tracing and repairing of faults on combi boilers. The use of these charts should help in tracing many of the common faults that occur. More complex problems associated with the boilers electronic system require a much greater degree of expertise. It must be remembered that the availability of the manufacturers instructions is an important factor in successful fault finding.

Problem

The customer complains that although her Potterton Myson Puma 80E Combination boiler is operating satisfactorily in the central heating mode there is no hot water supply.

Tracing the fault

Before attempting to trace the fault it is essential that the engineer understands how the boiler operates, below is some basic information about this boiler.

Special features

This boiler has a cross flow heat exchanger, an integral frost thermostat (situated on the CH flow pipe close to the CH thermistor) and a wax capsule type diverter valve. From a cold start the wax capsule will only allow water to circulate around the boiler, once the water temperature exceeds 38° C the wax capsule opens and allows the heated water to leave the boiler and enter the CH system. Because of this the closed boiler circuit will become hot before circulation to the CH system occurs. There is also a wax capsule device at the DHW outlet which restricts the flow of water to the draw off taps until it is hot.

Sequence of Operation Domestic Hot Water (DHW)

Mains light illuminated. Boiler switch is ON. DHW thermostat in full on position. System low pressure warning light not illuminated. Lockout light not illuminated.

1. Open a hot water draw off tap, the domestic hot water flow switch makes. (diaphragm type fitted between the cold water inlet and the domestic hot water outlet)
2. Domestic cold water flowing over the wax capsule operator in the diverter valve closes off the CH port and opens the DHW port to the pump manifold.
3. Pump starts after a seven second delay, unless the boiler is hot, in which case it will start immediately. (check by listening or viewing through the pump vent plug)
4. Fan starts to run at full speed and changes over the air pressure switch providing power to the ignition Printed Circuit Board (PCB).
5. The ignition PCB provides a ten second spark, pilot gas valve and flame proving period, after this the main 240 VAC sole-

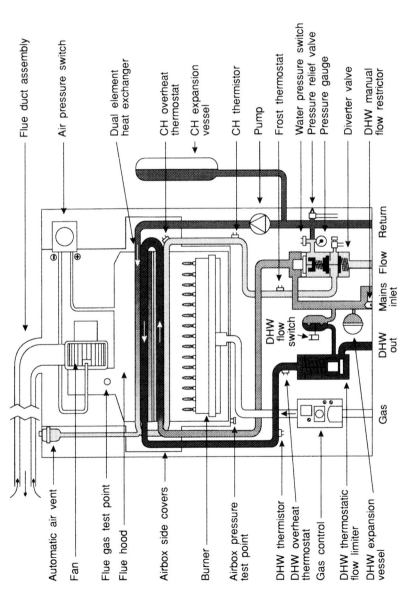

Flue duct assembly

Air pressure switch

Dual element
heat exchanger

CH overheat
thermostat

CH expansion
vessel

CH thermistor

Pump

Frost thermostat

Water pressure switch

Pressure relief valve

Pressure gauge

Diverter valve

DHW manual
flow restrictor

Return

Flow

Mains
inlet

DHW
out

Gas

DHW
flow
switch

Automatic air vent

Fan

Flue gas test point

Flue hood

Airbox side covers

Burner

Airbox pressure
test point

DHW thermistor

DHW overheat
thermostat

Gas control

DHW thermostatic
flow limiter

DHW expansion
vessel

Fig. 33   Potterton/Myson Puma operational diagram

noids will open and allow full gas rate to the burner. If no flame is detected during the proving period the boiler goes to lockout. (Light illuminates on facia panel). **Lockout is manually reset, and can only be reset when there is a call for heat at the time of pressing the lockout button**, this must be held in for at least ten seconds.

6.   The boiler comes on under the control of the DHW thermistor in conjunction with the customer thermostat, these modulate the burner to keep a constant temperature as the water flow rate varies. (45° C – 65° C).

7.   The DHW limit stat will cut the gas off altogether if the hot water temperature exceeds 75° C, it will automatically reset itself when the water temperature falls to about 60° C.

Sequence of Operation Central Heating (CH)

Mains light illuminated. Boiler switch and CH switch ON. CH thermostat full on position. System low pressure warning light not illuminated. Lockout light not illuminated.

1.   External controls (clock/room thermostat) call for heat. The diverter valve will be in the CH position.

2.   Pump starts. (check by listening or viewing through the vent plug)

3.   Fan starts to run at full speed and changes over the air pressure switch providing power to the ignition Printed Circuit Board (PCB).

4.   The ignition PCB provides a ten second spark, opens the gas valves to low rate and senses the flame, after this the modulating gas valve will allow full gas rate to the burner. If no flame is detected during the proving period the boiler goes to lockout. (Light illuminates on facia panel). **Lockout is manually reset, and can only be reset when there is a call for heat at the time of pressing the lockout button**, this must be held in for at least ten seconds.

5.   The boiler comes on under the control of the CH thermistor in conjunction with the customer thermostat, these modulate the boiler. When the system gets up to temperature and the boiler shuts down there is an anti cycling device controlled by the CH thermistor and the PCB, which prevents the boiler from cycling by holding off the burner for a period of approximately four minutes.

For both modes of operation the overheat thermostat will cut off the gas should the boiler temperature get too high, this is indicated by the lockout light illuminating, press the button to reset.

Also for both functions the pump has an overrun time of four to nine minutes depending on how long the boiler has been on, this is controlled by the PCB.

Equipment required

Basic tool kit. Electrical test meter. Fault finding charts.

Preliminaries

The nature of the customers complaint in that the CH is operating satisfactorily allows the engineer to make some deductions.

1. The system water must be satisfactory.
2. The pump itself must be in order.
3. The gas valve itself must be in order.
4. There is no boiler overheat situation.

Procedure 3 (from earlier in this chapter) can now be used to attempt to locate the source of breakdown.

Procedure 3
Symptom – No Domestic Hot Water/Central Heating Operating

1. *Carry out preliminary electrical safety checks.*
   The method of carrying out these checks will be found in Chapter 13.

2. *Switch the main electrical supply off for 20 seconds and then switch back on. This will clear any lock out or central heating anti cycling devices. (Some boilers may have a lock out re-set button).*
   This boiler has a lock out indicator, but because the boiler is operating in the CH mode lock out cannot have occurred.

3. *Check the mains water supply.*
   The minimum mains water pressure to this boiler is 1 bar flowing at least 2.5 litres per minute.
   This boiler reduces the flow of water to the taps until the temperature rises so it was necessary to test the supply at an adjacent tap where it was found to be 2 bar flowing at 8 litres per minute, which is satisfactory. A slight possibility exists that the supply to the boiler is being restricted by a blocked mains water inlet strainer, this could be checked later if the fault is not traced.

4.   *Check the diversion medium is in the correct position and all switches have operated.*
The diverter valve is a wax capsule type and has no moving parts that are visible from outside, the switch in this case that changes the boiler from CH to DHW is the DHW flow switch located on the mains inlet pipe. It is a diaphragm operated switch that makes when the mains water flows through the boiler. To test this switch, isolate the power to the boiler, disconnect the electrical leads, set the electrical test meter on the Ohms range and connect across the switch. High resistance or infinity should be the result. Now the draw off tap is opened to allow water to flow through the boiler, the reading should change to low resistance or continuity.
The switch is found to be operating correctly.

5.   *Check the pump.*
In the DHW mode there is a delay of approximately seven seconds before the pump starts.
The pump is operating correctly.

6.   *Check the hot water over heat thermostat.*
This is found on the DHW outlet pipe and is a normally closed, automatic re-set switch. Test by isolating the power supply, disconnecting the electrical leads and connecting an electrical test meter set on the Ohms range across the terminals. The result should be low resistance or continuity.
The switch is found to be operating correctly.

7.   *Check the hot water thermostat or thermistor.* (see thermistor guide)
This boiler is controlled by two identical thermistors, the DHW thermistor is located on the DHW outlet pipe. The thermistor testing guide (earlier in this Chapter) states that the Puma 80E boiler thermistor is a type C test and that its resistance should be between 7k Ohms when hot and 15k Ohms when cold. To check the thermistor resistance the power must be isolated, then remove the thermistor electrical leads and replace with an electrical test meter set on the Ohms range. As the boiler is cold then a reading of approximately 15k Ohms should be found. In this case a reading of 27k Ohms was found. As this is outside the limits the thermistor must be replaced with a spare. After draining down the boiler and fitting a new thermistor the boiler operated satisfactorily.

Condensing Boilers

By adding a second heat exchanger, or increasing the surface area of a single heat exchanger in a boiler, heat normally lost in the flue is recovered. This type of boiler is called a condensing boiler. Condensation occurs when the temperature of the flue gases drops below 55° C, this is known as the 'dew point'.

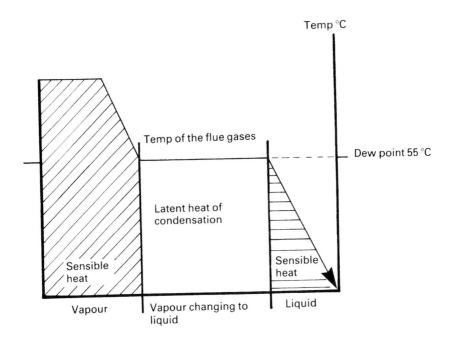

Fig. 34   Graph showing heat given off in a condensing boiler

Two types of heat are transferred to the water in a central heating system, Fig. 34:

(i)   *Sensible Heat*. This is the heat transferred as the temperature of the flue gases drops to 55° C.

(ii)   *Latent Heat of Condensation*. This is the heat transferred as the water vapour in the flue gases condenses to a liquid. It should be noted that the heat is given off as the vapour changes to a liquid and that its temperature remains constant.

Condensing boilers can attain an efficiency of 94%. They can be floor standing or wall-hung and all have room sealed, fan assisted flues. They are suitable for both new installations or as a replacement

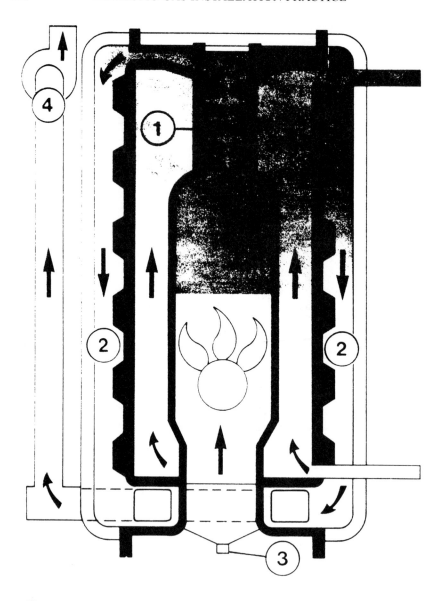

1   Primary heat exchanger   3   Condensate drain

2   Secondary heat   4   Fanned flue
       exchanger

Fig. 35   Condensing boiler with two heat exchangers. Burner firing upwards

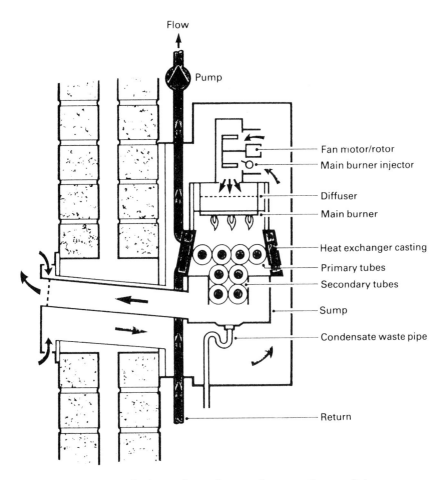

Flow

Pump

Fan motor/rotor

Main burner injector

Diffuser

Main burner

Heat exchanger casting

Primary tubes

Secondary tubes

Sump

Condensate waste pipe

Return

*Fig. 36   Condensing boiler with two heat exchangers. Burner firing
downwards*

boiler. To gain the maximum advantage of the increased efficiency,
the heating system should be designed to give a flow/return temperature difference of 21° C (11° C on a normal boiler installation). Figs.
35 and 36 show boilers with two heat exchangers and Fig. 37 a boiler
with one heat exchanger.

Discharge of Condensate

Where possible the condensate should discharge into an internal
stack pipe; the minimum pipe diameter is 22 mm and it should be
fitted with a trap (75 mm condensate seal). It is permissible to
discharge via a sink waste through a condensate syphon. Connection

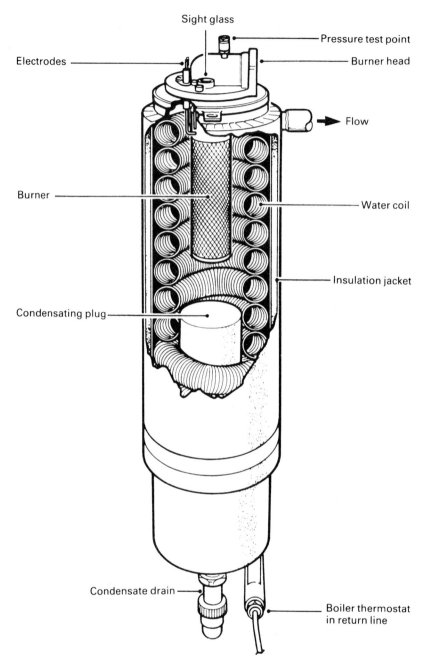

Fig. 37   Condensing boiler with one heat exchanger. Burner firing downwards

should be made downstream of the sink waste trap. If connection is only possible upstream, then an air break will be required between the two traps. Where it is not possible to terminate internally, the condensate discharge pipe may be terminated externally, Fig. 38.

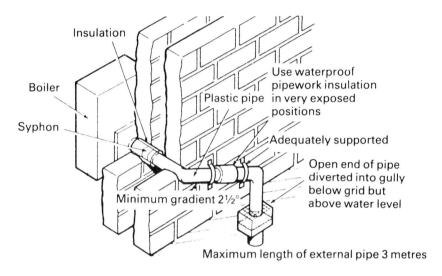

Insulation

Boiler

Syphon

Use waterproof pipework insulation in very exposed positions

Plastic pipe

Adequately supported

Open end of pipe diverted into gully below grid but above water level

Minimum gradient 2½°

Maximum length of external pipe 3 metres

Fig. 38   Condensate pipework, terminated externally

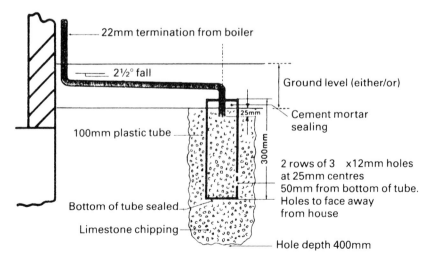

22mm termination from boiler

2½° fall

Ground level (either/or)

25mm

Cement mortar sealing

100mm plastic tube

300mm

2 rows of 3 ×12mm holes at 25mm centres 50mm from bottom of tube. Holes to face away from house

Bottom of tube sealed

Limestone chipping

Hole depth 400mm

Fig. 39   Condensate absorption point (soakaway)

Any external pipe run is susceptible to freezing, which on some models of appliance could cause it to lock out. To avoid this the pipework should be installed to dispose of the condensate quickly, with as much of the pipework as possible run internally before passing through the wall. From here the pipework should run to an external drain using the minimum number of joints and bends; the pipework should be insulated where necessary.

If none of the above options are possible, then the condensate pipe could be terminated at a condensate absorption point, Fig. 39. This should be sited as close to the boiler as possible, ensuring that no services (gas, water, electric etc.) are in the vicinity.

The latest addition to the condensing boiler market is the condensing combination boiler such as the latest development from Vaillant.

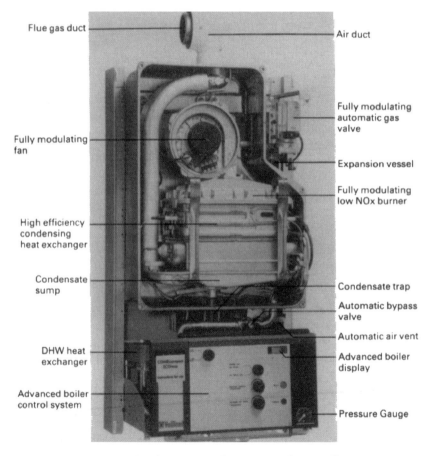

Flue gas duct

Air duct

Fully modulating automatic gas valve

Fully modulating fan

Expansion vessel

Fully modulating low NOx burner

High efficiency condensing heat exchanger

Condensate sump

Condensate trap

Automatic bypass valve

Automatic air vent

DHW heat exchanger

Advanced boiler display

Advanced boiler control system

Pressure Gauge

Fig. 40   Condensing Combination Boiler (Vaillant)

This combines all the advantages of a combination boiler with the efficiency of a condensing boiler. (see Fig 40)

This boiler also incorporates a multifunctional display to give information on the state of the boiler at any time. It operates in four modes.

Normal mode –   this displays the boiler flow temperature.

Status mode –   for a quick functional check during commissioning the display shows where the boiler is in its operating cycle. (e.g. S.4 – normal central heating operation)

Fault mode –   highlights system problems at a glance. (e.g. F.21 – no gas supply)

Diagnosis mode –displays key component operating information. (e.g. burner pressure can be checked directly from the display)

This boiler highlights the determination of manufacturers to upgrade the technical sophistication of their appliances to increase efficiency and user friendly operation in an attempt to increase their share of what is now a highly competitive market. As boilers and controls become more and more complicated it is left to the engineer to solve any subsequent problems. At least in this case the manufacturers have made an attempt to assist the engineer to deal with problems that might arise later.

European Directives

These Directives are binding on member states and as explained in Chapter 6 the Gas Appliance Directive is implemented in the UK by the Gas Appliance (Safety) Regulations which came into force in 1995

Fig. 41   Normal mode

Fig. 42   Status mode

Fig. 43   Fault mode

Fig. 44   Diagnosis mode

with a transitional period until January 1996. There is some overlap between the Gas Appliance Directive and five other directives.

One of these is the Boiler Efficiency Directive, which came into force in January 1995 with a transitional period until January 1998, it sets a minimum efficiency level at full and at part load. For example a 10kW boiler with a 1-star rating would have a full load efficiency of

86% and a part load efficiency of 83%. A 10kW boiler with a 4-star rating would have a full load efficiency of 95% and a part load of 92%.

The other Directives are: Construction Products Directive, this covers products incorporated within buildings and civil engineering works.

Electro-Magnetic Compatibility Directive, covering electrical and electronic appliances and equipment.

Low Voltage Directive, covering voltage ranges 75–1500 volts DC and 50–1000 volts AC.

Machinery Safety Directive, this covers performance of moving parts.

Other Directives and regulations are currently being drafted and include: CO emission limitation; general product safety (2nd-hand appliances); use of the CE mark: energy labelling; and measuring instruments (meters).

Heat Emitters

The heat emitters used with hot water circulation systems may be categorised as follows:

- radiators
  - pressed steel or cast iron
  - panel or column type
- convectors
  - wall or skirting convectors
  - natural or fanned convection
- pipework
  - pipe runs or emitter connections

Radiators (Fig. 45)

The type of radiators commonly used on the older gravity systems were cast iron column radiators which transmitted a large proportion of their heat by convection. These have been superseded on current small bore and micro bore systems by panel radiators, usually made from pressed steel, which emit about 50% of their heat by convection and 50% by low temperature radiation. Panel radiators can be made from copper and occasionally aluminium. They are available as single, double or treble panels and can be specially supplied curved or angled to fit bay windows or corner locations.

Because they have a high convection output, fitting radiators under shelves or in recesses can reduce the total heat emission by from 5 to 20%. Painting radiators with ordinary paints or enamels of any

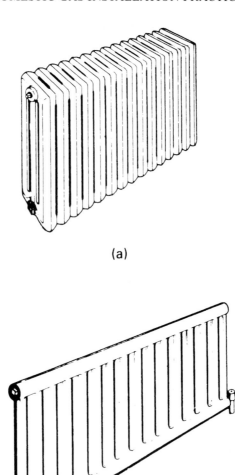

(a)

(b)

Fig. 45   Radiators: (a) cast iron column; (b) pressed steel panel

colour does not significantly affect the radiant output. Metallic paints may, however, reduce the heat emission by up to 15% (Vol. 1, Chapter 10).

Table 34 gives the values for the heat emission from typical steel panel radiators in a smallbore system.

TABLE 34 Heat Emission From Typical Steel Panel Radiators

Flow temperature   82° C
Return temperature   71° C
Radiator average temperature   77° C

Type of Radiator	Room Temperature			
	21° C	18° C	16° C	13° C
	W/m²	W/m²	W/m²	W/m²
Single Panel	605	648	685	726
Double Panel	520	556	587	625
Treble Panel	460	493	520	556

Convectors

Convectors consist essentially of water-to-air heat exchangers made from banks of fins surrounding the hot water circulating pipes. Air may flow upwards, over the fins, by natural convection. The pressed steel case has louvres at the top and bottom and the warm air output is usually controlled by a damper at the top louvre.

The skirting heater, Fig. 46, is a smaller version of the wall convector and is fitted along the bottom of the wall, replacing the skirting board, or slightly recessed. It gives an even temperature gradient throughout the room and is unobtrusive.

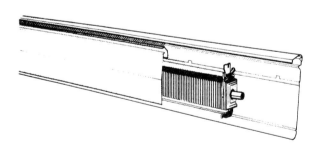

Fig. 46   Skirting heater

Fan convectors, Fig. 47, are considerably smaller than radiators of equivalent output. They may be usefully installed in large rooms where otherwise the radiators would occupy a large area of wall surface. They can have thermostatic controls which switch the fan on and off to maintain a selected room temperature. The thermostat can be internal or fitted remote from the convector on an internal wall.

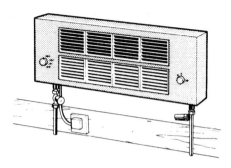

Fig. 47   Fan convector

TABLE 35 Heat Emission from Copper Pipes

Size of Pipe	Room Temperature			
	21° C	18° C	16° C	13° C
	Heat emission			
	W/m	W/m	W/m	W/m
8 mm	17	20	21	22
10 mm	21	23	24	25
15 mm	26	28	30	32
22 mm	35	38	40	42
28 mm	44	45	48	52
	Unheated Spaces, Temperature			
	10° C	4.5° C	− 1° C	
8 mm	23	24	26	
10 mm	26	29	31	
15 mm	34	44	47	
22 mm	44	60	64	
28 mm	54	74	79	

For 2 pipe runs multiply by 0.95   For painted pipes multiply by 0.55
For vertical pipes multiply by 0.8   For insulated pipes multiply by 0.2

Because the fan increases the efficiency of heat emission, the temperature drop across the convector is greater than that across a radiator. It is advisable, therefore, to supply the convector by a separate circulation and not to connect it directly into the same circuit with ordinary radiators.

Pipework

Pipes fitted at skirting level can provide uniform heating conditions and, at one time, were commonly used in churches, halls and schools. Unfortunately they are unsightly and they collect dust and cause wall staining so they are now rarely used as heat emitters. Pipes are, however, used to distribute hot water to other emitters and the amount of heat they emit may make a significant contribution to the heat requirement of the room. This must be taken into account when sizing the heat emitter.

Table 35 gives the heat emitted from pipe runs of bare, unpainted copper pipes to BS 2871, Table 'X'. The pipes are run singly and adjacent to the wall.

The heat emitted from radiator connections is given in Table 36. These are average values for the connections to a radiator fitted immediately above concealed horizontal flow and return pipes.

TABLE 36 Heat Emission from Radiator Connections

Size of Pipe	Room Temperature			
	21° C	18° C	16° C	13° C
	Heat emission			
	W	W	W	W
8 mm	20	21	23	24
10 mm	24	26	27	31
15 mm	29	32	34	36
22 mm	39	43	45	47

Cylinders

Hot water storage cylinders on a combined heating and hot water system may be:

● indirect double feed
● indirect single feed
● direct with immersion calorifier or coil.

Indirect Double Feed

This is the conventional cylinder with an annular or coil calorifier which has its own separate cold feed, Fig. 48. On sealed systems the coiled type only must be used.

Indirect Single Feed

These cylinders dispense with the separate cold feed and expansion cistern and its pipework and are consequently less expensive to install.

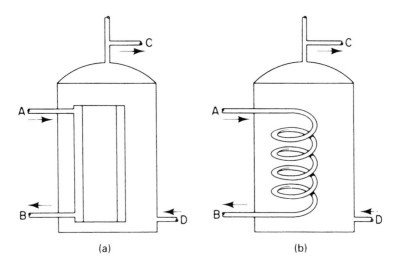

Fig. 48   *Indirect double feed cylinder: (a) annular; (b) coil. A: Primary hot water; B: Primary cold water; C: Secondary hot water; D: Secondary cold water*

The calorifier consists of two hemispherical chambers, shown at B and D in Fig. 49, connected by a tube, C. The two chambers are contained in an annular cylinder which has an air vent G in the form of an inverted 'U' tube.

When the system is being charged with water, the water enters through the cold feed at A. As the level in the cylinder rises, the water enters chamber B and spills over the top of the tube C down into the chamber D. From here it flows down the return E, so filling the boiler and the entire primary and heating system.

The air which is displaced is vented via the flow pipe and the 'U' tube G. When both primary and secondary systems are full, the primary and secondary water are kept separate from each other by two air locks which have been left. One is in the top of the upper chamber D, above the top of the tube, C. The other is in the 'U' tube G.

When the primary system is heated and the water expands, it raises the level slightly in the 'U' tube G and also displaces the air in D down the tube C and into the lower chamber B. When the system cools the air seals return to their previous positions.

The correct size of cylinder must be selected to accommodate the expansion from the entire primary and heating circuits. If the air seals are to remain intact the sizing must be based on the water capacity of the boiler, heat emitters and the circulating pipework.

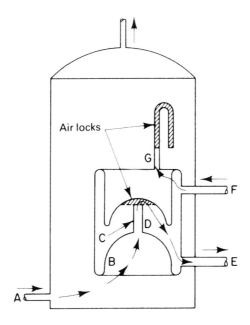

Fig. 49   Indirect single feed cylinder. A: Cold feed; B: Lower chamber;
C: Tube; D: Upper chamber; E: Return; F: Flow; G: Air Vent

Direct Cylinders

A direct cylinder must not be used in conjunction with a heating circuit but it may be converted to an indirect cylinder usually by means of an immersion calorifier. Cylinders should only be adapted if they are in good condition and provided that the conversion does not invalidate any maker's current guarantee.

Immersion calorifiers can generally be fitted into the top entry immersion heater boss on a standard cylinder. They are then connected to the primary flow and return pipes. Immersion calorifiers generally require a pumped primary circuit and may be used on sealed systems. An illustration of a typical calorifier is shown in Vol. 1, Chapter 10, Fig. 14.

All cylinders should be lagged with a glass fibre jacket, 75 mm thick, or with factory-applied insulation.

Pumps

The pump used on a central heating system is a centrifugal pump with an impeller revolving at the rate of about 1700 to 3000 r.p.m. The impeller may be an open or a closed type, Fig. 50. The closed type has a higher efficiency.

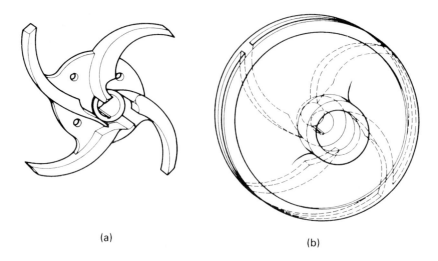

(a)

(b)

Fig. 50 Pump impellers: (a) open; (b) closed

Impellers consist of a series of curved vanes which pick up water at their centre inlet and throw it rapidly outwards to the outlet. As the velocity decreases, the pressure rises, so creating the pump head which forces the water round the system.

Although the open impeller is fitted very closely within the two halves of the pump casing, some leakage of water from the pressure side to the suction side is possible. The closed impeller has its vanes enclosed between two metal discs and fits against a renewable ring at its outer circumference.

Figure 51 shows an exploded view of a pump. The motor is a single phase type with the condenser producing an out-of-phase current for the starter windings. The starter and starter windings are sealed from the waterways by the stainless steel rotor cover.

The rotor has an iron core with short-circuited copper bars embedded in it. The changing magnetic fields in the windings induce a current in the bars of the rotor and so produce the force to rotate the rotor and the impeller shaft.

Stainless steel is used for a number of the components because it resists scale deposition. Bearings are often ceramic and tungsten carbide.

Pumps are generally made to close tolerances and so may be affected by scale deposits or by impurities in the water which can settle in horizontally mounted pumps if they are shut down during the summer period.

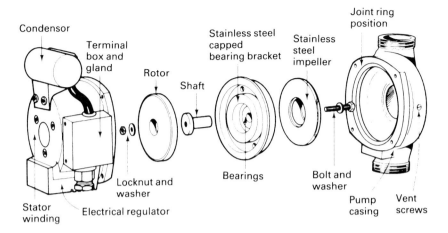

Fig. 51   Centrifugal pump

INSTALLATION

When locating and installing wet central heating systems, the following points should be considered.

Boilers

Boilers should be installed to BS 6798 : 1987 : Installation of gas-fired hot water boilers of rated output not exceeding 60 kW.

General Considerations

Boilers may be installed in a variety of locations.
   In all cases the following notes apply.
   Manufacturer's instructions must be followed.
   Wherever practicable, a room-sealed boiler should be used. Where this is installed in a bathroom, any electrical control must not be within reach of anyone using the bath or shower.
   Open-flued boilers must not be fitted in:

● bedrooms
● bedsitting rooms
● any room or space containing a bath or shower.

Fitting open-flued boilers in toilets and cloakrooms is not recommended but may be acceptable if air vents to outside air are provided.
   With all boilers there must be:

- an effective flue system to discharge the products of combustion to the outside atmosphere (see Chapter 5)
- adequate air vents to the room or the compartment in which the boiler is sited (see Chapter 5)
- adequate space around the boiler for air circulation and to permit servicing or exchanging components.

If fitted in a cellar or low-lying boiler house an appliance burning 1st or 2nd family gases should be protected against flooding or water seepage. Appliances burning 3rd family gases should not be installed in cellars or basements.

Boiler Compartments

No boiler should be installed in an understairs cupboard if the premises are more than two storeys high.

Compartments were dealt with in Chapter 5. The main requirements are:

- the compartment must be a permanent, rigid structure
- internal surfaces should be of non-combustible material or situated at least 75 mm from the boiler
- makers' instructions must be followed for the protection of the floor or wall on which the boiler is to be mounted
- adequate air vents must be provided for ventilation or cooling and, if required, for combustion air. With open-flued boilers the air vents must not communicate with a garage, bathroom or bedsitting room
- adequate space must be provided for servicing and the door must permit withdrawal of the boiler
- a notice should be attached warning against its use for storage.

Airing Cupboards

These may be used provided that they satisfy the recommendations for compartments and, in addition, the airing space is divided from the boiler compartment by a perforated partition.

The perforations must not be larger than 13 mm; expanded metal mesh is suitable.

Where any flue pipe passes through the airing space it must be protected by a guard giving an air space of at least 25 mm around the pipe.

Roof Spaces

In addition to the general requirements the following additional points must be considered:

- a sufficient area of insulated flooring, strong enough to support the boiler and its accessories and adequate for access and servicing, must be provided
- a permanent means of easy access to the roof space is necessary
- there should be a safety guard around the opening
- any articles stored in the space must be guarded against contact with the boiler or its flue
- fixed lighting must be installed
- sufficient headroom is required to give an adequate head of water at the feed and expansion cistern, alternatively a sealed system should be used
- a means of electrically isolating the boiler should be provided in the roof space
- the user should be provided with the means to shut down the boiler without having to enter the roof space
- the positions of the cold feed and vent connections are critical
- it may be necessary to incorporate thermal delay on the pump to avoid overheating
- where an existing brick chimney is used, the appliance should be installed to BS 5440 (see Chapter 5).

Garages

The installation must conform to the requirements of local bylaws, fire regulations and insurance companies.

Frost Protection

When an appliance is fitted in an exposed position, e.g. roofspace, garage, a frost thermostat should be fitted to operate at a temperature of approximately 4° C and sited to manufacturers' instructions.

Fireplaces

At the planning stage, ensure that the dimensions of the opening can accommodate the appliance selected. Before installation the chimney must be swept and lined (Chapter 5).

Where the gas fire is to be wall mounted, it may be necessary to raise the boiler to bring it into line.

Any pipework within the opening must be protected against corrosion or damage by soot or debris by wrapping with PVC tape.

External Locations

Boilers may be:

- designed to be fitted in an external location without the need for any additional protection
- installed in a purpose-made enclosure.

Where an enclosure is required it should comply with the recommendations for boiler compartments and, in addition:

- it should be made from weatherproof materials
- it should contain a waterproof, fused double-pole switch to isolate the boiler electrically
- the low-level air vent should be at least 150 mm above floor level
- any openings should be not wider than 16 mm to keep out birds or rodents
- the openings should be at least 6 mm wide to prevent them becoming blocked
- the system must be protected from freezing by a frost thermostat or by the addition of antifreeze.

Where antifreeze is to be used, ethylene glycol is recommended in conjunction with an inhibitor. Usually a 25% solution, which gives a freezing point of at least $-12°$ C, is adequate. The mixture should contain $2\frac{1}{2}$ parts of antifreeze to 1 part inhibitor.

A label should be tied to the boiler drain cock stating the quantity and type of antifreeze and the date it was introduced.

Table 37 gives the quantities of inhibitor and antifreeze for domestic systems.

TABLE 37 Quantities of Antifreeze Required

Type of Premises	Number of Radiators	Approximate Water Content of System litres	Corrosion Inhibitor litres	Antifreeze Required litres
3 bedroom house or bungalow	4 to 7	45 to 90	4.5 to 9	11 to 23
4 bedroom house	6 to 8	90 to 114	9	23

Gas Supply

The gas supply pipes and the meter must be capable of meeting the total gas load.

The meter should be a credit meter, a prepayment meter would cause considerable inconvenience to the customer.

The supply to the boiler should be fitted with a union cock to allow the boiler to be disconnected without interfering with the gas supply to other appliances. The installation must comply with the requirements of BS 6891 for natural gas and BS 5482 for LPG and the Gas Safety (Installation and Use) Regulations 1998.

Heat Emitters

Radiators

Radiators are generally located under windows or adjacent to doors so that draughts are eliminated. Although fitting the radiator below the window has some advantages, the amount of heat lost through an uninsulated cavity wall is quite considerably more than would have been lost if the radiator was installed on an internal wall.

However the problem can be overcome by fitting a panel of reflective material (foil) to the wall behind the radiator.

Radiators should be fitted with a minimum clearance at the back of 40 mm and a space of at least 100 mm below, to allow for cleaning.

In situations where dust or tobacco smoke entrained in the convection current can discolour the wall decorations, shelves should be fitted. The shelf should project about 75 to 100 mm at each end and 25 to 50 mm at the front. It should be sealed to the wall surface, usually with a strip of plastic foam. The shelf should have a valance at each end but this must not obstruct any control or air release valves.

Radiators have commonly been supplied with four points for connecting to pipework or controls. Some are available with one or two connections situated at the centre or at one end. The use of the four connecting points is illustrated in Fig. 52. Most radiators are fitted with:

- control valve, manual or thermostatic
- lockshield valve, for balancing and isolating the radiators
- air release valve
- Control and lockshield valves are obtainable, fitted with an integral drain-off device (Fig. 53).

Combined control and balancing valves are available and should be fitted in conjunction with an on-off valve on the return connection to enable the radiators to be isolated when decorating or servicing.

Convectors

Convectors should be sited in the same positions as radiators. With fanned convectors, direct draughts on to the occupants should be

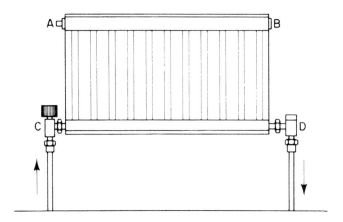

Fig. 52   Radiator connections. A: Air release valve, 6 mm; B: Plug, R½; C: Control valve, 15 mm; D: Lockshield valve, 15 mm

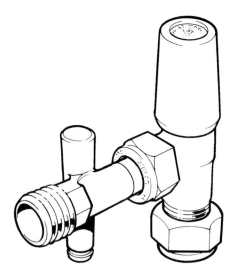

Fig. 53   Radiator valve fitted with an integral drain-off device

avoided and a supply of electricity must be available. The casings of all convectors must be sealed to the wall to avoid staining.

Cylinders

Where the primary circulation is by gravity, the cylinder must be fitted above the boiler at a height which will provide an adequate circulating head. This is usually a minimum of about 1 m. The flow and return pipes should be at least 28 mm diameter and should preferably be connected to independent tappings on the boiler.

With pumped primaries the cylinder may be fitted in any position although the cylinder is still normally fitted near to the boiler. This location has the advantage of allowing some gravity circulation to take place even when the pump fails.

The flow and return pipes are usually 15 to 22 mm diameter. A balancing valve should be fitted in the return from the cylinder when a single pump supplies both the cylinder and the heating circuit.

When towel rails are fitted they may be connected to the cylinder primary flow and return so that they continue to operate when the heating circuit is shut down.

A drain cock must be fitted to provide a means of emptying the secondary water from the cylinder.

The cold feed pipe must supply the total demand of the draw off taps and should be not less than 22 mm diameter. It should be fitted with a gate valve for isolating purposes. The open vent should run from the top of the cylinder rising continuously to terminate over the cold feed cistern. It should be not less than 22 mm diameter and must not have any valves fitted to it.

The other requirements of the hot water system should be as detailed in Chapter 9.

Pumps

Pumps should be fitted in readily accessible positions and in accordance with the manufacturer's instructions. Isolating valves should be fitted at each side so that the pump may be removed for cleaning without draining down the system. Some pumps have integral valves.

The pump should be installed with the shaft in a horizontal, or slightly above horizontal position. If the pump is fitted with the shaft below an horizontal position it will cause rapid wear to the bearing and shaft. Fig. 54 shows the right and wrong way of installing a pump.

The position of the pump in relation to the vent pipe and the cold feed is critical, two problems may occur:

• air may be sucked into the system
• water may be pumped out of the open vent.

Both these faults cause inefficient heating and corrosion. To avoid this happening the following rules should be complied with when locating a pump in an open system:

RIGHT   WRONG

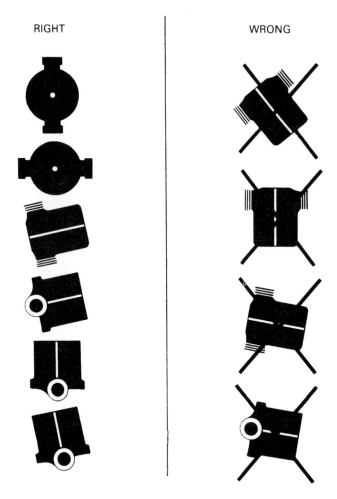

Fig. 54   Siting of pumps (Grundfos)

- the cold feed pipe should be connected to the return side of the boiler and should be free of all valves
- if possible, there should be no valves between the feed cistern and the boiler, any intervening pump should have full-way isolating valves
- the cold feed pipe should normally never be connected into the return pipe between the pump and the boiler, if it is, most of the system will be under suction
- the open vent should be connected to the flow side, near to the boiler and should be free of all valves

- if the vent is connected to the suction side of the pump, the water level in the vent pipe must never fall so low that air can be drawn into the system
- if the vent is connected to the outlet side of the pump the water level in the vent pipe must never rise so high that water can be discharged into the feed and expansion cistern
- at no time should any part of the system be subjected to pressure below atmospheric pressure to avoid the possibility of drawing in air.

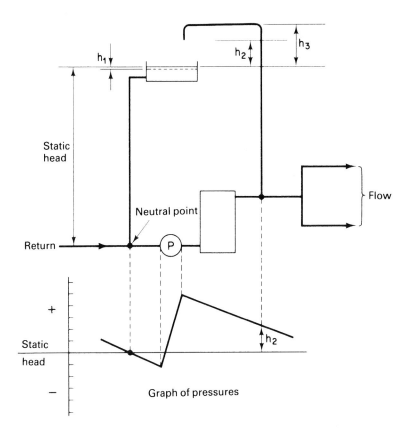

Fig. 55   Effect of pump fitted on return pipe
h_1 : drop in cistern water level
h_2 : rise in vent water level, equal to pump pressure at this point
h_3 : height of vent above static water level
The system operates satisfactorily provided that h_3 is always greater than h_2

A pump has the effect of raising the pressure on its outlet and lowering the pressure on its inlet. This can have the effect of raising or lowering level of water in an open vent by a considerable amount. However, the change in the volume of water in the cistern is comparatively small and has very little effect on the water level in the cistern. So the point at which the cold feed joins the circuit is a 'neutral point' at which the only pressure is the static head of the system.

Figure 55 shows the effects of fitting a pump on the return pipe, between the cold feed and the boiler. The vent is on the outlet side of the pump.

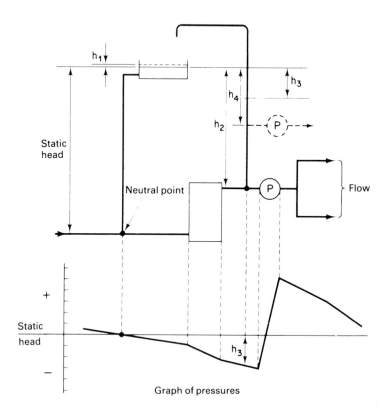

Fig. 56   Effect of pump fitted on flow pipe
  h_1 : *rise in cistern water level*
  h_2 : *head at vent connection into system*
  h_3 : *drop in vent water level, equal to pump suction at this point*
  h_4 : *head at higher vent connection*
  The system operates satisfactorily provided that h_2 is always greater than h_3

The graph indicates the pressures prevailing at the various points on the circuit when the pump is running. At the neutral point, the pressure is equal to the static head. Between the cold feed connection and the pump inlet the pipework is under suction and the pressure then rises sharply through the pump. From then onwards the resistance of the boiler, pipework and radiators cause a gradual decrease in pressure until the returning water reaches the neutral point again, completing the circulation and a return to static pressure.

The diagram shows the small drop in the cistern water level caused by the suction of the pump and the increase in the vent water level caused by the pump pressure. Most of the system operates at pressures above the static head so air cannot be drawn in. However, water can be pumped out of the open vent unless the vent is taken to a height above the cistern (h_3) which is greater than the rise in the vent water level (h_2).

Figure 56 shows the effects of fitting a pump on the flow pipe. The vent is now connected on the suction side of the pump, so the vent water level drops when the pump is running and the cistern water level rises slightly.

The graph shows the pressures falling from the neutral point to the pump inlet, due to the pump suction. Most of this pressure drop is due to the resistance of the boiler itself and it results in an equal drop in the vent water level (h_3).

With this system water cannot be pumped out of the open vent. However, air can be sucked in through the vent, if the pump suction causes the vent level to drop to the point at which the vent connects to the flow (h_2).

While this is unlikely to occur on most house systems, three points must be considered.

Firstly, if the vent connection is made at a higher point in the system, as shown by the broken lines, the head at the connecting point is now h_4, which is only half h_2.

Secondly, low water content boilers have a high flow resistance which may be of the order of 1.2 m. This will result in a drop in the vent water level of 1.2 m.

Thirdly, when the pump is suddenly switched on, the pump suction causes an increased fall in the vent water level to occur momentarily.

So a combination of a high connecting point and a high boiler resistance can result in air being sucked into the system each time the pump is switched on. To avoid this it is necessary to ensure that the head at the vent connection (h_2) is at least $1\frac{1}{2}$ times the boiler resistance (h_3).

Heating Circuit

Feed and Expansion Cistern (Fig. 57)

The cistern combines the jobs of providing a supply of make-up water for the system and accommodating the additional volume of water produced by expansion when the system is heated. It must not be used to supply water for any other purpose.

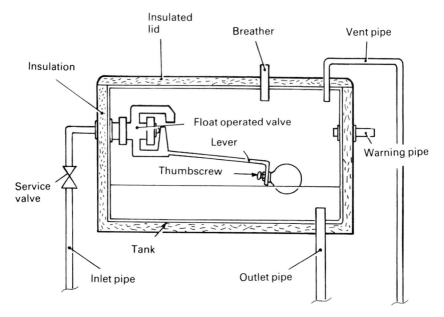

Fig. 57   Feed and expansion cistern

The cistern should:

- be made from a suitable material
  – high density polythene to BS 4213
  – galvanised mild steel to BS 417
  – copper or glass fibre may also be used
- have a volume of not less than 18 litres
- be fitted with a ball valve capable of withstanding a temperature of 100° C adjusted so that, when the system is hot, the water level is not less than 25 mm below the overflow level. (The expansion is approximately $\frac{1}{20}$ of the total water content of the system)
- be fitted with a warning pipe or overflow of 22 mm diameter installed below the level of the cold water inlet

- be fitted with a cover and be suitably insulated against freezing
- have a cold water supply to the ball valve of 15 mm, fitted with a stop cock or valve to BS 1010
- be fitted with a cold feed pipe of 15 mm copper or 22 mm galvanised mild steel, connecting it to the circulating pipework
- be fitted at least 1 m above the highest point of the circulating system or to the manufacturer's instructions.

The vent pipe should terminate over and into the feed and expansion cistern. The vent should be 22 mm in diameter and should be sufficiently high to prevent water being pumped over under all normal conditions.

The cold feed pipe from the feed and expansion cistern should not be fitted with a valve unless it is a requirement of the local water company.

Pipework

Pipes must be laid so that air can be removed through the vent pipes or the air release valves on heat emitters. Any other high points must also be fitted with a vent or an air release valve.

Pipes must be securely supported but the support must allow movement for expansion or contraction. The distance between supports should not normally exceed 1.8 m or, on skirtings, 1.2 m. Where practicable, pipes should be fitted clear of wood joists and floor boards. Where this is not possible the pipes should have pads of vermin-proof insulating material between themselves and the woodwork.

Notches in joists should allow sufficient clearance for expansion and contraction.

When passing through brickwork or masonry, pipes should be suitably sleeved.

Provision for draining the entire system must be made.

Cold feed and vent pipes in open systems should be fitted as shown in Figs. 58 and 59.

The arrangement shown in Fig. 58 A is suitable for a system with a gravity primary circuit or with an assisted gravity circuit by injector. There is no risk of vent overflow or air entrainment.

Fig. 58 B shows the arrangement is advised with pumped primary circuit to an indirect cylinder. Again there is no risk of vent overflow or air entrainment. Heating and domestic hot water controls should be fitted with this system.

In Fig. 58 C the feed is on the inlet side of the pump. Unless the height above the water level equals 'h', there is a risk of the vent overflowing.

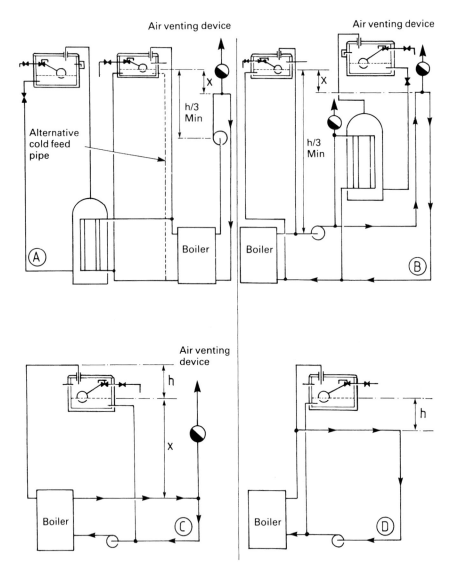

Fig. 58   A, B, C and D. Arrangements for vents and cold feeds, 'h' is the maximum head developed by the pump and 'x' is at least 1 m above the highest point of the circulating system or such lesser height recommended by the manufacturers

In Fig. 58 D the feed is on the outlet side of the pump. There is a risk of air entrainment unless the head above the circuit is equal to 'h'.

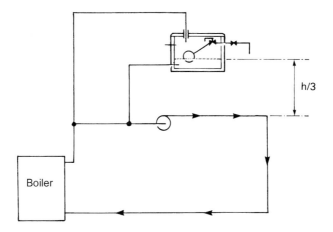

*Fig. 59   Vent and cold feed arrangement for high hydraulic resistance boiler.
'h' is the maximum head developed by the pump*

Fig. 59 shows the vent and cold feed arrangement for a high
hydraulic resistance boiler, e.g. a wall mounted boiler, unless any
special instructions are issued by the manufacturers.

Pipes should be insulated when heat emission is not required, that
is, under floors, in roof spaces, outbuildings or wash houses. Insulation is not necessary in intermediate floors.

Safety Valves

Safety valves are not absolutely necessary on open systems, although
they may be required for insurance purposes. Where they are used,
they should be connected directly to the boiler or to the flow pipe
immediately adjacent to the boiler. On sealed systems a safety valve
must be fitted and should be set to operate at 3 bar.

Commissioning

Open Systems

The procedure for commissioning an open system is as follows:

- flush out the water system with all the valves open, at least twice,
  once cold and once hot, preferably with the pump removed for
  the first flush out
- check all joints and make good any water leaks, recheck after
  heating up
- fill the system with water and release all the air

- add the corrosion inhibitor
- test the gas installation for soundness and purge
- check for adequate ventilation
- light the boiler in accordance with the manufacturer's instructions
- check the burner pressure, allow the boiler about 15 minutes to heat up and recheck
- check the flame picture and, if necessary, check the gas rate by the meter test dial
- check that the pilot flame is adequate and stable
- set the pump adjustment to give the required design temperature difference between flow and return pipes
- check for noise and/or pumping over at the vent
- balance all the circuits to give the design temperature drop
- on open-flued boilers check for spillage at the draught diverter
- check the controls
  - flame supervision device
  - boiler thermostat
  - any other controls
- if the boiler has a bypass, carry out manufacturer's setting instructions or open the valve until kettling is just eliminated
- leave the installation and servicing instructions at the meter and the operating instructions at the boiler
- instruct the customer on the operation of the boiler, system and controls and on the need for regular servicing.

Sealed Systems

These should be commissioned in a similar way to the open systems with the exception of the following points:

- fill the system at the filling point either from the mains water supply using a temporary connection approved by the water authority, Fig. 14 or by a sealed system filler pump with a break tank
- after flushing and refilling
  either – release water from the safety valve until the level in the top-up bottle falls, then refill the bottle
  or – if no bottle is fitted, introduce or release water until the desired pre-pressurisation pressure is achieved
- check that the safety valve operates at a pressure of 3 bar, as shown on the pressure gauge, ± 0.3 bar
- check the operation of the boiler high temperature cut-off.

Servicing

The servicing should be carried out annually and should include the following operations:

- question the customer and examine the boiler for faults
- isolate the boiler electrically
- inspect the heat exchanger and clean the flueways
- clean pilot and main burners
- check gas taps and ease if necessary
- check gas soundness
- check ventilation
- test the ignition device, light the boiler
- check burner pressure and flame picture
- check pilot flame correctly located and stable
- check controls, including, as fitted
  - flame supervision device
  - boiler thermostat
  - cylinder thermostat
  - room thermostat
  - frost thermostat
  - pump delay thermostat
  - modulating thermostat
  - overheat (high temperature cut-off) thermostat
  - safety valve
- check ball valve in feed and expansion cistern
- on a sealed system, check system pressure
- check pipework and valves for water leaks
- check pump and electrical controls including clocks and zoning valves
- check flue operation
  - open flues for spillage
  - room-sealed for flue seals
- check that the appliance is securely fixed and stable.

When an appliance performance tester is used, (Vol. 1 Chapter 14) and results are favourable the engineer may not need to strip down and disturb equipment that is operating safely and efficiently.

System Faults

Many of the common faults which give problems on central heating installations are simple faults on electrical components. These are dealt with in the last two chapters.

Faults which may occur on the heating system itself may be due to:

- excessive gas formation in the system
- formation of sludge or bacterial growths
- water leakage
- design or installation faults.

The first three causes are related to corrosion which is associated with composition of the water, the amount of dissolved oxygen and the chloride content. Generally the soft waters which are neutral or acidic, are the more corrosive or 'aggressive'. Corrosion is usually accelerated by the presence of dissolved oxygen or by a high chloride content, resulting from chlorination of the water.

The effect of these substances on metals in a central heating system is as follows.

Cast Iron

This is generally satisfactory with most waters. Acidic waters can cause large growths or 'tubercules' to form on the surface with deep pitting under them.

Steel

Both bare and galvanised mild steel are satisfactory in hard waters but have little resistance to soft, corrosive waters. The protection of the zinc is satisfactory at low temperatures even when there are breaks in the coating. At temperatures above 65° C, however, the zinc becomes more positive than the iron and corrosion is accelerated.

Stainless Steel

This has a high corrosion resistance due to the continual formation of an invisible oxide film on its surface. When the film breaks down it is reformed by contact with oxygen. If access to oxygen is limited, as in tiny cracks or crevices on the surface, the film will not be reformed and the steel will corrode. This occurrence may take place on riveted or bolted seams, on bolts and washers and under debris or oil or grease films. The steel is also attacked by chloride which can cause pitting and possible cracking.

Copper

Copper is generally more resistant to corrosion than iron or steel. It can, however, be dissolved by some acidic waters. While copper itself rarely fails, the dissolved copper in the water can accelerate the corrosion of iron and aluminium.

Brass

Brasses and bronze have a high resistance to corrosion generally. However, in some soft waters the brass is subject to 'dezincification'. This is caused by the brass being dissolved and the copper being reprecipitated, without the zinc. In this form, the copper is spongy, porous and mechanically weak. Fittings to which this occurs generally leak and may disintegrate. When dezincification is taking place, patches of the pinkish copper may be seen on the yellow brass. Arsenical alpha 70/30 brass is more resistant to dezincification.

Aluminium

This is liable to rapid pitting in waters which contain dissolved oxygen, calcium, bicarbonate and chloride ions and copper.

Gas and Sludge Formation

Excessive amounts of air or gases in a system can cause 'air locks' which prevent circulation either in radiators or in whole circuits. This results in the radiators being cold. It can cause noise if the air is trapped in a pump. The remedy is to remove the air at the air release valve.

Care must be taken because, although it may only be air which has been sucked into the system, it could also be hydrogen, formed by the corrosion process. This is, of course, highly inflammable and could easily cause damage. The hydrogen is formed when the reaction of the iron and water produces ferrous hydroxide (Fe $(OH)_2$) and hydrogen (H_2). Then the ferrous hydroxide is slowly converted to magnetite (FE_3O_4) and more hydrogen. The equation is:

$$3 \text{ Fe } (OH)_2 \rightarrow Fe_3 \text{ } O_4 \downarrow + 2H_2O + H_2 \uparrow$$

The magnetite forms a black sludge which settles in the low points of the system and can block pumps and pipework. Pumps have narrow clearances and can easily become jammed. They are generally fitted with a knob or a screw for freeing a jammed rotor. Any other blockages may need to be flushed out of the system.

Bacterial Growths

Bacteria can grow in the water of central heating systems when corrosion inhibitors have been added. The growths can clog the ball valve or the outlet in the feed and expansion cistern. They can impede the water circulation in radiators or pipework and may jam the pump.

Growths can be prevented by the addition of a 'biocide' to the water, when the system is filled, to kill the microorganisms. It is not a good idea to add a biocide to a system which is already infested. In addition, the feed and expansion cistern should be covered to keep out dust, dirt and animal droppings which contain microorganisms and can cause the trouble. Cisterns and covers should be opaque to discourage the growth of algae, which needs light for its development. They should be made of plastic or metal and cellulose-based materials must not be used. Epoxy resin paints should only be used.

Corrosion Inhibitors

Inhibitors are produced to reduce the rate of corrosion, they cannot stop it completely. Recommended inhibitors contain the following substances:

- sodium benzoate
- sodium nitrite
- sodium dodecamolybdophosphate
- benzo triazole
- dichlorophene.

Of these the first four are 'anodic' inhibitors, they slow down the passage of metal ions from an anode. The fifth is a general biocide.

Inhibitors are often in liquid form in 4.5 litre bottles. A domestic system requires two or three bottles. The liquid should be added in two stages. Two-thirds should be added at the feed and expansion cistern as the system is being filled. The remaining one-third should be added when the cistern is nearly full of water.

Water Leakage

This may be due to corrosion, mechanical failure of pipes, joints or packing glands. Failure of soldered joints may be due to:

- use of corrosive fluxes
- soldering temperatures too high
- unsatisfactory preparation of the jointing surfaces.

Design or Installation Faults

A system which has been designed and installed by a reputable and competent firm is unlikely to be faulty. Unfortunately not all systems are faultless and there may be a need to check the adequacy of the boiler, heat emitters and circulating pipework. Any investigation needs to be based on a calculation of the actual heat requirements and the size of equipment to supply them.

In central heating, as in life generally, you get what you pay for.

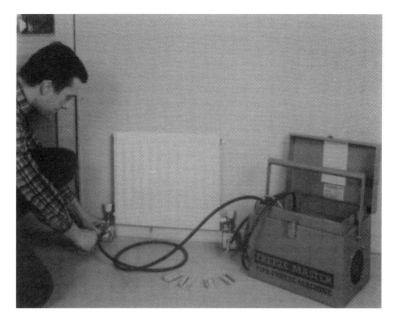

Fig. 60(a)   Portable pipe freezing equipment (Freeze Master): Exchanging a radiator and valves

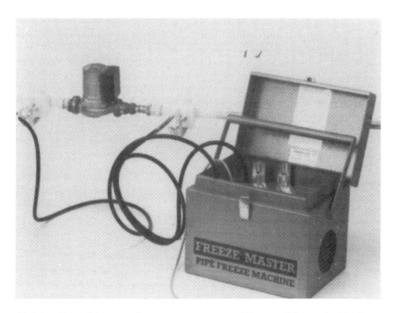

Fig. 60(b)   Portable pipe freezing equipment (Freeze Master): Exchanging a pump

Portable Pipe Freezing Equipment

Portable pipe freezing equipment makes it possible for water leaks to be repaired or parts to be replaced on heating or hot water systems without the need to drain down. It provides a quick and easy means of exchanging pumps, radiators or zone valves when isolating valves have not been fitted. It can save a considerable amount of time by using the equipment rather than searching for a stop tap. In addition it very much reduces the loss of water and/or inhibitor. The equipment freezes the water in the pipe on either side of the component to be disconnected.

Earlier pipe freezing equipment used carbon dioxide (CO_2) or in some instances Chlorofluorocarbons (CFCs). CO_2 was injected into jackets that were wrapped around the pipes to create the necessary ice plugs to stop the flow of water. The more recently developed equipment, illustrated in Figs 60 (a) and (b), uses mains electricity and a recently developed chlorine-free refrigerant gas from I.C.I., KLEA 134a.

This recently developed electrical equipment has the advantage of being less bulky than the earlier equipment which required a cylinder of CO_2 or CFC. It is also less likely to fail as the equipment is plugged into the electricity supply. With the earlier equipment it was difficult to tell when the gas cylinder was exhausted. Another big advantage of the Freeze Master equipment is the fact that it is environmentally friendly. The carbon-free refrigerant gas is enclosed within the unit and not allowed to escape into the atmosphere during the freezing process.

When using the Freeze Master equipment to stop the flow of water it is vital to ensure:

- the system is cold
- there is no flow of water
- the pipe(s) to which the freezeheads are connected is clean
- the freezeheads are the correct size for the pipe being frozen and secured with the clamps supplied
- where there is only one freeze point both freezeheads are attached at the point to be frozen
- a clean dry cloth is wrapped around the freezeheads to provide greater insulation from ambient temperatures and to prevent anyone from touching the freezehead which could cause frostbite
- the machine is plugged in to a 13 amp wall socket and switched to position 1 until the ice plug(s) is formed

- when the ice plug(s) has formed and no water can flow, the machine must be switched to position 2*
- the machine is positioned to blow away from the pipe being frozen.

The times taken to freeze will vary according to pipe diameter. Approximate freeze times for normal room temperatures and cold water pipes are:

8 and 10 mm – 8 to 10 minutes
15 mm – 15 to 20 minutes
22 mm – 20 to 25 minutes
35 mm – 30 to 35 minutes

using both freezeheads at one point will shorten the freeze time by approximately 50%.

---

* It should be noted that where the system contains anti-freeze the machine may be left in position 1 for the duration of the working time, to prevent the ice thawing while the machine is in the off-cycle.

Central Heating – Warm Air Systems

Chapter 11 is based on an original draft by B. J. Whitehead

Introduction

Warm air central heating is provided by an appliance in which air is forced over a heat exchanger by a fan and conveyed through ducts to the individual rooms. It has been popular in the USA and Canada for many years and began to be used in Britain for domestic heating in the early 1960s.

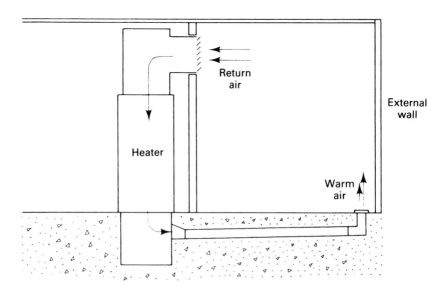

Fig. 1   Warm air system, perimeter heating

The systems may be designed to provide:

- whole house heating
- partial or selective heating.

A full heating system should produce even temperatures across the rooms by a gentle circulation of heated air. This is best achieved by introducing the air at points from which it can blanket the walls and windows which are the main source of heat loss. Warm air outlets at floor level, situated below windows or close to outside walls, are known as 'perimeter heating' and generally give the highest standard of comfort, Fig. 1.

Mounting the outlet registers on the internal walls at low level and blowing air towards the outer walls can produce good results. The cost of installation can be reduced by these 'low side wall registers' but the system design is more critical and they have some limitations. High level registers can be effective but their application is even more critical.

The air is returned to the heater by relief air openings over the doors in some of the rooms and then through return air grilles which are connected by ducts to the heater, Fig. 2. The return air grilles are often situated in a hallway which serves to collect the returning air.

When it reaches the heater, the air is filtered, its temperature is raised by about 50 deg C and it is forced back into the distributing ductwork again. So the air is generally recirculated and is supplemented with fresh air from the hall.

Selective heating is by means of short ducts and only three or four outlet registers. These are situated usually in the two main rooms and in the hall or landing. By opening and closing the registers the space to be heated may be selected. The whole output of the heater could be used for heating the lounge or main bedroom or provide background heating to the other spaces. Heating units have been used with outputs from 5 to 9 kW.

Larger heaters are available to cover all domestic and many commercial requirements.

The other main points about warm air heating may be summarised as follows:

- the system is easily controlled automatically
- it heats up quickly and has a fast response to changes in demand
- a gas circulator may be incorporated and use a common flue
- during hot weather some heaters can circulate unheated air
- a controlled amount of fresh air from outside can be introduced to the heater to provide ventilation by warmed air
- humidifiers and electrostatic filters may be added to the system
- some heaters can be used for air conditioning with the addition of extra equipment for cooling and humidity control.

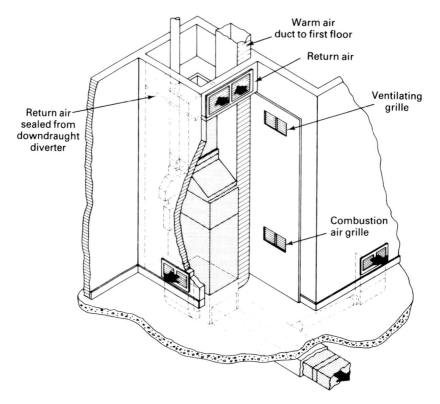

Fig. 2 Heater installation with low side wall registers and return air from hall

Warm Air Heaters

Warm air heaters or 'units' as they may be called, are made in a variety of types. They may be either:

- directly heated or
- indirectly heated

and are manufactured to BS 5258 Part 4: Specification for fanned-circulation ducted-air heaters, and Part 9: Specification for combined appliances: fanned-circulation ducted-air heaters/circulators.

Direct Heating (Fig. 3)

The majority of heaters are of this type. They have a gas-to-air heat exchanger, often consisting of tubes between two large manifolds. The lower contains the gas burner and the upper is connected to the flue. Air is blown by the fan around the heat exchanger, so becoming heated.

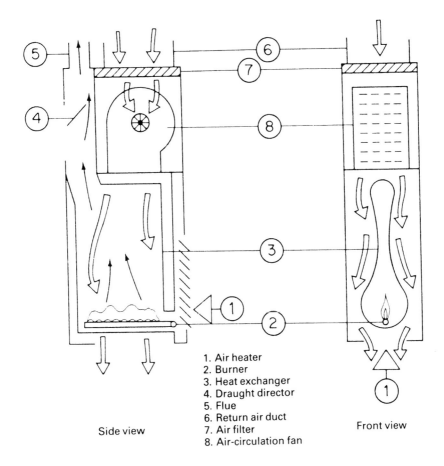

1. Air heater
2. Burner
3. Heat exchanger
4. Draught director
5. Flue
6. Return air duct
7. Air filter
8. Air-circulation fan

Side view

Front view

Fig. 3   Direct gas fired warm air heater

Indirect Heating (Fig. 4)

In this type, air is heated by a water-to-air heat exchanger, similar to a car radiator. The water is heated by an ordinary central heating boiler which can also be used to provide domestic hot water, Fig. 4.

Heaters are usually described by the direction of the air flow through them. They may be:

- upflow
- downflow

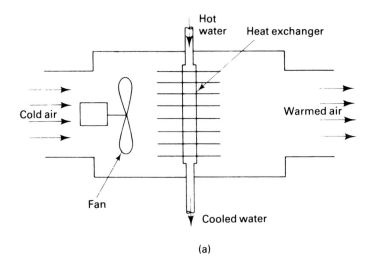

(a)

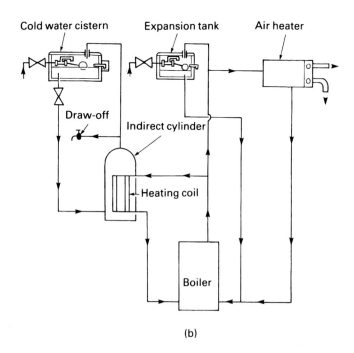

(b)

Fig. 4   Indirect heater: (a) section through heater; (b) heater connected to system

- horizontal
- basement (cellar).

Upflow Heaters

A typical heater is shown in Fig. 5(a). The warm air leaves the heater at the top and cooled air is returned to the bottom. This type is best located in a basement (cellar) and so it is less frequently installed in this country than the downflow heater. The construction of the heater is, however, similar to that of other types of heater, Fig. 5(b).

It consists of:

1. Heat exchanger, curved to give maximum heat transfer and reduce stress which can cause noise when heating up.
2. Draught diverter outlet.
3. Centrifugal fan or 'blower'. Air is drawn into the centre and thrown outwards, through the blades, into the casing. The pressure developed is from 0.5 to 2 mbar at the outlet of the heater.
4. Hammock-type glass fibre air filter, designed to present a large area to the incoming return air.
5. Cover, housing the automatic fan control and high and low limit temperature controls.
6. Multifunctional control with flame supervision device, regulator and solenoid valve.
7. Thermocouple and pilot.
8. Pre-aerated burners.
9. Electric motor driving the fan by means of pulleys and vee belt. The securing bolts are in slots so that the motor can be moved to adjust the belt tension to the 25 mm play required. Many heaters have motors directly driving the fan shaft.
10. Glass fibre and aluminium foil insulation.
11. Flue outlet.
12. Spigot for warm air outlet duct.
13. Return air connection in the base.

The natural buoyancy of the heated air assists its movement through the heater and gives an apparent increased velocity above that of the fan performance.

Down Flow Heaters

Also known as 'counterflow' heaters, these are the types most commonly used in Britain. They have the fan mounted above the heat exchanger, blowing the warm air downwards, or counter to its natural

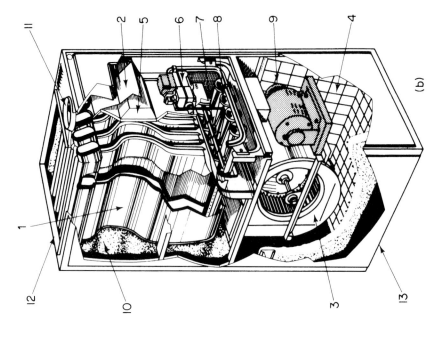

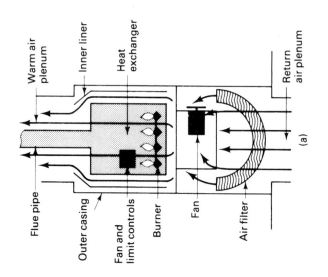

Warm air plenum

Inner liner

Heat exchanger

Return air plenum

Flue pipe

Outer casing

Fan and limit controls

Burner

Fan

Air filter

(a)

Fig. 5   Upflow heater: (a) heater; (b) construction

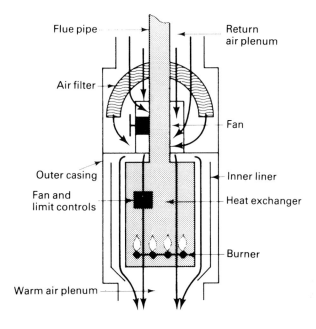

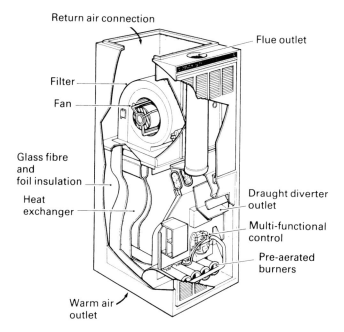

Fig. 6   Downflow heater: (a) heater; (b) construction

tendency to rise, Fig. 6. The heater is connected at floor level to a plenum box which collects the warm air and passes it to the ducts. The extra energy required to move the air downwards reduces the margin for error in duct sizing compared to an upflow heater.

Horizontal Flow Heaters

The horizontal heater has its fan and heat exchanger side-by-side, with the air discharged horizontally, Fig. 7. It is also called a 'cabinet' heater and may form part of a kitchen unit, being fitted beside the wall cupboards. It can also be installed beneath working tops. Examples of installations are shown in Fig. 8.

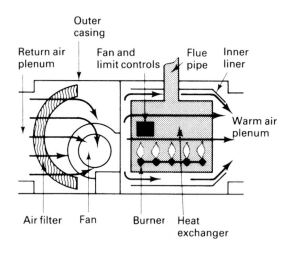

Fig. 7   Horizontal flow heater

'Basement' Heater

This is similar to the horizontal flow heater but it has its warm air outlet and return air inlet at the top, Fig. 9. It is particularly suitable for installation in basements, as its name implies.

Installation of Warm Air Heaters

The heater should be installed to BS 5864 : 1989. Installation in domestic premises of gas-fired ducted-air heaters of rated input not exceeding 60 kW.

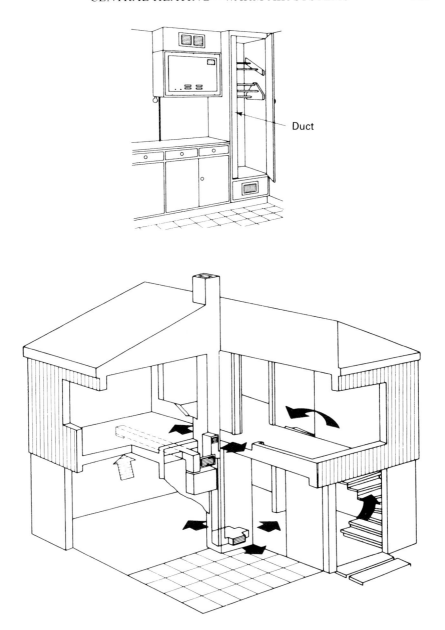

Duct

Fig. 8   Installation of horizontal flow heater

Location of Heater

One of the main considerations when deciding on a suitable location is the provision of a satisfactory flue. If a water heater is incorporated

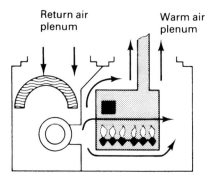

Return air plenum

Warm air plenum

Fig. 9   Basement heater

with the air heater then the route of the flow and return pipes to the hot water storage cylinder must also be considered.

Heaters should be sited so that the warm air and return air ducts are as short and direct as possible. Adequate space must be provided, in accordance with the manufacturer's instructions, for operating and servicing the appliance.

Care must be taken to seal the return air duct system to prevent vapours or products of combustion being drawn into the heater. In buildings of more than two storeys, special requirements given in Section 8 of BS 5864 : 1989 will apply.

Products of combustion must never be permitted to escape into a bathroom or bedroom. So open-flued heaters must not be located in bathrooms or bedrooms or compartments opening off these rooms. In exceptional circumstances room-sealed heaters may be used.

Care must be taken to ensure that the sound level from the heater does not cause annoyance. Heaters should not generally be located in living rooms or bedrooms because of the lower noise level expected in these rooms.

The final points to consider are the provision of gas and electricity supplies. The gas supply to an air heater with an integral water heater should be sized to allow for both appliances operating at the same time. The electric supply requires a switched and fused spur or socket outlet. Fuses should be to the maker's specification.

Installation of Heaters

Heaters may be either wall or floor fixing. When a heater is fitted to a wall the maker's recommendations with regard to the methods of securing the heater and the proximity of other materials must be followed. The wall should preferably be of solid construction and not a light studding partition. Where the latter cannot be avoided, it will be necessary to provide local protection against vibration.

When floor-standing, heaters should be mounted on a level, fire-proof base of concrete or similar material. Where necessary, the adjacent surroundings should be insulated to reduce noise. If the heater is on a suspended wood floor, resilient mountings or a padded concrete slab should be used. Some manufacturers provide a special base for suspended floor installations. If heaters burning 1st and 2nd family gases are installed in cellars or low-lying situations they must be protected against water seepage or flooding. Heaters burning 3rd family gases must not be installed in cellars or low lying areas.

All heaters should be protected from weather or excessive draughts but should have adequate ventilation in accordance with the recommendations in Chapter 5.

Heaters may be fitted in specially designed compartments. Such compartments should meet the requirement laid down in Section 9 of BS 5864 : 1989.

- they should have a half-hour resistance to an internal fire; the inner lining should be non-combustible, BS 476, Class 1 finish
- the door must have at least the fire resistance of the enclosure and should be a good fit
- compartments should have good sound insulating properties and preferably be made of brick or clinker block, plastered on at least one side
- they must be large enough to allow access for operating and servicing the heater but not be so large that they can be used as storage cupboards
- the door must be big enough for the heater to be removed
- compartments must be ventilated as recommended in Chapter 5, Tables 25 and 26
- where the compartment houses an open-flued heater and the door communicates with a bedroom or bedsitting room, the door must be self-closing and draughtproofed. It should also have a warning notice attached, stating that: 'Except when resetting the appliance controls, this door must be fully closed. The compartment must not be used for storage'.

When an air heater is fitted in an existing compartment which was not designed specifically for that purpose, then the compartment must be checked to ensure that it complies with the requirements listed, or modified as necessary.

'Storey-height' heaters are made as ceiling-height cabinets containing the heater, controls, flue and rising duct. They are only 200 to 260 mm in width and may be fitted in a recess or in the angle of two walls, Fig. 10. When fitted in a recess they are called 'slot-fix' units.

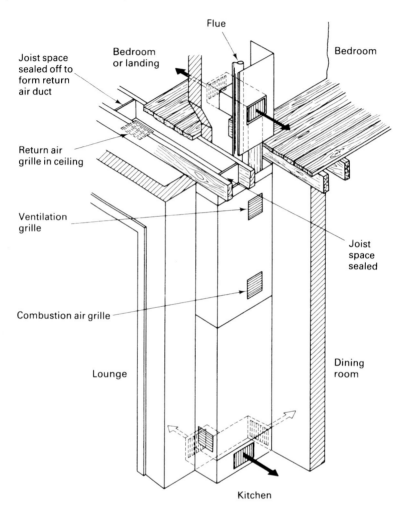

Fig. 10   Storey-height heater with stub ducting

Duct Systems

The duct system starts from the 'plenum' which serves to equalise the air pressure to the ducts. The connections should be taken from the sides of the plenum box and should not directly face the outlet from the heat exchanger. Plenums used with down flow heaters may support the weight of the heater as in Fig. 10.

The ducts which convey the warm air from the heater to the registers or diffusers may be run in a variety of ways. The main systems are: ● stub duct; ● radial duct; ● perimeter loop; ● extended plenum; and ● stepped duct.

Stub Ducts

These are normally used for partial or selective heating systems. They are generally installed in small houses and flats and provide heating for a living room and one or two other rooms. General background heating is obtained from a warm air outlet in the hall with heat spillage to the other rooms. The system may be operated selectively by opening or closing the registers.

Figure 10 shows a storey-height air heater with stub ducts from the plenum box at the base supplying warm air to lounge, kitchen and dining room. Within the heater case a duct is extended from the plenum box to wall registers in the bedroom and the landing. The return air path from the lounge is via a grille in the ceiling to the joist space above. This is lined with heat resisting material and blanked off at each end.

Radial Ducts

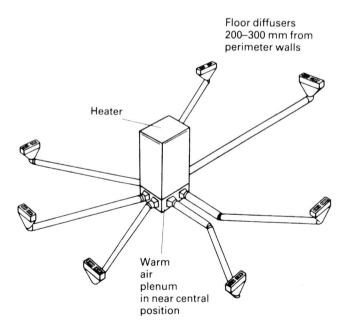

Fig. 11   Radial ducts

This system consists of individual ducts branching directly from the plenum box and run to the perimeter diffusers, Fig. 11. The ducting is often circular and embedded in concrete. The branch ducts must not be longer than 6 mm and should have not more than two bends.

Perimeter Loops

This system is not commonly used in this country. It has, however two main advantages:

- the duct system is easily allowed for in the site concrete
- the outlet pressure of the heater is more equally balanced throughout the system, like the pressures in an arterial mains system.

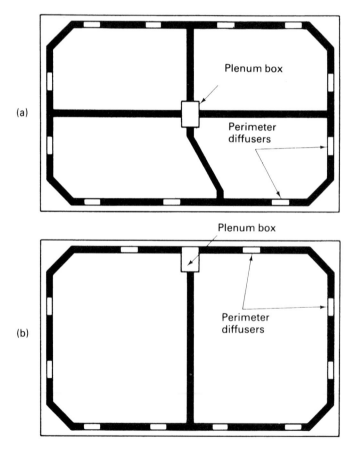

Fig. 12   Perimeter loop: (a) heater at centre; (b) heater on perimeter

Figure 12(a) shows the arrangement. Circular ducting is usual and floor diffusers are branched directly on to the ducts. The builder can leave a duct space around the perimeter and suitable spaces back to the plenum box, making sure that these do not join the perimeter at any proposed diffuser location.

Alternatively, the heater may be located at the perimeter, Fig. 12(b), when it becomes part of the loop.

Extended Plenum System

This is a simplified system in which the main duct is the same size throughout its length, forming an extension to the plenum box. The branches may be either rectangular or circular and are connected to the main duct by take-off fittings which are slanted in the direction of the air flow, Fig. 13.

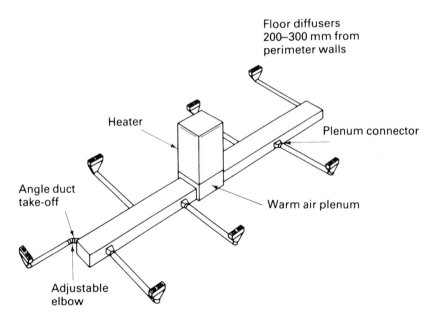

Fig. 13   Extended plenum

Main ducts should not be longer than 6 m and branch ducts should not be longer than 6 m and have no more than two bends.

Stepped Ducts

This system is shown in Fig. 14 and is frequently used when the heater is located on an outside wall. It consists of main rectangular ducts with either rectangular or circular branches. The cross-sectional area of the duct is kept roughly proportional to the volume of air passing through and the main duct is reduced as branches are taken off. The air velocity should not exceed 4 m/s.

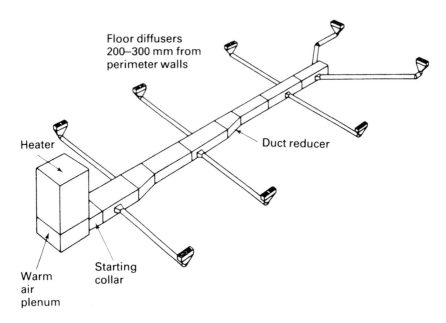

Fig. 14   Stepped ducts

   Some systems may be more complex than that illustrated, with two or more main ducts taken from either the plenum box or from another main duct.

General

Duct systems should be as short and as simple as possible. They should use the least number of fittings and those used should have a low resistance to air flow. Reductions in area should be tapered and take-offs should follow the direction of air flow. Branches must be taken from straight sections of ducting and never from the end of a duct.

Although ideally the system should be designed so that all the ducts have the same resistance to air flow, in practice this is not always possible. In order to maintain balanced pressures throughout the system, dampers are usually fitted. Balancing dampers may be positioned in the ducts or immediately behind the warm air registers. When set, (see Testing and Commissioning) they are usually locked in position.

Register and Diffuser Installation

Warm air terminal points are fitted with one or other of the following devices:

- grilles
- registers
- diffusers.

Grilles

Grilles have openings which cannot be closed. They are usually louvred and, although they were used on some of the early systems, they are now generally only used for ventilators or return air openings.

Registers

Registers have a bank of vanes behind the louvres, operated by a small lever to open or shut the outlet, Fig. 15. The vertical louvres are set at the angle required for air distribution. The term is now generally applied only to terminal devices which are located in walls. They are available in sizes from 150 × 100 mm to 300 × 200 mm.

The recommended air velocities through low side wall registers are 1 to 1.8 m/s. Earlier systems may have been designed for velocities up to 2 m/s.

If the register serves as the balancing adjustment, a screw is fitted in the slot below the lever to limit the extent to which the register may be opened. If the register will not be required to close, a removable handle, Fig. 16, may be used. Balancing is best achieved by a friction damper located in the stackhead, Fig. 18.

Low side wall registers on perimeter heating systems should be located to throw an air curtain across the outside walls. They should be at a distance of at least 150 mm from the outside wall to prevent staining or curtain movement and a minimum of 75 mm from the floor, Fig. 17.

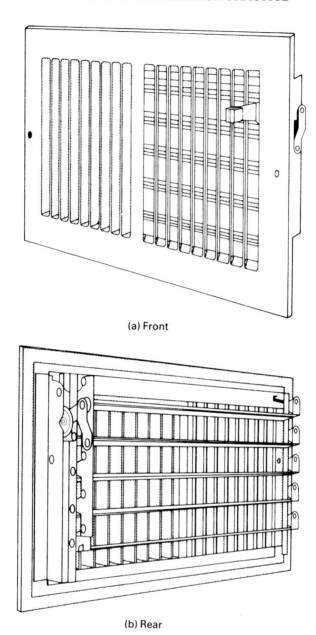

(a) Front

(b) Rear

Fig. 15   Register: (a) front; (b) rear

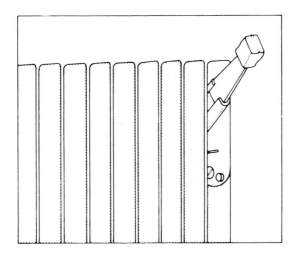

Fig. 16   Balancing adjustment

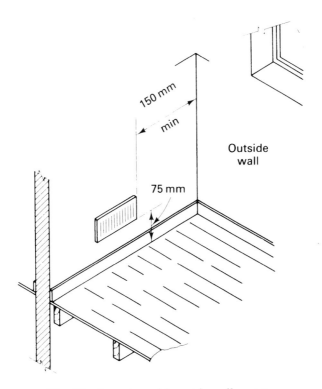

Fig. 17   Location of low side wall register

Details of the method of connecting the register to the duct are shown in Fig. 18. Wall registers should generally not be fitted on outside walls. If this cannot be avoided the back of the stackhead must be insulated.

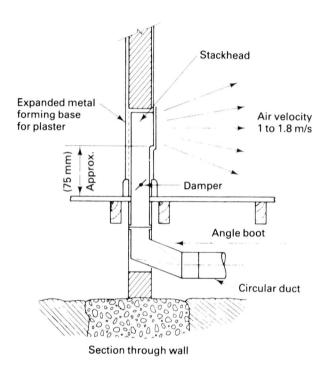

Section through wall

Fig. 18   Installation of wall register

Registers must be adequately sealed to the stackhead or the wall surface to avoid staining of the wall by warm air leakage. This is done by fixing a foam rubber seal around the inner face of the register flange, Fig. 15(b).

High side wall registers give a less satisfactory performance and their use is limited. The top of the register must be at least 100 mm from the ceiling and the bottom a minimum of 2 m from the floor. Recommended air velocities are 1.5 to 2 m/s.

All registers should emit air at an angle slightly below horizontal. High level registers should be set at about 15° below.

If registers are mounted on inside walls, facing the cold window walls, they should give adequate air distribution in rooms up to about 4.25 m in width.

Diffusers

Diffusers are similar to registers but are longer and narrower, Figs. 19 and 20. They are mounted in floors or ceilings and are available in sizes from 57 × 250 mm to 100 × 300 mm. The recommended air velocities are 1.5 to 3 m/s for floor diffusers and 2 to 4 m/s for ceiling diffusers.

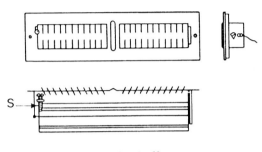

Fig. 19   Diffuser

Most diffusers incorporate a closing device which can be used for balancing but it is recommended that a separate damper should be fitted. The screw at S in Fig. 19 can be used to limit the degree to which the vanes may be opened if the diffuser vanes are used for balancing. The vanes are operated by the serrated quadrant shown at A in Fig. 20. This can be turned by a foot.

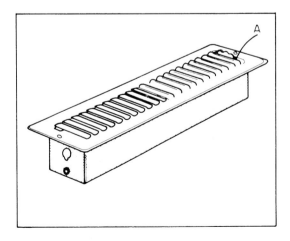

Fig. 20   Floor diffuser

Floor diffusers are usually located under windows so that they provide a curtain of warm air across the window and the exposed wall. More than one diffuser is used if the window areas are large or if there are windows on more than one wall. They should be positioned 200 to 300 mm from the wall to allow full-length curtains to be drawn without interfering with the flow of warm air. Where more than one diffuser is used they should be at least 1.8 m apart.

If a diffuser is positioned within 1 m of a side wall, the louvres should be adjusted so that the air will not impinge on the side wall and cause staining.

Details of the method of installation are shown in Figs. 21 and 22. If a balancing damper is fitted it would be sited in a boot extension between the diffuser and the universal boot.

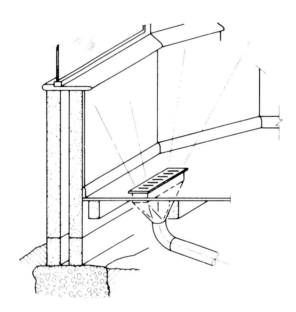

Fig. 21   Location of floor diffuser

A ceiling diffuser is shown in Fig. 23. This is similar to a floor diffuser except for a sealing strip between the flange and the ceiling to prevent air leakage and a lever to operate the vanes. The lever may be fitted with cords to bring it within reach.

The high velocity needed to get air down to near floor level can make these diffusers noisy. The downward flow of air is helped if the louvres are adjusted to direct the air straight down.

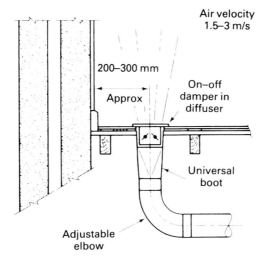

Fig. 22   Installation of floor diffuser

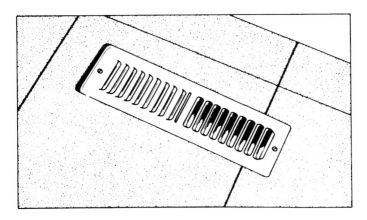

Fig. 23   Ceiling diffuser

Duct Installation

The essential requirements of a duct system are:

- ducting should be fire-resistant
- it should have a very low heat capacity
- the ducts and components should be easy to transport and assemble
- ducts should have only a low frictional resistance to the air flow, fan pressures being rarely much more than 1 mbar

- ducts in spaces where heat is not required should be insulated to have a minimum thermal resistance of 0.625 m² °C/W
- insulation must be non-combustible
- insulated ducting within 2 m of the heater must be stable up to 120° C
- ducting must be resistant to the entry of water
- ducts and components should be corrosion resistant
- ducts must have adequate strength to retain their shape.

No one single material meets all the above requirements and duct installations are frequently made from galvanised sheet steel duct-work, suitably insulated and waterproofed.

Ducts in Concrete Floors

Close co-operation with the architect and builder is essential to ensure good drainage, insulation and care of ductwork. Where channels are to be provided in the floor structure, the builder must be supplied with adequate drawings.

The site for the building must have good drainage and not hold surface water. Ground that does hold surface water has high capillary attraction and will hold water on the underside of the site concrete. Even though the ducts may be well sealed from moisture and also insulated, this can lead to high running costs. The top screed should preferably be 150 to 200 mm above outside ground level at least, to prevent water being held in the hardcore.

On completion of the foundation walls up to damp course level, the site hardcore should be excavated and replaced with an underslab fill of 100 mm of coarse gravel. This will prevent water being drawn to the underside of the concrete slab by capillary attraction. If the site hardcore is already of a uniform grade, similar to gravel, there is no need to replace it. It must not, however, contain any ash, sand, rock with fines or rock which can disintegrate into fines.

A suitable moisture or 'vapour' barrier, such as a thick polythene sheet, must be positioned between the underslab fill and the site concrete. This keeps the concrete dry and presents a smooth face to the fill, so reducing heat transfer by keeping the area of contact to a minimum. Overlapping joints in the barrier should be at least 100 mm wide and be sealed to prevent capillary attraction. The vapour barrier must be carried up the inside of the foundation wall and overlapped on to the damp proof course, Fig. 24.

A considerable amount of heat is lost at the perimeter of a solid floor where the site concrete is in contact with the inner foundation wall, it is therefore usual to insulate the edge of the concrete with

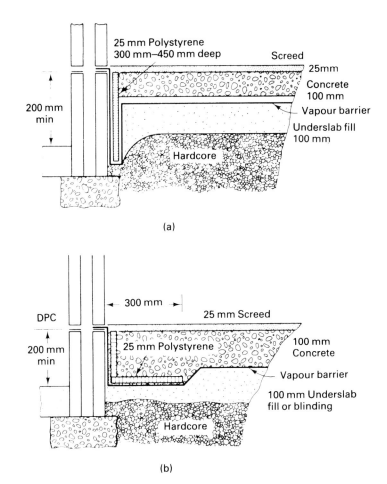

Fig. 24   Insulating perimeter of solid floor: (a) normal method; (b) for loop ducting

polystyrene slabs, 25 mm thick and about 300 to 450 mm deep, Fig. 24(a). If a loop system of ducting is installed, the arrangement shown in Fig. 24(b) is preferable.

Figure 25 shows a duct laid in concrete. The use of a vapour barrier around the duct itself is not theoretically necessary provided that:

- there is a reasonable thickness of concrete below the duct, about 64 mm
- there is a satisfactory vapour barrier below the concrete
- the duct insulation is waterproof.

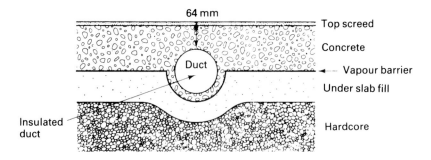

Fig. 25   Duct laid in concrete

Because ducts have a tendency to float when the site concrete is poured, it is necessary to anchor them down, Fig. 26. The duct is wired, at intervals, to bricks or pegs driven into the hardcore. Alternatively, the duct may be covered with a quantity of dry-mix concrete or 'haunched', before the main concrete is applied. The site concrete should be a very wet mix to avoid the need to use shovels around the duct.

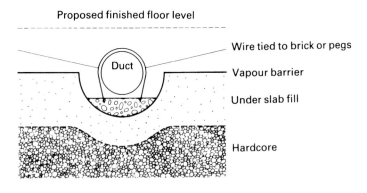

Fig. 26   Method of anchoring ducting

Figures 27 and 28 show ducts which have been installed in pre-formed channels. The ducts are inserted and covered by concrete slabs or tiles before the floor screed is laid.

A simplified method is shown in Fig. 29, where the duct is laid with half its depth in the underslab fill or hardcore and directly on to the site vapour barrier. A similar installation of a rectangular duct is illustrated in Fig. 30. The top of the duct is protected by corrugated iron, reinforcing rods or expanded metal.

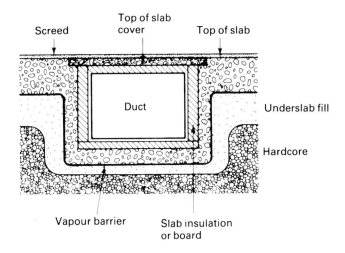

Fig. 27   Rectangular duct in preformed channel

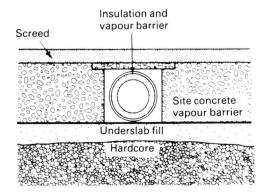

Fig. 28   Circular duct in preformed channel

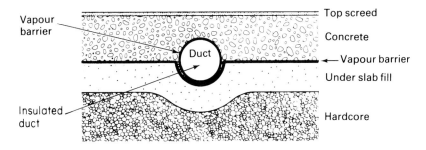

Fig. 29   Circular duct halfway in hardcore and concrete

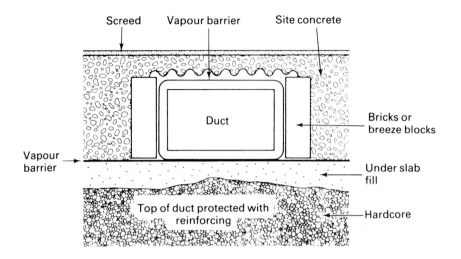

Fig. 30   Rectangular duct in site concrete

Ducts in site concrete are usually insulated with 50 mm of glass fibre or 30 mm of sprayed urea formaldehyde or polyurethane. The latter is waterproof so that only the joints need to be sealed. Glass fibre is usually resin bonded so that it is not easily crushed and the insulated duct is wrapped in a polythene sheet. Care must be taken to seal all the joints in the polythene.

The warm air outlets or 'boots' for floor diffusers must be positioned before the concrete is poured. They must be secured against movement out of position either to the side or downward, as concrete is poured round them. The boots should be protected by wooden blocks to prevent the entry of concrete and possible distortion of the opening. The blocks should be well painted to make them waterproof. Stackheads must also be correctly located and covered before concrete is poured.

The plenum box should be installed so that it will project about 25 mm above the top of the screed. This allows the top edge to be folded outwards into an angle frame to receive the base of the heater. The box must be positioned accurately, particularly if the heater is located in a cupboard where there is little margin for error. The top of the box must be capped to prevent the entry of debris or vermin. The angle frame is fitted when the heater is installed.

Builders' workmen must be made aware of the need to take care not to puncture the vapour barrier or move any duct outlets from their position. Attention must be drawn to avoiding treading on lightweight ducting or using ducts to support planks for wheelbarrows. It must be emphasized that any flattening or distortion of ducting must· be reported immediately, while the fault can be remedied. If possible, someone should be on site to observe when the concrete is being laid.

Ducts under Suspended Wood Floors

Where joists are used as duct routes, they must be suitably lined with fire-resisting material. Ducts below suspended floors should be supported at about 1 m intervals by metal straps secured to the joists. The ducts should be insulated with glass fibre, 50 mm thick. Polyurethene should not be used without the agreement of the local fire officer.

The insulated duct should be wrapped with a polythene vapour barrier. Particular care should be taken when the duct is in contact with the site concrete to ensure that it is fully waterproofed. Figure 31 shows the duct installation.

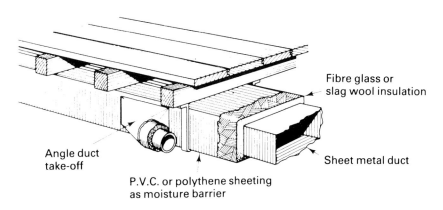

Fibre glass or slag wool insulation

Angle duct take-off

Sheet metal duct

P.V.C. or polythene sheeting as moisture barrier

Fig. 31   Duct installation below suspended wood floor

Plenum boxes installed into suspended floors must be insulated from contact with any part of the floorboards or joists. The heater can be mounted on a purpose-made base which fits on top of the plenum box so that the base and the bottom side of the heater are insulated from the floorboards. The design of these bases varies from one manufacturer to another. Figure 32 shows a heater standing on a plenum box which is mounted on a concrete plinth. Details of the installation are shown in Fig. 33.

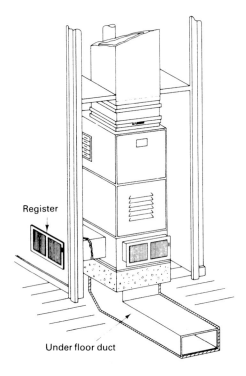

Fig. 32   Heater fitted on plenum mounted on concrete plinth

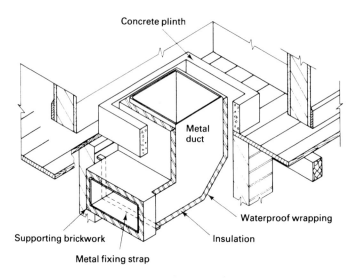

Fig. 33   Detail of underfloor adaptor installation

Return Air

The warm air from the registers and diffusers must have an unrestricted return path to the heater. The output of warm air is equal to the volume of air returned. So restriction on the return path will result in a restricted output and possible overheating and cut-off by operation of the overheat device.

The return air grilles and relief air openings should be capable of passing a volume of air equal to that supplied to the room. The air velocity through the grille should not exceed 2 m/s.

Grilles are available in sizes from 200 × 100 mm to 750 × 300 mm. Relief air openings should have a free area of 8800 mm^2 for each kW of heat input to the room. The free area of a grille is usually 75% of the louvred area.

The return air path must be provided from all heated rooms with the exception of:

- bathrooms
- kitchens
- toilets.

It is usual to site the return air grilles in the main collecting areas, i.e. the hallway or landing and convey the air to the heater by a common duct. The air from the rooms is returned to the collecting area through relief air openings, usually located over the doors, Fig. 34. The openings must be positioned so that accidental blockage is not likely, undercutting a door is unacceptable.

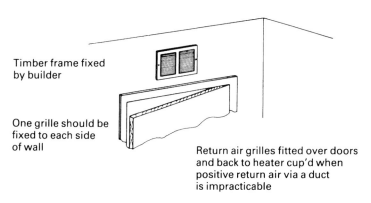

Fig. 34   Return air grilles over door

Large rooms should have their own individual return air grilles, situated in the ceiling and ducted to the main return air duct. Large houses require a separate return air grille on each floor. A typical return air system is shown in Fig. 35.

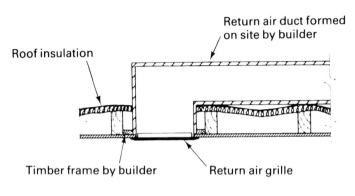

Section through return air duct
in roof space

Fig. 35   Return air duct in roof space

Return air arrangements should assist the circulation of warm air in a room. The return or relief opening should be at the opposite end to the register or diffuser. If warm air is supplied at high level, the return air should be taken from low level and vice versa. Avoid creating draughts by locating low level grilles at least 1 m from areas where people normally sit.

Although a single return air grille immediately above a heater fitted in a hallway is adequate for a small house, it can allow the noise from the fan to become a nuisance. This can be overcome by ducting the return air through the ceiling from a suitable position at a little distance from the heater. There should be at least one bend and the duct may be lined with sound-absorbing material.

The ideal system would be to have a return air grille in the ceiling of each room, ducted back to the main return air duct and with an additional grille in the hall for fresh air, equal in volume to that not returned from the kitchen, bathroom and toilet. Alternatively, with fully ducted systems it is possible to connect a duct to outside air to draw in this fresh air. The amount required is about 10 to 15% of the heater output.

With open-flued heaters the return air must be arranged so that there can be no interference with the operation of the flue by the heater fan. The return air must be ducted directly back on to the heater, Fig. 36. Room-sealed appliances do not require a direct connection, but the return air path must not be capable of being obstructed.

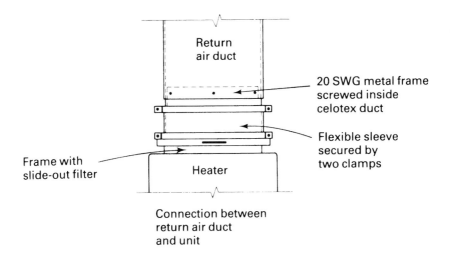

Fig. 36   Return air duct direct connection to heater

Return Air Supply (Diluted)

Where a supply of outside air is induced by the appliance circulating-air fan, a minimum flow of 2 m/h must be drawn into the air chamber for every 1 kW of appliance input rating. Provision must be made for the adjustment of the flow of induced air by means of a lockable damper or other suitable control.

An acceptable method is shown in Fig. 37. It shows a duct from a ventilated roof space providing an air supply to the return plenum. Where this method is employed, reference should be made to the appliance manufacturers for guidance. Where the air supply is taken from the roof space, a bird guard must be fitted to the duct inlet.

Following the report of the Fire Research Organisation, No. FROSI 9514, April 1972, the following recommendations have been adopted to prevent the spread of smoke or fire.

In blocks of flats or maisonettes of three storeys or over, all pressure relief, air transfer and return air grilles should be fitted at a height of not more than 450 mm, measured from the finished floor level to the top of the grille. Return air should be ducted directly to the heater and relief air grilles should not be sited in any wall on an escape route.

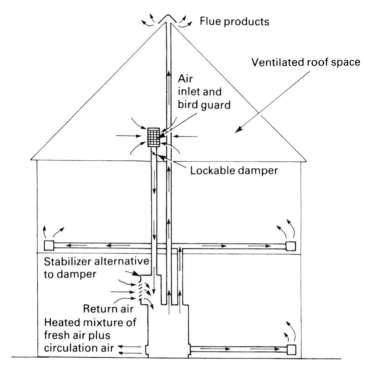

Fig. 37   Air supply from a ventilated roof space

Duct Materials

Duct materials, including jointing and insulation must be non-flammable and dimensionally stable up to an internal air temperature of 120° C. Joints and seams must be securely fastened and then made air tight by applying heat-proof, non-flammable tape, usually 50 mm wide.

Galvanised sheet steel is commonly used for duct work and 18 swg is recommended for commercial installations. Domestic ducting is usually 24 swg. Fully assembled ducting is difficult and expensive to transport so this has resulted in the development of 'snaplock' ducting also known as 'button lock' ducting. The rectangular ducting is delivered in two halves and joined as in Fig. 38. The snaplock joint is pressed together along its whole length, starting at one end and working forward. Using a mallet speeds up the operation.

Circular ducting is supplied in bundles with the lengths nesting inside each other. They are joined along their length by the same snaplock joint, Fig. 39.

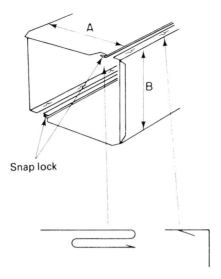

Fig. 38   Snaplock joints in rectangular ducting

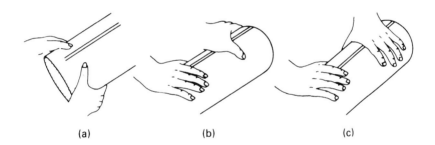

Fig. 39   Snap-lock circular ducts.
(a) Starting from crimp, engage pipe edges. Work forward with a slight jolting movement with left hand closing seam with gripping motion.
(b) Work forward, joining seam as you go. Always keep hands behind where edges first lock.
(c) Be sure edges are solidly joined all the way to the end of the pipe. Release downward pressure and pipe is then snapped in position.

Lengths of rectangular ducting are joined together by S and D cleats, Fig. 40. The S cleat is cut and fitted to the exact width on the top and bottom of the duct lengths. The D cleat is then slid over the two side projections with an overlap of 40 mm at the top and bottom.

The overlaps are bent over and hammered down on top of the S cleats. The joint is made air tight with 50 mm heat resistant and waterproof tape.

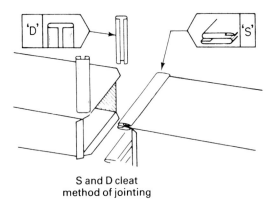

S and D cleat
method of jointing

Fig. 40   S and D cleats

Pre-insulated rectangular duct is delivered fully assembled and can be obtained with spigot and socket joints.

Lengths of circular ducting are crimped at one end and plain at the other. The crimped spigot is wedged tightly into the plain end and the joint is made secure by inserting a self-tapping screw on each side, Fig. 41.

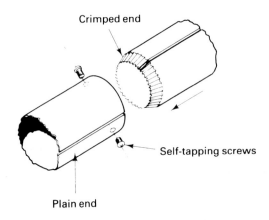

Crimped end

Self-tapping screws

Plain end

Fig. 41   Joint in circular ducting

Ducting is available in 1 m and 1.5 m lengths. Rectangular ducting is in sizes from 150 × 200 mm increasing in increments of 50 mm. Circular ducting is from 100 mm diameter increasing in increments of 25 mm for sizes up to 250 mm and then in 50 mm increments.

Tools

The principal tools required for cutting and jointing sheet metal ducting are shown in Fig. 42.

Monodex Cutter (a)

This is used to cut openings into ducting so that take offs may be fitted. The hole is first marked out with the scratch awl scriber. The marked area is then drilled in one corner with a drill of sufficient size to allow the smaller lower jaw of the cutter to enter. The cutters are operated to remove a narrow strip of metal all round the scribed line.

Tin Snips or Shears

Three types are illustrated

    (b) Bow shears for cutting in straight lines
    (c) and (d) Left and right hand snips for cutting circular openings.

Scratch Awl Scriber (e)

This is simply a hardened steel pin, sharpened at the end and fixed in a plastic handle. It is used, with a steel rule, to mark out metal for cutting.

Hand Crimpers (f)

Squeezed together, as shown, these can be used to crimp the end of a cut length of round duct to produce a spigot in order that it may form a joint with another plain end.

Duct Fittings

A range of duct fittings is shown in Fig. 43 and the way in which these may be used to make up a duct system is illustrated in Fig. 44.

Fittings which have a rectangular end, for connecting into a rectangular duct, are provided with a stop flange and a fixing flange with 'fishlock' edge, Fig. 43(2). The fishlock is a projecting edge which has been cut into small lengths so that it can easily be bent over. An opening is made in the duct through which the fixing flange

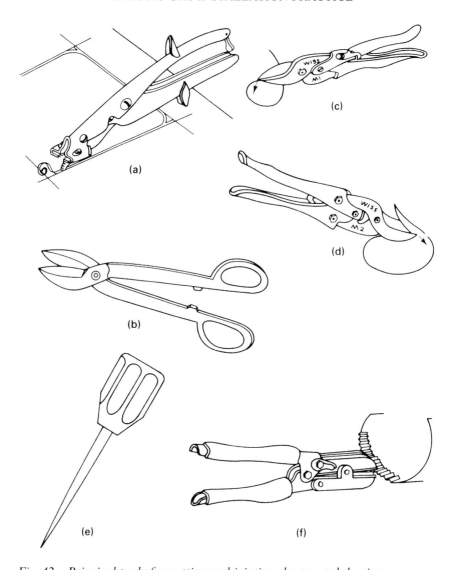

Fig. 42 Principal tools for cutting and jointing sheet metal ducting:
(a) Monodex cutter; (b) Bow shears; (c) Left hand snips;
(d) Right hand snips; (e) Scratch awl scriber; (f) Hand crimpers

can pass so that the stop flange bears against the wall of the duct.
The fishlock edge is then turned over by finger pressure to trap the
wall of the duct against the stop flange. The joint is finally sealed with
tape.

The purpose of the fittings illustrated is as follows:

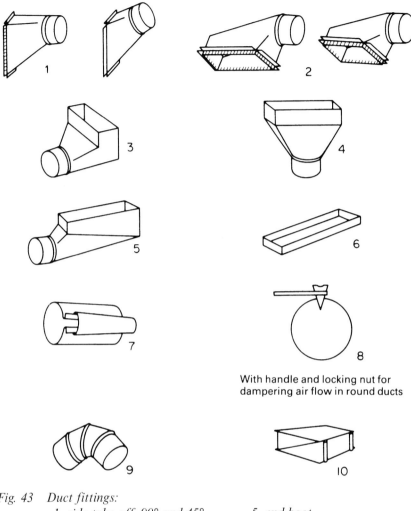

Fig. 43   Duct fittings:
   1. side take off, 90° and 45°   5. end boot
   2. top take off, longway and   6. boot caps
      shortway   7. wedge connections
   3. angle boot   8. circular damper
   4. universal boot   9. adjustable elbow
                                         10. reducer

1. Side Take Off 90° or 45°

To join circular branch ducts to the side of rectangular ducts. The graduated change from rectangular to round section keeps the pressure loss to a minimum. The equivalent length of duct having the same pressure loss is:

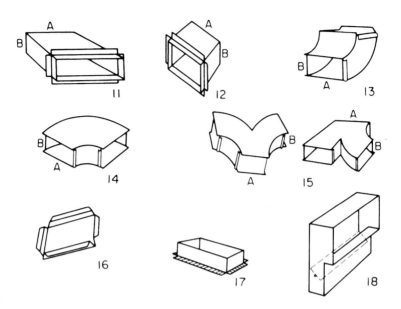

Fig. 43   *(continued)*

11. *side take off*   15. *Y fitting, branch fitting*
12. *starting collar*   16. *rectangular duct end*
13. *flat elbow*   17. *fishlock collar*
14. *side elbow*   18. *stack head*

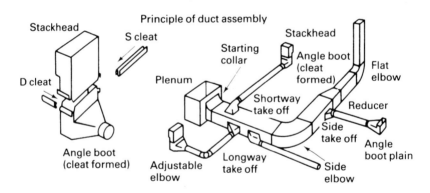

Fig. 44   Fittings assembled into duct system

$$90° = 10 \text{ m}$$
$$45° = \ 7 \text{ m}.$$

2. Top Take Off, Longway or Shortway

Similar to the side take off but used for top or bottom connections. It may be used at the side where the circular duct is to continue in the same direction as the rectangular duct is leaving. This saves using a side take off and an elbow. When connected to the top or bottom of a rectangular duct it offers a means of rising or dropping into a joist run. Equivalent length,

$$\text{Longway} = 8 \text{ m}$$
$$\text{Shortway} = 14 \text{ m.}$$

3. Angle Boot

All boots may be obtained as:

- plain, to accept a diffuser
- cleat formed, for connection to a stackhead.

Angle boots are used at the end of a circular branch and are usually fitted to floor diffusers. Equivalent length, 8 – 12 m.

4. Universal Boot

Used to terminate a vertical duct or, with an elbow, used where an angle boot would not reach the floor surface. Equivalent length, 2 – 6 m.

5. End Boot

Similar to an angle boot but air enters on the short side. It does not give a good air distribution although it is usually adequate for a small room. Equivalent length, 8 – 15 m.

6. Boot cap

To seal boots and ducts against entry of foreign matter on the building site, prior to completion.

7. Wedge Connections

To join two circular fittings or lengths of duct with plain ends. May be used as a make-up piece to adjust for length.

8. Circular Damper

With handle and locknut, for balancing in circular ducts.

9. Adjustable Elbow

Usually made in four pieces which can be adjusted in relation to each other to form angles from 90° to 180°. It can also form offsets. The equivalent length may vary from 4 to 10 m.

10. Reducer

For changing from one duct size to a smaller size, in a straight line. Equivalent length depends on the reduction, usually about 10 m.

11. Side Take off

A fitting providing a graduated connection between two ducts at right angles. Equivalent length, 8 – 15 m.

12. Starting Collar

To join the plenum box to a rectangular duct. The equivalent length is included with that of the plenum and might be 3 m on up-flow and 10 m on downflow.

13. Flat Elbow

To turn a rectangular duct vertically. Equivalent length, 1 to 7 m.

14. Side Elbow

To turn a rectangular duct horizontally. Equivalent length 4 to 6 m (4 m on widths up to 300 m). Although a flat and a side elbow for a square duct appear to be identical, they are not. The difference is that the cleats are on opposite sides, with the D cleat always at the side, or vertical.

15. Y Fitting, Branch Fitting

To divide a rectangular duct into two. The equivalent lengths are 4.5 m on the curved branches and 1.5 m on the straight.

16. Rectangular Duct End

To block the end of a duct.

17. Fishlock Collar

Used to connect registers or diffusers to a rectangular duct or wall registers to a box base plenum. May be rectangular or circular.

18. Stackhead

To connect a boot to a wall register. It includes a balancing damper which can be adjusted when the register is removed. Equivalent height, 8 – 15 m.

The above list of fittings is representative of those in common use. There is, however, a number of angles, elbows, offsets and adaptors which may be employed. In addition, special fittings and ducts may be obtained from sheet metal workshops.

Controls

The controls fitted to warm air heating systems are described in detail in Chapter 12. However, it is necessary to introduce them at this point so that the testing and servicing of the appliance can be dealt with later.

The controls normally included are:

- clock
- room thermostat
- main gas cock
- regulator
- limit thermostat or overheat control
- automatic fan control or fan switch
- gas valve or solenoid
- flame supervision device
- ignition device
- fan speed control.

The location of these devices on the heater is given in Figs. 45 and 46. These show a typical storey-height heater with a close-up view of the control panel.

Of these controls, some were dealt with in Vol. 1, Chapter 11 and the following require some explanation.

Fan Switch

This starts up the fan only when the heat exchanger has warmed up and the air is warm enough for distribution. Then, when the burner has been shut off by the clock or room thermostat, it allows the fan to continue to run until the useful heat has been taken from the heat exchanger and the air is beginning to cool down. The temperatures at which the switch cuts in and cuts out can be adjusted to suit the particular system. The usual temperatures are approximately:

- cut-in at 58° C
- cut-out at 38° C
- differential between the two, 20° C.

Limit Thermostat

Like the fan switch, this is another thermally operated switch. It shuts off the main burner when the warm air temperature becomes too

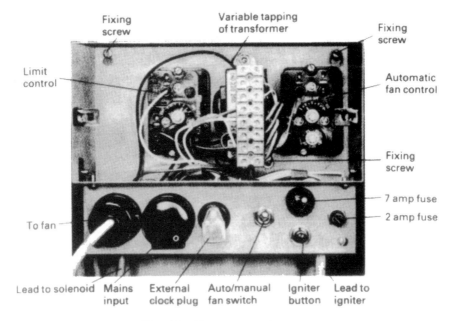

Fig. 45   Heater control panel

high. So the gas supply will be shut off automatically if all the registers are closed or if the fan does not operate for any reason. Ideally, heaters should also be fitted with an overheat control (see Chapter 12). The limit thermostat may be preset or adjustable and the usual setting is 93° C. It has a fixed differential of 14° C. So it will shut off the gas when the air at the outlet of the heat exchanger reaches 93° C and switch it on again when the air has cooled to about 79° C.

Fan Speed

The speed of the fan must be set to give the required air temperature, that is, 66° C to 76° C. It should be set as low as possible, consistent with an acceptable temperature.

The adjustment may be by:

- variable tappings on a transformer
- a variable resistance in series with the motor
- an adjustable pulley on the motor shaft.

It is a simple matter to alter the transformer or the variable resistance and they should be initially set in accordance with the manufacturer's instructions.

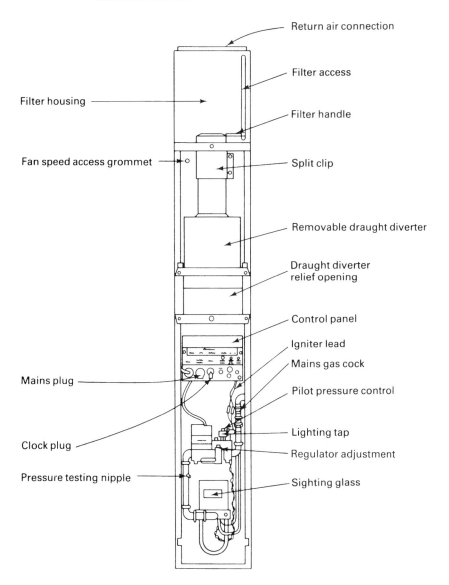

Fig. 46   Location of controls on typical heater

To adjust the pulley, remove the fan belt and slacken the Allen screw on the inside of the pulley, nearest to the motor.

Turn the inside half-pulley:

- clockwise (closing) to decrease speed
- anticlockwise (opening) to increase speed.

the direction being viewed from the end of the shaft. The half-pulley must be turned in either complete half-turns or full-turns to ensure that it will lock in position.

Retighten the Allen screw making sure that it grips on to a flat on the shaft. Replace the belt and adjust the tension until there is about 25 mm play, by means of the belt tensioning screw.

Some modern appliances such as the Johnson and Stanley Modair-flow use a modulating system to control the fan speed. This is an electronic system that uses a thermistor in place of the fan switch to control the fan. More information on this system can be found in Chapter 12.

Pulse Combustion

Up flow, down flow and horizontal warm air appliances using 'pulse' combustion units are manufactured by Lennox Industries Ltd. Pulse combustion units have a totally sealed combustion system, with air taken from outside the building by means of a 50 mm PVC pipe. The products of combustion are discharged, also through a 50 mm PVC pipe which is fitted with a plastic tube to drain off condensate. The vent (flue) pipe can be run to outside, horizontally through a side wall, vertically through a roof or at any angle in between. It can also be fitted with a number of 90° bends. Pulse combustion gives efficiencies between 91 and 96%.

Process of Combustion

The process of pulse combustion begins as gas and air are introduced into the sealed combustion chamber with the spark plug igniter, Fig. 47. Spark from the plug ignites the gas/air mixture, which in turn causes a positive pressure buildup that closes the gas and air inlets. This pressure relieves itself by forcing the products of combustion out of the combustion chamber through the tailpipe into the heat exchanger exhaust decoupler and on into the heat exchanger coil. As the combustion chamber empties, its pressure becomes negative, drawing in air and gas for ignition of the next pulse of combustion. At the same instant, part of the pressure pulse is reflected back from the tailpipe at the top of the combustion chamber. The flame remnants of the previous pulse of combustion ignite the new gas/air mixture in the chamber, continuing the cycle. Once combustion is started, it feeds upon itself allowing the purge blower and spark plug igniter to be turned off. Each pulse of gas/air mixture is ignited at a rate of 60 to 70 times per second. Almost complete combustion occurs with each pulse. The force of these series of ignitions creates

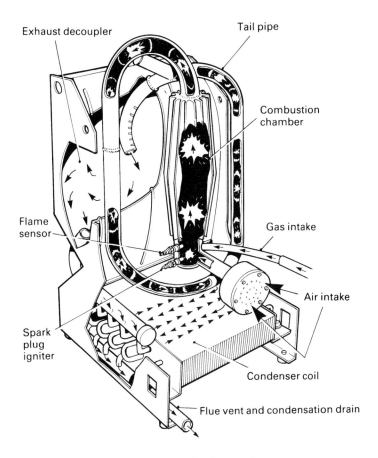

Fig. 47 Process of pulse combustion

great turbulence which forces the products of combustion through the entire heat exchanger assembly resulting in maximum heat transfer.

Figs. 48 (1) to (5) shows in sequence how pulse combustion works.

(1) Air and gas are introduced into the combustion chamber through two valves.

(2) A spark creates combustion which in turn causes a positive pressure buildup that closes off the air and gas valves.

(3) This pressure forces products of combustion down a tail pipe.

(4) As the combustion chamber empties its pressure becomes negative, drawing in air and gas for the next ignition. At the same instant part of the previous pulse is reflected back from the tail pipe.

(5) This re-enters the combustion chamber causing the new gas/air

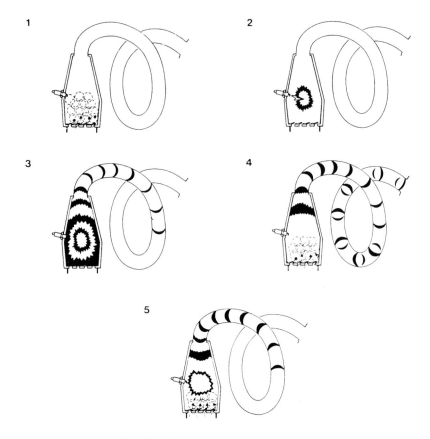

Fig. 48   How pulse combustion works

mixture to ignite and continue the cycle. Once combustion is started it feeds upon itself and the spark igniter becomes redundant.

Fig. 49 shows the section of an up flow pulse unit and Fig. 50 a horizontal pulse unit.

Testing and Commissioning

When an installation has been completed, the following points should be checked:

- ventilation and combustion air requirements have been met
- the return air path is clear and adequate
- all controls have been set in accordance with the manufacturer's instructions

Fig. 49   Up flow pulse unit (Lennox Industries)

● fan and motor work satisfactorily
  – rotating in the right direction
  – fan belt correctly tensioned and aligned ⎱   when fitted
  – adjustable pulley set and locked   ⎰
  – transit packing removed from the mountings
● air filter is clear
● all registers and diffusers are open.

Light up the heater in accordance with the manufacturer's instructions. A typical procedure is:

● make sure the controls are in the 'off' position
● turn on the main switch, clock control and room thermostat
● turn on main gas cock

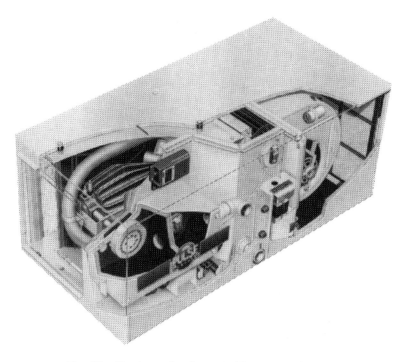

Fig. 50   Horizontal pulse unit (Lennox Industries)

- turn on and ignite the pilot; this is usually associated with the flame supervision device and must be maintained by manual pressure until the device is energised
- turn on gas to main burner.

When the heater has been lit, continue to check the following:

- the fan switch should start the fan about 3 to 4 minutes after ignition
- the gas rate should be checked by checking the pressure or by reading the meter test dial
- flue operates satisfactorily with no spillage from the draught diverter
- controls operate satisfactorily
  - flame supervision device
  - limit or overheat control
  - clock
  - room thermostat.

When this has been done, the system should be balanced. This is carried out by setting the fan speed to give the required temperature rise across the heater. Then measure the air velocity at each of the warm air outlets in turn and adjust the dampers to give the desired velocity.

The procedure is as follows:

- check that all dampers and outlets are open
- check that the fan speed is set for the required temperature rise, usually 50 to 55° C
- place thermometers in the main air duct near the plenum (or in nearest air outlet) and in the return air duct, (or at the filter)
- turn on and allow the heater to operate for about 20 minutes so that the temperature can stabilise
- measure the temperature rise and if it is more than 6° C too high or too low, readjust the fan speed
- measure the air velocities at each register or diffuser with a velometer, starting at the outlet nearest to the heater and ending with that furthest away
- set the dampers to give velocities indicated on the drawing of the installation and lock the dampers in position
- if the air velocities at several outlets are too low, increase the fan speed and balance from the beginning
- re-check the temperature rise and, if necessary, adjust the fan speed and re-check the registers.

A velometer is preferable to an anemometer because it allows for the change in air density with temperature rise. Velocity measurements may be made more accurately from an opening of known area in the neck of a cone placed over the register.

When balancing is completed, turn off the gas by the room thermostat, letting the fan continue to run. Watch the thermometer in the warm air duct and check that the fan switches off when the thermometer reaches 6° C above room temperature. If necessary alter the differential adjustment. If the differential is too small, the fan will cycle on shut-down and on start-up.

Finally, the customer should be instructed in lighting, operating, including cleaning filters and shutting down procedures. The manufacturer's instructions should be left on site and the need for regular servicing should be stressed.

Servicing Warm Air Heaters

The installation should be inspected and serviced at least once a year. However, the filter needs attention frequently, possibly at monthly intervals, depending on local conditions.

Filters are usually made from impregnated glass fibre or rubber foam material supported in a frame or hammock of wire mesh. Access to the filter varies with the type of heater. It may necessitate turning off gas and electric supplies and removing the fan compartment front panel.

Air filters must be kept clean. If they become clogged with dust or lint the heater cannot pass the required volume of air. This decreases the heater efficiency, so increasing fuel bills while the building will not be heated to the temperature required. In extreme cases it can cause overheating and shutting down by the limit control. A completely loaded filter should be replaced. Customers should be instructed to follow the manufacturer's instructions for cleaning filters.

When carrying out a routine service, the procedure to be followed should include:

- check the operation of the appliance
- isolate the heater electrically, switch off the supply, remove the plug or fuse
- turn off the gas supply
- remove and clean the filter
- disconnect and remove the fan, or fan and motor unit
- clean the fan, remove dust from the blades, check for damage, distortion or loose balancing weights
- clean the outside of the motor and lubricate if necessary, check for wear on the bearings
- replace fan and motor and check any anti-vibration pads on the mountings
- if a belt is fitted, check the cleanliness and alignment of the pulleys and the belt tension
- replace the filter
- disconnect and remove the gas burner assembly
- clean the flueways and combustion chamber
- examine the heat exchanger for corrosion or damage due to thermal fatigue
- clean dust and lint from the burner, injector and any gauzes
- clean the pilot burner and examine any probes or thermocouples
- replace the burner assembly
- examine the electrical wiring visually, wires should be properly insulated, connections clean and sound
- replace the fuse or plug and switch on the electrical supply
- turn on the main gas cock; grease cock if required
- check for gas leaks with detection fluid
- test the ignition device, adjust spark gap or orientation of glow coil if necessary

- check the pilot flame and test the flame supervision device
- check that the ventilation grilles are free, correctly sized and unobstructed
- check the gas rate and adjust the pressure if necessary
- check the operation of the controls
    - fan switch
    - overheat or limit thermostat, by closing all registers
    - clock
    - room thermostat
- test flue pull with a smoke match and inspect the flue joints adjacent to the appliance
- leave the appliance in working order and clear up the work area.

Fault Diagnosis and Remedy

Faults in the controls or in the electrical circuits are dealt with in more detail in Chapters 12 and 13. The faults which occur from other causes are as follows.

Fan Operates but Main Gas Cuts In and Out at Intervals

This may not be an actual fault condition. It could be due to some or all outlet registers being closed, causing the limit thermostat to operate. It might also be caused by a blocked filter, fan speed too slow, limit thermostat not set correctly, or a perforated heat exchanger.

Heater Cycles On and Off Frequently in Cold Weather

This could be due to a blocked filter as well as to the room thermostat heater setting being incorrect (see Chapter 12).

Air Outlet Temperature Too Low

May be due to gas input rate too low. This could be an incorrect pressure setting or a fault in the supply or the gas controls. It could also be due to the fan speed being too fast or the limit thermostat incorrectly set.

Insufficient Air Flow

This might be caused by a choked filter or by some obstruction in the return air path; check and clear all return air ducts and grilles.

Noise

1. At the heater. Noise generally originates from the fan and motor unit. So the fan speed should be set as low as possible for the heater to deliver the required volume of warm air. The fan may cause vibration if it is out of balance. However this is the least common cause. Impellers are balanced and run on test at the factory and are not likely to get out of balance unless damaged. Vibration is usually due to bent shafts, worn bearings, dirt on impellers or damaged mountings. On older units with belt-driven fans, vibration may also be caused by unbalanced pulleys, misalignment, or worn or slack belts.

Check the fan for signs of damage, bent blades, build up of dirt on the blade, particularly on one side, and for clearance between the blades and the casing.

Make sure that it is not loose on the shaft.

Rotate the fan by hand and listen for scraping or clicking sounds which indicate contact with the case or worn bearings. By using a screwdriver as a stethoscope the source of the noise can often be found.

Check the mounting for worn antivibration pads or metal-to-metal contact due to overtightening.

Give the fan several short spins and check to see whether it always stops at the same point. If so, it may be out of balance or the shaft may be bent. In both cases it is better to exchange the unit; balancing is not easily done on site.

On belt-driven units check the alignment of the pulleys with a straight edge or hold a taut string up to the pulley faces to check that they are both parallel to the string when it just touches both pulleys.

Rotate the pulleys slowly and hold a pencil or piece of chalk rigidly and just close enough to touch the pulley. It should make a continuous line around the pulley if the shaft and the pulley are not bent.

Check for dirt on the belt or in the grooves of the pulleys. Check that the groves are smooth and concentric.

Belts must neither be too loose nor too tight. About 25 mm play half-way between the pulleys is about right.

Some a.c. solenoids may be a source of vibration and the appliance panels should also be properly secured.

2. At the grilles or registers. Noise in the ducts may be due to high air velocities and to sharp edges or narrow slots which can set up turbulence. Register dampers should give a good seal to avoid whistling noises when closed.

Noise from the return air grille may be minimised by an acoustic lining to the duct or by attaching it to the heater spigot by means of a flexible sleeve connection.

Customers' views on what is, or what is not, an acceptable noise level vary considerably. Measurements of the sound level in 'decibels' (dB) may be taken using an electrical meter. But this task is perhaps best left to the experts. The range of sound levels is from 0 dB at the threshold of hearing to deafening noise at the threshold of pain at about 100 to 120 dB. The average house has a level of about 40 to 50 dB.

Testing Heat Exchangers

During conversion to natural gas it was discovered that some warm air heaters had defective heat exchangers. These components had developed small cracks caused by metal fatigue, corrosion or faulty welding or swaging.

Under certain circumstances this may be dangerous. If, as is most likely, air leaks into the combustion chamber, it may adversely affect combustion. This, combined with spillage from an open flue, could result in toxic products being circulated through the premises.

In the less likely case of products of combustion leaking into the air system the final result may be the same if the leakage exceeds 5% by volume and combustion is incomplete.

Because of the possibility of leakage occurring, all heat exchangers should be examined when the appliance is being serviced. The procedure for testing is as follows:

- examine the heat exchanger for corrosion or damage when cleaning the flueways
- turn on and light the main burner
- check the flames before the fan switches on, ensure that flames are even and burning correctly
- check that the gas rate is correct
- observe the flame picture when the fan switches on; interference indicates a leakage.

The leakage may be on the heat exchanger or on the section of primary flue which passes through the fan chamber. All joints or gaskets should be checked. Where leakage is suspected, it can be proved to exist by a smoke match test as follows:

- light the heater and leave on for 10 minutes
- switch off the appliance

- insert a smoke pellet into the back of the combustion chamber and wait for it to burn out
- open the register nearest to the appliance and close all other outlets
- switch on the fan
- check the heat exchanger and the open register for traces of smoke.

If smoke is discovered at the register, the heat exchanger or the flue should be examined. On open-flued heaters the flue should be checked for spillage at the draught diverter.

Where there is any doubt the combustion should be tested with a flue gas analyser.

Faulty heat exchangers should be exchanged.

Heating Control Systems

Chapter 12 is based on an original draft by J. Norwebb

Introduction

Gas controls were originally dealt with in Vol. 1, Chapter 11. This chapter now covers the whole system of controls for heating and hot water systems, including the electrical components.

Controls are necessary to:

- ensure that the heating unit operates safely
- provide and maintain the desired level of comfort
- provide automatic operation of the system, at the discretion of the user
- ensure that the unit operates efficiently and economically.

The devices which make up heating control systems may be categorised as follows:

- unit controls
- system controls
- temperature controls
- time controls.

Unit Controls

The essential controls on any heating unit include:

- main control cock or valve, to shut off all gas for servicing purposes or for when the unit may need to be disconnected
- pressure control, by means of a regulator, to maintain the correct heat input rate (this may be at the meter)
- flame supervision device, to shut off all gas in the event of flame failure
- limit thermostat, to control or limit the temperature in the heating unit, usually by means of a solenoid or relay valve
- ignition device, to provide safe and easy ignition

- burner control cock, to shut off the main gas manually but leave the pilot on and all the controls operational.

As was shown in Vol. 1, these devices may be combined, in various groups, into multifunctional controls. It is then usual for the flame supervision device to be the first in line, followed by the regulator and the gas valve controlled by the thermostat.

In addition to the essential controls, the heating unit may also be fitted with a fan delay switch or a pump over-run thermostat and sometimes a modulating thermostat. All of these devices are integral with the unit.

System Controls

Both wet central heating systems and warm air heating systems have controls/valves that turn on or off the source of heat to system components and consequently to parts of the system and zones of the building. Some controls cause the heat to be diverted to areas of priority or share the heat between components.

The operation of these controls is generally determined by a temperature control (thermostat etc.) or a time control (clock or programmer).

The following ancillary controls may also be included in the system:

- room (air) thermostat
- overheat thermostat
- frost thermostat
- outside temperature sensor
- thermostatic radiator valves (as an addition or alternative to a room thermostat)
- zone valves
- diverter valves
- mixing valves.

Hot Water Controls

The temperature of the domestic hot water in the cylinder is generally controlled by a cylinder thermostat. This may be:

- mechanical – operating a valve in the primary return pipe
- electrical – controlling either a pump and/or zone valve or diverter valve.

Time Controls

All units have some form of timer. This may be integral with the unit or fitted immediately adjacent to it. The clock control may range from the old original clockwork operated gas valve up to the electronic programmer with a digital clock.

Clocks give on/off control usually once or twice a day; programmers offer, in addition, a variety of combinations of time and either heating and/or water heating.

Set-back devices are available which switch the room thermostat to a low setting for a selected period.

Unit Controls

Gas Controls

Many heating units incorporate the control devices or multifunctional controls already described in Vol. 1. Some units are fitted with full sequence automatic controls. These controls bring the heating unit into operation simply by switching on the electric supply.

The control turns on gas to the pilot burner and, at the same time, operates the ignition device. A flame supervision system checks that the pilot flame has been established and, if so, turns on the main gas and shuts off the igniter. If the flame is not established, some of the more sophisticated devices will shut off completely. Some full sequence controls will switch on a combustion fan and allow a purge period before attempting ignition.

Full sequence controls may be operated by either thermal or electronic systems.

Thermal Controls

A thermal system is shown in Fig. 1. The sequence of operation is as follows.

When the electrical supply is switched on, current flows to the pilot solenoid and, via the 'cold' contact of the thermal switch, to the ignition transformer and the heater of the bimetal switch.

So the pilot valve opens, gas flows to the pilot and is lit by the glow-coil.

If the pilot flame is established, it heats up the phial of the thermal switch, the diaphragm expands and moves the switch over to the 'hot' contact. Current now flows to the main solenoid. This allows gas to pass to the main burner where it is lit by the pilot. At the same time the ignition transformer and bimetal switch heater are both shut off.

If the pilot flame is not ignited, no current can flow to the main valve. In addition, the bimetal switch continues to be heated and finally opens, breaking the neutral connection from the solenoid valves and the ignition transformer.

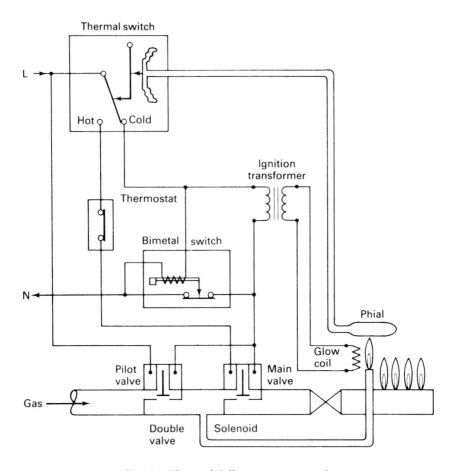

Fig. 1   Thermal full sequence controls

The unit cannot then be lit until the main supply is switched off and remains off until the bimetal switch cools down and reconnects the neutral.

Thermal systems are not very common. They are generally less expensive than electronic systems and are suitable for the smaller, more simple appliance usually without fan-assisted combustion. Generally the thermostat or other controls operate the main gas valve, leaving the pilot always burning.

Electronic Controls

The schematic layout shown in Fig. 2 consists of:

- control box

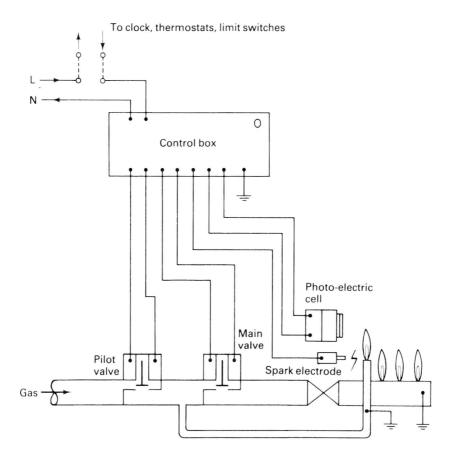

Fig. 2 Electronic full sequence controls – visual type

- double solenoid valve
- igniter
- flame detector.

The control box contains the circuits which determine the timing and the sequence in which the purge, ignition, flame detection and lock on or lock out operations are carried out. It will house an ignition transformer or pulse generator and may rectify the current for the solenoids.

Domestic appliances often use double solenoid valves rather than separate solenoids, an example is shown in Fig. 4. Large heating units use a pilot solenoid and two main valves, one of which is designed to give slower turn on for the main burner.

The igniter may be either spark or glow-coil and the flame detector is usually flame rectification or a visual type as shown. In both cases

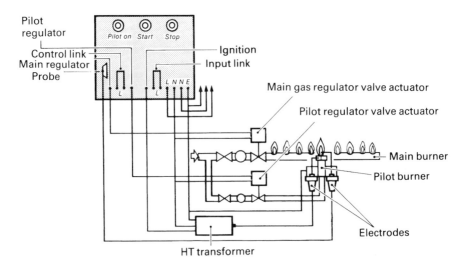

Fig. 3　Electronic full sequence controls – flame rectification

Fig. 4　Double solenoid valve (Johnson-McLaren)

the detector will activate the control to switch on the main gas and switch off the ignition when the pilot flame is established.

Electronic controls may be linked to a combustion fan and will be arranged to switch the fan on before ignition to purge the combustion chamber.

Because the thermostat and other controls are fitted in the main electric supply line before the control itself, the pilot does not remain lit all the time. It is extinguished at the end of each heating cycle and relit automatically each time there is a call for heat. This has the advantage of conserving energy.

Wet Central Heating System Controls

Figure 5 shows a simple control layout for a domestic central heating system. The boiler thermostat controls the main gas solenoid valve and the room thermostat switches the pump on and off. The clock controls both solenoid and pump. The boiler will, of course, also be fitted with some form of pilot and flame supervision device so that it comes into operation when the solenoid valve opens.

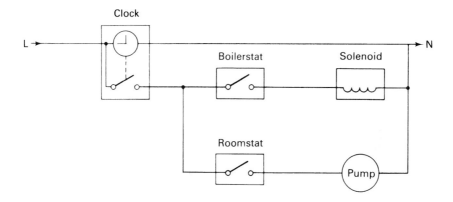

Fig. 5   Simple central heating control layout

The sequence of operations is as follows:

- clock switches on
- water in the boiler is cold, so boilerstat is closed, or 'calling for heat'
- air temperature is lower than roomstat setting so roomstat is closed, or calling for heat
- solenoid valve opens, burner lights up
- pump starts circulating heated water

- rooms reach roomstat setting temperature, roomstat switch opens
- pump stops, water stops circulating
- boiler water temperature increases until boilerstat switch opens
- solenoid valve shuts, burner is turned off
- room temperature falls to below roomstat setting, roomstat switch closes
- pump starts and cool water from the radiators enters the boiler
- boiler water temperature falls to below boilerstat setting
- boilerstat closes
- solenoid valve opens, burner relights.

The sequence continues until the clock switches off, when both boiler and pump stop working together.

If the central heating is combined with a domestic hot water system having a gravity primary circulation, then the same simple control layout can be used. It is, however, an advantage to control the hot water temperature separately. This is because the boiler thermostat in a combined system is usually set at 82° C, whereas the domestic hot water should not normally be more than 60° C to avoid scaling up in hard water areas and the possibility of scalding.

Modulating Thermostats

On conventional boilers the heat input to the circulating system is controlled by the room thermostat and the boiler thermostat. Because the boiler efficiency is lowered if the gas rate is reduced, an on/off method of control is used. Some conventional boilers do, however, operate efficiently across a range of heat inputs.

With low thermal capacity boilers, which retain their efficiency at lower gas rates, modulating thermostats may be used.

One type of modulating thermostat uses a vapour pressure bellows system with the sensing phial mounted in the hot water outlet from the boiler. The bellows operate a gas valve in the supply to the main burner, modulating the gas flow down to about one-third of the full-on rate. If the water temperature continues to rise, the valve snaps shut. When the water temperature falls, the valve snaps partly open and then modulates the gas rate up to fully on. This maintains a constant temperature in the hot water circulation to the radiators.

Modulating thermostats help to prevent overheating in low thermal capacity boilers. They may be used in conjunction with room thermostats or thermostatic radiator valves for full control of the system. Many combi boilers use modulating thermostats. One of the most common being the Honeywell 'Modureg' valve, Fig. 6.

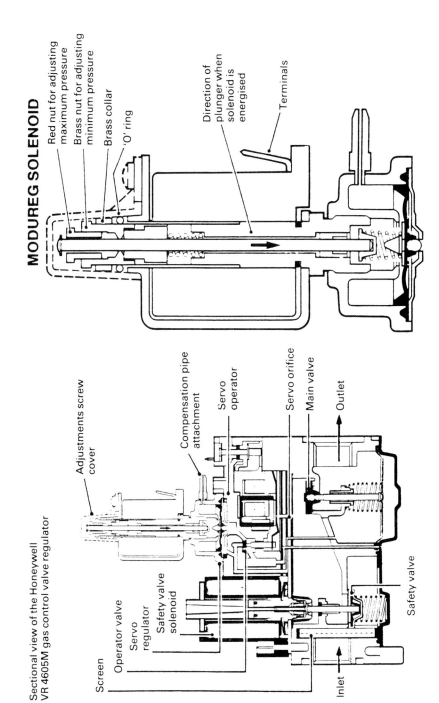

MODUREG SOLENOID

Red nut for adjusting maximum pressure

Brass nut for adjusting minimum pressure

Brass collar

'O' ring

Direction of plunger when solenoid is energised

Terminals

Adjustments screw cover

Compensation pipe attachment

Servo operator

Servo orifice

Main valve

Outlet

Sectional view of the Honeywell VR 4605M gas control valve regulator

Screen

Operator valve

Servo regulator

Safety valve solenoid

Safety valve

Inlet

Fig. 6   Modureg valve (Honeywell)

This current controlled regulator varies the outlet working pressure to the burner in response to demand from a sensor (usually a thermistor). The minimum input to the appliance is controlled through a solenoid valve and the modulating (variable) part of the input is controlled by the Modureg (current controlled) device. The sensor (thermistor) works on the principle of heat effecting the electrical resistance of a material and as the temperature of the water in the appliance increases, the electrical resistance of the sensor changes and this in turn controls the electric current flowing through the Modureg controller. The Modureg controller reduces the gas supply as the water temperature increases and increases the gas supply as the water temperature decreases. The sensor (thermistor) is connected through an electronic circuit containing a potentiometer (variable resistor), usually in the printed circuit board (PCB), to the Modureg controller. This potentiometer is used to set the maximum gas rate supplied to the appliance.

A further method of modulating (varying) the gas supply is used in the Chaffoteaux Celtic combi boiler. The gas supply to the burner is controlled through three solenoid valves, Fig. 7 (a) and (b). The solenoid valve (2) supplies $\frac{1}{3}$ of the appliance maximum gas rate and is opened when there is a call for either central heating Fig. 7 (a) or domestic hot water Fig. 7 (b). The solenoid valve on the left, (1) opens when there is a call for central heating and the volume of gas passing through this valve can be varied between $\frac{1}{3}$ and $\frac{2}{3}$ by setting the adjuster (4), thus controlling the total heat input to the appliance when supplying central heating. The boiler thermostat controls the temperature of the water between 50° C and 82° C by opening and closing these solenoid valves.

When there is a call for domestic hot water the solenoid valve (2) again opens and supplies $\frac{1}{3}$ of the appliance maximum gas rate and the solenoid (3) also opens to supply the other $\frac{2}{3}$ of the gas rate. A limiting thermostat prevents the hot water temperature from exceeding 60° C, by cycling solenoid (3) without interrupting the flow of water. There are also overheat devices, (10) and (11), to prevent the appliance from operating unsafely.

Boilers which have modulating thermostats are usually also fitted with an overheat thermostat. This may be in the form of a thermally operated switch which cuts off the gas if the modulating thermostat valve fails to shut off completely. The overheat thermostat has usually to be manually reset.

One form of overheat device operates an interrupter in the thermocouple lead.

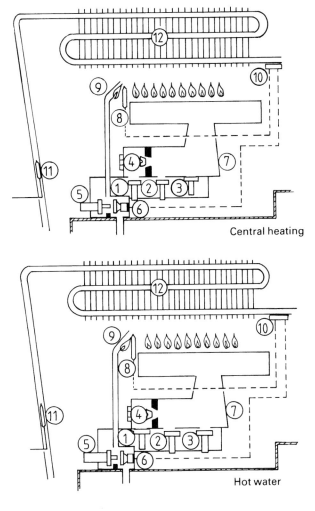

Central heating

Hot water

1 Solenoid valve	6 Thermoelectric valve	11 Boiler thermostat
2 Solenoid valve	7 Gas section	12 Heat exchanger
3 Solenoid valve	8 Thermocouple	
4 Gas adjuster	9 Pilot	
5 Igniter button	10 High limit thermostat	

Fig. 7   Modulating gas supply to combi boiler (Chaffoteaux):
(a) central heating; (b) domestic hot water

Room Thermostats

A room (air) thermostat, frequently abbreviated to 'roomstat', regulates the amount of heat emitted by the system. It must be fitted in a position which is at approximately average room conditions. This is usually:

- on an inside wall
- not near a direct heat source
- not in a draught
- clear of doors or windows
- about 1.5 m above floor level.

Where only one roomstat is used it is normally located in the living room or hall.

Roomstats are generally bimetal switches, Fig. 8. The type shown consists of a bimetal strip A, wound into a spiral and attached at one end to a lever, C, pivoted at B. The other end is attached to the moveable contact of the switch D.

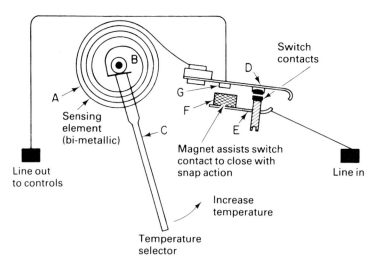

Fig. 8   Room thermostat

When heated, the bimetal spiral tends to unwind, finally carrying the moving contact D away from the fixed contact, E. The temperature at which this occurs will depend on the extent to which the temperature selector lever, C has initially tightened the coil. The tighter the coil, the higher the temperature required to unwind it.

In order to prevent sparks and arcing on the contacts, a magnet F is attached to the fixed contact E. This attracts an armature G, mounted on the moving contact D. When they are sufficiently close, the magnet causes the switch to close suddenly, so avoiding arcing. The switch stays closed until the bimetal strip is able to overcome the pull of the magnet when the switch snaps open again.

Some thermostats use sealed contacts as shown in Fig. 9. A small quantity of mercury is sealed into a glass capsule which has two contacts at one end. Tilting the capsule tips the mercury onto, or away from, the contacts.

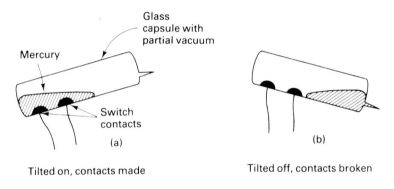

Fig. 9   Sealed switch contacts: (a) on; (b) off

There is always some lag between the air temperature in a room and the temperature sensed by the thermostat. Also, the system is full of heat when the thermostat opens so the temperature of the room continues to rise slightly above that required. These factors cause thermal lag which can be compensated for by an 'anticipator' or 'accelerator'. Fig. 10 demonstrates the effect an accelerator has on the temperature of the room or space being controlled. This is a small heating element which is fitted in the thermostat near to the bimetal strip. The accelerator raises the temperature of the air surrounding the bimetal strip so that it reaches its set point slightly before the air temperature and not after it, as previously. When the thermostat switch opens, the accelerator is cut off and cools down.

Accelerator heaters may be connected in either series or parallel.

Series Accelerators

Thermostats which control low-voltage equipment generally have the accelerator in series, Fig. 11. Since the heater now passes the total

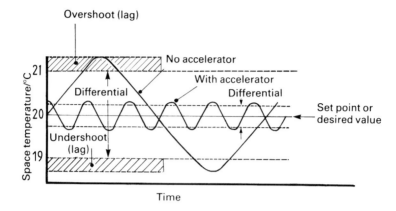

Fig. 10   The effect when an accelerator is fitted to a room thermostat

load current of the circuit being controlled by the thermostat, it must be matched to this value if it is to give the right amount of heat.

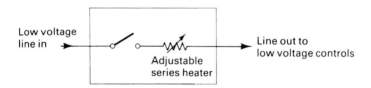

Fig. 11   Accelerator in series

The accelerator is usually adjustable over a range of about 0 to 1 amp and must be set to the load current of the control served. For a 24 V a.c. solenoid valve, this is about 0.4 amp. The heating effect is proportional to the square of the current ($Q \propto I^2$) so a small change in current can cause a much greater change in the heat output.

If the thermostat is controlling a warm air installation there is very little residual heat in the system and it may be desirable to lengthen the 'on' cycle of a low-voltage roomstat by adjusting the heater. This is done by raising the current rating of the accelerator heater.

Parallel Accelerators

Also known as 'shunt accelerators' these heaters are independent of the current taken by the control load, Fig. 12. The heater has a fixed value accurately matched to the thermostat and is connected directly to the neutral supply. The shunt accelerator is used on mains voltage roomstats.

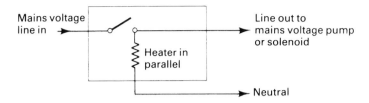

Fig. 12   Accelerator in parallel

Integral Thermometer

Some roomstats incorporate a small, bimetal thermometer. If this becomes inaccurate it may mislead the customer and it should be checked against an accurate mercury-in-glass thermometer if it becomes suspect.

Frost Thermostats

A frost thermostat is fitted on a wet central heating system when the boiler is installed in an external, unheated location, for example, a

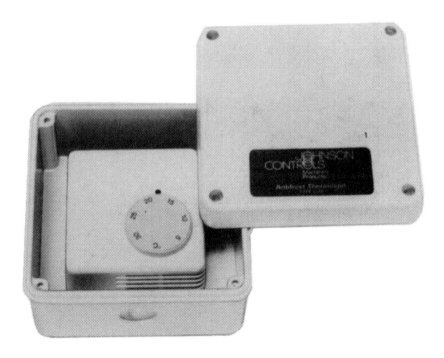

Fig. 13   Froststat (Johnson-McLaren)

garage. The froststat protects the boiler from damage by freezing in excessively cold conditions. It is similar to a roomstat but designed to operate at lower temperatures. The temperature range may be from 0 to 20° C and the stat would be set to operate at about 3° C. A typical example is shown in Fig. 13.

The froststat is wired into the control system so that it overrides the clock and can turn the boiler on at any hour of the day or night. It is located adjacent to the boiler to sense the temperature in the immediate vicinity.

Some roomstats are capable of being set to about 4° C and have been used as frost thermostats.

External Temperature Control

This provides a means of controlling the heat output from a heating circuit in response to temperature changes both inside and outside the building. It consists of:

- outside air thermostat
- room thermostat
- electric controller
- three-port mixing valve.

The layout is shown in Fig. 14. The mixing valve controls the temperature of the circulating water by mixing the hot water from the boiler with some of the cool water from the return.

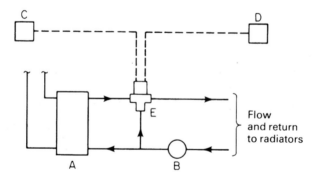

Fig. 14   External temperature control. A: Boiler; B: Pump; C: Outside air thermostat; D: Room thermostat; E: Controller and mixing valve

The temperature is raised or lowered in response to the outside air temperature and the selected temperature set on the room thermostat. The roomstat has priority over the outside air thermostat.

The more sophisticated external controllers were originally used on larger, usually commercial, installations which detected changes in solar gain and wind velocity as well as the air temperature. Increasing awareness of the need for economy and energy conservation plus the desire for more automatic control has lead to the development of electronic domestic heating control systems. The centre point of these systems is the controller which collects information from strategically positioned sensors – outside the building, at the hot water cylinder, in the flow pipe and in the main living room. The controller then manages the boiler, pumps and zone valves to achieve the desired comfort levels. Fig. 15 shows a typical layout of this type of heating control system.

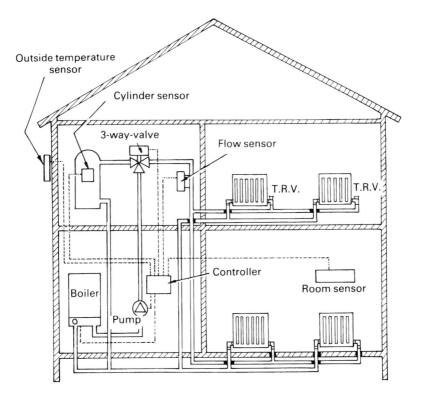

Fig. 15   Environmental heating control system

The information from the outside as well as the inside of the building is fed to the microprocessor to determine when to switch on the heating system to achieve comfortable conditions when the occupants need them e.g., at the start of the day or arriving home from work etc. Outside temperature is also used to adjust the flow temperature from the boiler. The ratio of flow temperature to outside temperature is called the heating curve and the steeper the curve the higher the flow temperature required. Many of these controllers have battery back-up to retain the memory if there is a power failure. This battery back-up also allows some controllers to be removed from their mountings to be programmed. Many of the cables used in the installation of these controls carry only a low voltage and a simple telephone cable or bell wire can be used. Self-diagnosis of faults is a further advantage of several of these modern electronic systems.

Thermostatic Radiator Valves

These provide a form of 'zone control' since they effectively control the temperature within a room or space. They may be fitted to radiators in a system or used selectively in rooms with intermittent or variable usage, in addition, or as an alternative to the conventional roomstat.

Thermostatic radiator valves are mechanically operated and do not require connecting to an electrical supply. Examples are shown in Fig. 16. They are usually liquid expansion thermostats in which the water supply to the radiator is reduced by the valve when increasing room temperature expands the bellows. The temperature is selected by the

(a)

Fig. 16   Thermostatic radiator valve: (a) radiator thermostat

(b)

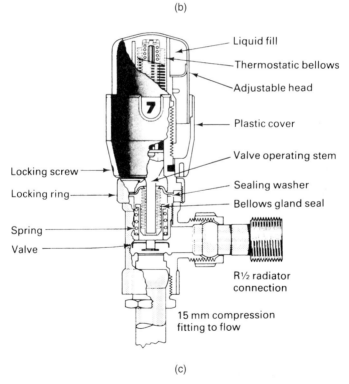

Liquid fill

Thermostatic bellows

Adjustable head

Plastic cover

Valve operating stem

Locking screw

Sealing washer

Locking ring

Bellows gland seal

Spring

Valve

R½ radiator connection

15 mm compression fitting to flow

(c)

Fig. 16   (b) remote sensor; (c) Thermostatic radiator valve: cross section

customer and the valve is controlled by the bellows so that, when this temperature is reached, it will pass just sufficient hot water to maintain the set temperature. The thermostat has, therefore, a modulating control without the lag or overshoot associated with air thermostats.

Radiator thermostats should be fitted to either the inlet or outlet connections to the radiator, in accordance with the manufacturer's instructions.

Because these positions are not always the best ones in which to sense the room temperature, models are available with remote sensing element, Fig. 16 (b).

Zone Control

This is generally applied to larger domestic installations in order to provide heating to specific parts of the building at the particular times and temperatures at which it is required.

Zone control is effected by controlling the heat input to large rooms or groups of rooms separately.

Fig. 17   Two-port motorised valve

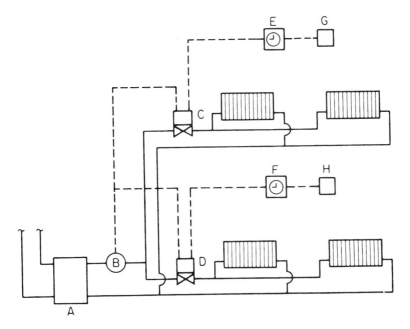

Fig. 18   Zone control by two-port valves, two storey installation:
A: Boiler; B: Pump; C: Upper floor zone valve; D: Lower floor
zone valve; E, F: Clock controls; G, H: Roomstats

The need for zoning arises from the following factors:

- the use of the various rooms varies considerably; heat is not generally required continuously in all parts of the building
- the heat requirements of different parts of the building will vary, depending on the aspect and the time of day, due to changes in wind velocity and solar gain
- differences in structure will affect heat requirements, for example large areas of glass, rooms over unheated spaces.

With all these variables it follows that a single room thermostat cannot satisfactorily provide the varied temperatures required and one time control cannot programme heating to satisfy the different requirements. Controlling different zones separately can result in improved comfort conditions and considerable economies in fuel.

Zone control is only possible if the pipework layout divides up the areas to be controlled. It should, therefore be planned at the time that the heating system is designed. In the average house, top and bottom floors have separate circulations and may be separately controlled.

Control is effected by means of 2-port motorised valves, Fig. 17, each controlled by a separate clock control and room thermostat. The general arrangement is shown in Fig. 18. The pump continues to run until both zones are shut off.

It is also possible to control a zone by modulating either the volume or the temperature of the circulating water using a three-port motorised valve, Fig. 19. This is more frequently applied to large commercial installations. Two methods are shown in Fig. 20.

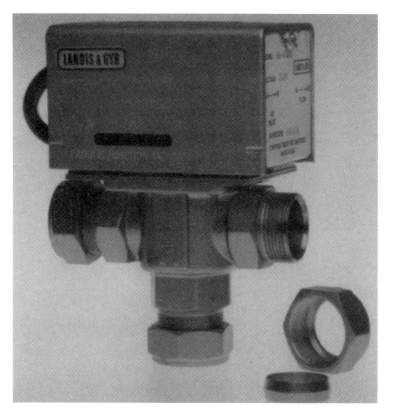

Fig. 19   Three-port motorised valve

In both the layouts illustrated, the valve is used as a mixing valve, that is, with two inlets and one outlet. One inlet is hot water and the other, cool water. The outlet is mixed water.

The valve can also be used as a diverter valve, that is, with one inlet and two outlets.

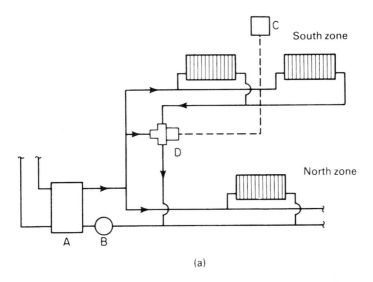

(a)

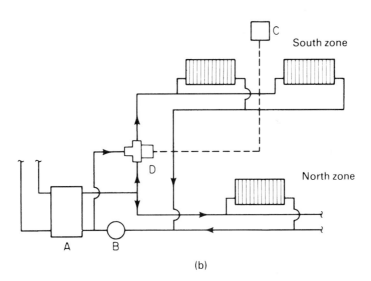

(b)

Fig. 20   Zone control by three-port valve: (a) variable volume at constant temperature; (b) variable temperature at constant volume: A: Boiler; B: Pump; C: Roomstat; D: Zone valve

Domestic Hot Water Control

In combined heating and water heating systems the boiler thermostat may be set at temperatures up to 82° C. This is unnecessarily high for domestic hot water which needs only to be at about 60° C. A cylinder thermostat can reduce heat losses and minimise scale formation.

Mechanical Thermostats

The simplest form of cylinder thermostat is the liquid expansion type which is similar to a radiator thermostat. The valve is opened and closed by the contraction or expansion of a bellows in response to the temperature of the heated water. The thermostat is fitted in the return pipe from the cylinder. Its temperature-sensitive element may be the bellows itself or a separate phial connected by a capillary tube and located in the stored water. Figure 21 shows a typical thermostat with a remote sensor. The temperature required may be selected by turning the knob of the valve.

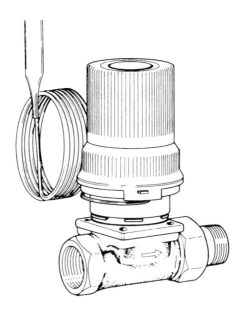

Fig. 21   Cylinder thermostat, liquid expansion

Electrical Thermostats

These are usually strapped on to the outside of the cylinder about a third to a half of the way up from the bottom. A typical thermostat is shown in Fig. 22.

There are several methods of wiring up the thermostat to different controls including:

- overriding the boiler thermostat
- controlling a separate pump
- controlling a two-port zone valve
- controlling a three-port diverter valve.

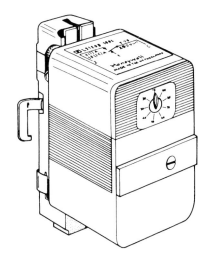

Fig. 22   Cylinder thermostat, electrical

Boiler Thermostat Override

When the domestic water is heating up, the boiler thermostat operates normally.

When the water reaches the temperature set on the cylinder thermostat, this overrides the boiler thermostat and shuts off the gas to the burner. The cylinder thermostat continues to control the boiler until the roomstat calls for heat and switches the cylinder thermostat out of circuit. The boiler thermostat then becomes the temperature control for the boiler until the roomstat is satisfied.

Separate Pump

If the heating circuit and the water heating circuit each have their own pumps, the cylinder thermostat can control the temperature of the domestic water by switching the second pump.

Two-port Zone Valve

The use of these valves to control separate heating zones was described earlier in this chapter. The water heating circuit can be regarded as another zone and controlled by the cylinder thermostat, Fig. 23.

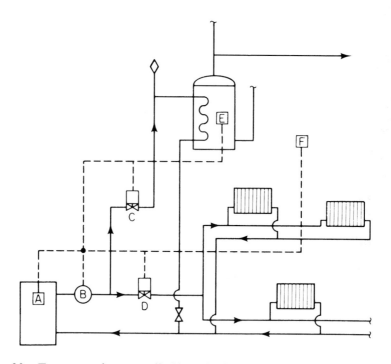

Fig. 23 *Two-port valve controlled by cylinderstat. A: Boiler thermostat; B: Pump; C: Hot water control valve; D: Heating control valve; E: Cylinder thermostat; F: Room thermostat*

The pump must be kept running while either the cylinderstat or the roomstat are calling for heat.

Three-port Diverter Valve

This system was introduced briefly in Chapter 10. When the primary hot water circuit is pumped, a three-port motorised valve can be used as a diverter valve. The layout is shown in Fig. 24.

A three-port diverter valve is fitted at the point where the primary flow divides to feed the domestic hot water calorifier (cylinder) and the heating circuit (radiators).

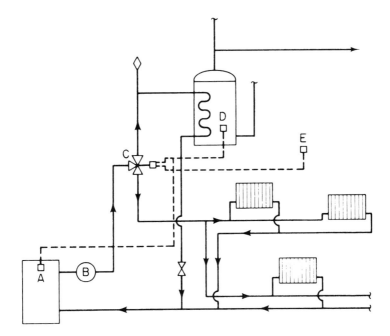

*Fig. 24   Three-port diverter valve controlling cylinder and heating circuits.
A: Boiler thermostat; B: Pump; C: Diverter valve; D: Cylinder
thermostat; E: Room thermostat*

The valve has one inlet port and two outlets. On earlier models of
diverter valve a motorised ball plug was driven to close off either one
or other of the outlet ports dependent upon which of the thermostats, cylinder or room, was calling for heat. When both thermostats
were calling, domestic hot water generally took priority although it
was possible to modify the wiring circuit to give priority to the
heating circuit. With this type of system it was necessary to set the
cylinder thermostat at least 17° C below the setting of the boiler
thermostat in order to ensure that the domestic water heat up period
was not prolonged to the detriment of the central heating service.

More recent 'diverter' valves have the facility to 'share' the flow of
heated water between the two circuits. The position of the motorised
valve is determined by the demand from the controls (usually clock,
cylinder thermostat and room thermostat). Calls for heat from all
three controls result in the valve being driven to a mid position and
the heated water to be shared between the two circuits. If calls for
heat come from only the clock and room thermostat then the valve is
driven to close off the outlet to the calorifier and all of the heated
water flows through the central heating circuit. When the clock and

cylinder thermostat are the only controls calling for heat the valve is driven to close off the outlet to the heating circuit and all of the heated water then passes through the calorifier. Preferably a programmer should be fitted (see later section), thus giving the customer the facility to select the time periods for domestic hot water and central heating.

Time Controls

Clock Controllers

Many of the older clocks in use are driven by small synchronous motors which run in step with the frequency of their a.c. supply, 50 Hertz. They may be either mains voltage 240 V a.c. or low voltage, 12 to 24 V a.c.

The clock motors are connected so that they run continuously when the main isolating switch is on.

The clock dial is divided into 24 hours with each hour subdivided into 15 minute sections. The early models were marked 'Day' and 'Night', with 12 hours in each sector. Examples are shown in Fig. 25. The clock motor turns the dial round, completing one revolution in 24 hours. The dial is set to the correct time of day against the indicator on the case by turning it in the direction of the arrow.

There are two or four tappets. Each tappet is marked with 'on' or 'off', or with a number. With numbered tappets 1 and 3 are 'on', 2 and 4 are 'off'. The tappets are set to the times when it is required to turn the heating on or off and they rotate with the dial.

On the clock shown in Fig. 25 (a), the tappets are set by pressing down the black central triangular time indicator and sliding the tappet to a new position. The central triangle is spring loaded and fits into slots at the back of the dial.

On the clock shown in Fig. 25 (b), the tappets are locked into position by the central bakelite knob. This is a left-hand thread and must be turned clockwise to release the tappets. When they have all been moved to the required times, the knob is turned anticlockwise to lock them into position.

The minimum duration of an 'on' or 'off' period is about 1½ hours.

Timers are generally made with some form of manual override or 'advance' device. This operates directly on the switch mechanism to advance the switch sequence to the next operation. So if the timer is 'on', the advance device will turn the switch off, or vice versa. The override only affects the switch sequence until after the next tappet operation, when the normal programme is resumed.

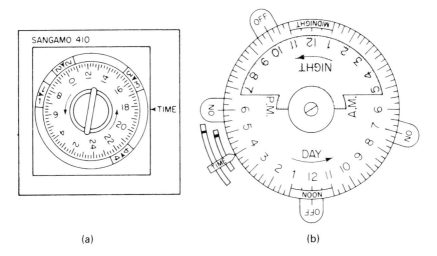

SANGAMO 410

Fig. 25   Clock controls: (a) 24 hour; (b) day/night

Some clocks incorporate a 'day omitting device'. These are seldom used on domestic installations but are useful in shops or offices where no heating is required from lunch time on Saturday until Monday morning.

There are several different types and the one shown in Fig. 26 is a 14-point star wheel with screwed pegs inserted into it. To omit a day, remove the pegs from that particular sector of the wheel. On the example shown, the system will shut down at the first 'off' operation on Saturday and remain off all day Sunday. It will resume the normal daily sequence beginning with the first 'on' tappet on Monday morning. The star wheel is usually rotated by one of the 'on' tappets, which must be set for the first operation of the day. On some clocks, days are omitted by inserting screws or by turning screws through 90° C.

Because it may be necessary, at some time, to run the system continuously, most timers have some facility for manually overriding the clock switch. This is usually a separate switch, mounted on the clock panel, which provides a continuous electrical supply to the system.

Timers may have either three or four terminals, Fig. 27. The three-terminal models have an inner link between the inlet connection to the clock motor and the inlet of the clock switch. So the motor is at the same voltage as the switched supply.

Four terminal clocks may be used with the clock motor at the same voltage as the switched controls but they could have different voltages.

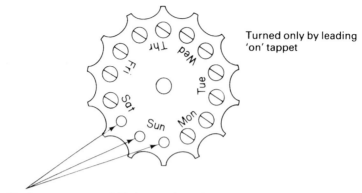

Turned only by leading 'on' tappet

Stud screws removed. No heat at these times.

Fig. 26   Day-omitting device

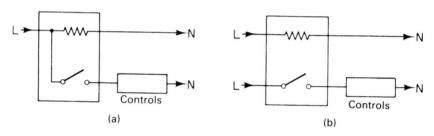

(a)   (b)

Fig. 27   Timer wiring: (a) three-terminal; (b) four-terminal

A low-voltage clock motor can switch mains current to a pump or fan. Alternatively a mains voltage clock can control a low-voltage control system. Care must be taken when exchanging a clock to ensure that the resulting installation is completely safe, see Chapter 13 for details.

Programmers

There are many different programmers in use. Basically they all consist of a clock unit and a switch with multiple contacts, or a bank of multiple switches, each covering a different time period. Figure 28 shows two typical programmers. These range from the simple model with five programmes up to the electronic type with a digital clock and several selections.

Programmers, like timers, may be integral with the heating unit. If not, they must be located in an easily accessible position.

To set the conventional five-programme type:

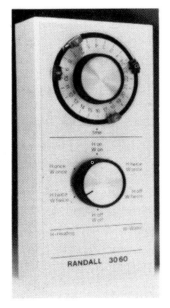

Fig. 28   Programmers

- set tappets A and C to the two 'on' times, A is the first 'on' of the day
- set tappets B and D to the two 'off' times
- set the dial to the correct time of day
- turn the rotary switch to the number appropriate to the programme required.

There is usually an off position which cuts off all the electrical output from the programmer. This leaves the clock still running and the froststat, which overrides the programmer, still operative.

So, any boiler with fully automatic control will be shut down completely and other boilers will have the main gas off but the permanent pilot still on. It is generally not possible to alter the rotary switch if the clock has started a switching operation. If it is necessary to do this, first turn the clock dial to complete the switching operation, then alter the rotary switch and finally reset the dial to the correct time.

Until the 1980s most central heating programmers were electromechanical and this limited the functions to relatively simple on/off clocks or time switches. In more recent years almost all new central heating programmers have been electronic and these vary considerably. They fall into the following categories:

- clocks/time switches for controlling the boiler only – these have no separate hot water or central heating functions
- gravity domestic hot water and pumped central heating control – on this system the selections are limited to (i) hot water only, or (ii) central heating plus hot water
- fully pumped systems giving independent control of hot water and central heating or both. The programmes can then be further subdivided depending upon the number of programmable days:

(i)   One day operation only – i.e., all days are the same
(ii)   Separate modes for weekdays – Monday to Friday – and week-ends – Saturday and Sunday.
(iii)   Independent control for each of the seven days.

A variety of time periods can be available e.g., 2, 3 or 4 periods a day. Electronic programmers also display the time of day. They are generally delivered with a built-in pre-set programme with options to re-set to the customer's needs. There is usually the protection of a battery back-up facility that protects the set programme in the event of failure in the mains electricity supply and in some instances allows the programmer to be removed for re-setting. In addition to the set functions many programmers have the extra facilities to override the selected programme by advancing or boosting the settings.

There are many manufacturers of electronic programmers and Fig. 29 shows the control layout of the programmer.

Another type of control is the Honeywell AQ 6000 Boiler Management System. The system components are shown in Fig. 30 and the system wiring is shown in Fig. 31.

Control of the central heating is through a combination of sensors monitoring:

- outside air temperature
- flow water temperature
- room temperature.

Provided that a diverter valve, or equivalent, is fitted to separate the central heating and domestic hot water circuits, a further sensor may be fitted on the hot water cylinder allowing the domestic hot water to be fully controlled. Information from the sensors is fed through to a control unit which also serves as a wiring centre. Programming the system is done through the room temperature unit. The control can be switched up to six times each day and any one of three temperature levels – comfort, activity or economy – can be selected. There is also built-in protection against frost when the unit is in an off position. The system also has a built-in pump overrun

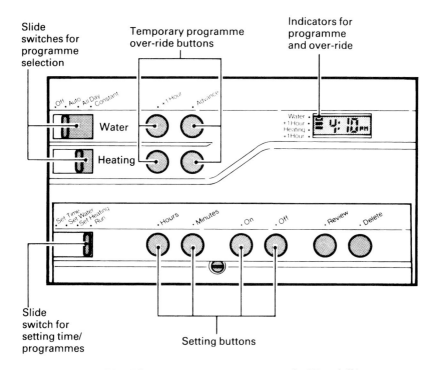

Fig. 29 Electronic programmer controls (Randall)

Fig. 30 System components (Honeywell)

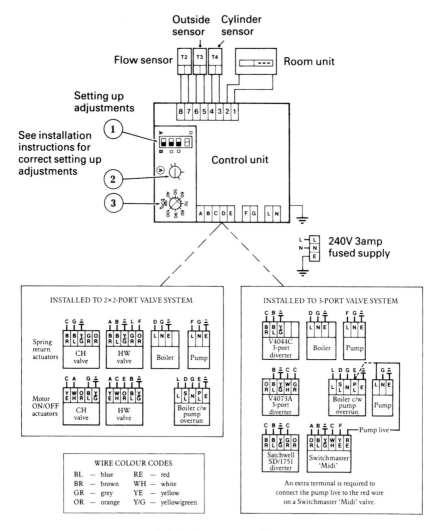

Fig. 31   System wiring

facility and also exercises the pump and valves during periods of inactivity. Any five days of the programme can be set up as 'work-days' while the other two days can be set up as 'rest days'.

Programmable Room Thermostats

These electronic controls are tending to replace the traditional pro-grammer and room thermostat method of control. They are most compatible with combi boilers since these only require control of the space heating element as the domestic hot water is supplied on an instantaneous basis. They can, however, still be used on an appliance and system supplying space heating and stored domestic hot water. The device would then require an additional component for on/off control of the domestic hot water.

The traditional method of control through a programmer and room thermostat relies upon the programmer to operate the system components (boiler, pump, zone valves etc.) at pre-set times while the room thermostat controls the temperature at which these components are switched on or off. The programmable room thermostat (PRT) measures temperature and then instructs the components of the system to act on that measurement. This enables the device to control at the varying temperatures required, when the occupants are carrying out different activities. Typical temperature zones and corresponding activities are:

- high setting – low level of activity, e.g. watching TV, reading etc.
- medium setting – occupants active, e.g. cleaning, cooking etc.
- low setting – dwelling temporarily unoccupied
- frost setting – dwelling unoccupied but heat required to prevent frost damage, condensation etc.

Many of these controls have the facility for up to six space heating periods with a different temperature zone for each of the selected periods. A separate programme can often be selected for each day of the week. Fig. 32 illustrates a typical programmable room thermostat and the lay-out of the control.

Warm Air System Controls

Many of the control devices on warm air systems are the same as those used on wet systems. They include:

- main gas control
- regulator
- flame supervision device ⎫
- ignition device ⎬ or multifunctional control
- gas solenoid valve ⎭
- overheat control
- limit thermostat
- fan delay switch
- winter/summer switch
- fan speed control
- timer or programmer.

If the heating unit has an integral gas circulator there may also be:

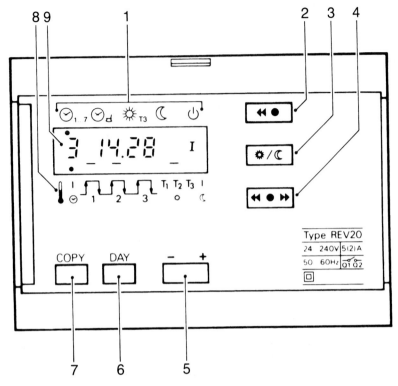

1   Indication of the mode of operation; a dot
     indicates the selected mode of operation
     symbol:

     ⊙, ,   Automatic operation according to
            weekly programme
     ⊙d   Automatic operation according to
            exception programme
     ☼ ₁₃   Continuously comfort temperature
     ☾   Continuously energy saving
            temperature
     ⏻   Standby
2   Mode-of-operation selector button
3   Override button
4   Function selector button
5   Button for changing numerical values
6   Button for selecting the day of the week
7   Copy button, used for copying a daily
     switching programme for the next day

8   Setting the operating functions; a dot
     indicates the selected function:

     ⌇   Display of the room temperature
     ⊙   Display of day and time
     ⌐⌐⌐   Entering of the switching
     ₁ ₂ ₃ times for the comfort periods
            — Switch-on (start) of a
            comfort period
            — Switch-off (end) of a
            comfort period
     T₁   Set value for the first comfort period
     T₂   Set value for the second comfort
            period
     T₃   Set value for the third comfort period
     ☾   Set value for the energy saving
            temperature
9   LCD display field.

*Fig. 32   Programmable room thermostat: Layout of controls (Landis &
Gyr)*

- circulator flame supervision and ignition devices
- cylinderstat
- circulator solenoid valve.

On a warm air system the room thermostat and the timer, control the gas solenoid valve and not the fan. So there is no need for the controls to carry mains voltage current. Consequently use is made of low voltage controls, usually 24 V a.c. A simple layout is shown in Fig. 33.

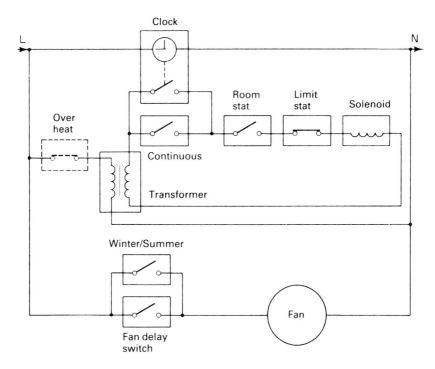

Fig. 33   Warm air control layout

This has a mains voltage clock switching a low voltage supply to the solenoid, through the thermostats. An override switch allows continuous running, when required.

An overheat control could be fitted, as shown, to cut off the gas and leave the whole low voltage circuit dead.

The fan is supplied directly through the delay switch. A winter/ summer switch is included to override the fan switch and allow the fan to run in the summer when the unit is not heating, to give a circulation of cool air to the rooms.

Figure 34 shows a similar layout, now modified to include a circulator. When the water heating switch is on, the circulator operates continuously under the control of the cylinderstat. Only the warm air unit is controlled by the timer.

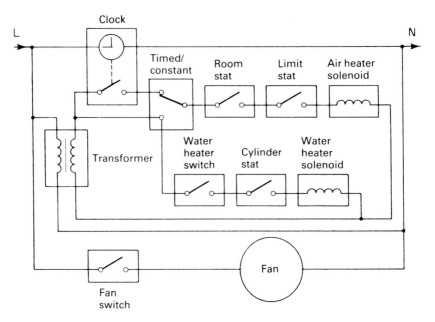

Fig. 34   Warm air and circulator controls

Overheat Control

The heat exchanger of a warm air unit can very quickly become overheated if air is not circulating over the heated tubes. It would be badly damaged if the gas burner was allowed to remain alight after the fan had failed or the duct or filter became blocked.

So there is a need for a cut-off device to turn off the gas to the main burner when overheating occurs. The maximum temperature is set by BS specification at 110° C for the mean air temperature at the inlet and outlet of the warm air unit (BS 5258, Part 4).

Control of overheating may be by means of an overheat or a limit thermostat or both.

The overheat is added, as an extra control usually to down-flow heaters. This is because, in these models, the filter unit at the top of the heater can quickly rise to a high temperature if the fan is slow in switching on.

A device is shown in Fig. 35. It consists of a bimetal element acting on a switch. The switch is wired in series with the gas solenoid and is normally closed. The device is set to open at about 93° C and on some devices, will close again automatically, when the temperature falls by about 14 deg C. A number of overheat controls have to be manually reset.

(a) Overheat control (manual reset)

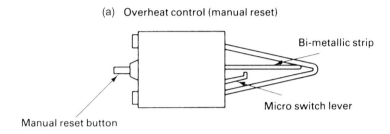

(b) Overheat control (auto reset)

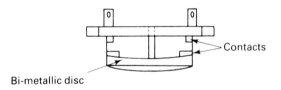

Fig. 35 Overheat Control: (a) Manual Reset; (b) Automatic Reset

Limit Thermostat

This is very similar to the overheat control and on many heaters it is the only temperature limiting device fitted. The limiting temperature is set by BS specification at 95° C for the mean temperature of the air at the outlet. The bimetal element is usually located in the outlet of the heat exchanger and the switch is wired in series with the gas solenoid. The switch is preset to cut out at about 93° C and usually has a differential of about 14 deg C. So the heater will be turned on again when the temperature falls to about 80° C.

Because the limit thermostat is automatically reset, it will cause the heater to cycle on and off every few minutes when an overheat condition occurs. An adjustable limit thermostat is shown in Fig. 36.

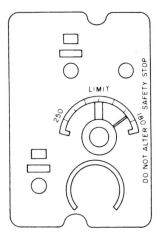

Fig. 36   Limit thermostat

Fan Switch

If the fan and the gas solenoid valve were both switched on at the same time, the fan would start to blow air into the rooms before the heat exchanger had had time to heat up. So there would be an influx of cold air which would make the rooms even more uncomfortable before the temperature started to rise.

To avoid this, most units are fitted with a thermally operated switch which delays the fan from running until the heat exchanger has reached an acceptable temperature, Fig. 37.

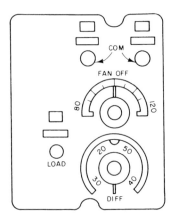

Fig. 37   Fan switch

The switch is wired in series with the fan and the contacts make on temperature rise. It is usually operated by a bimetal element located in the outlet of the heat exchanger. This element may be the same one that also operates the limit thermostat on a combined control device, Fig. 38.

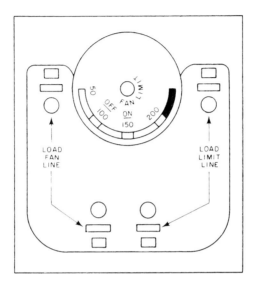

Fig. 38 Combined fan and limit switch

Most fan switches have two adjustments.
These, with their usual temperature settings are:

● fan 'off' at about 38° C
● fan 'on' at about 58° C or
● 'differential' of about 20 deg C.

In operation, the fan switch will switch on the fan when the air in the heat exchanger has reached about 58° C. After the gas solenoid has closed, the fan will remain running until the temperature of the air leaving the heat exchanger has fallen to about 38° C. In this way the residual heat from the unit is usefully employed.

Fan Delay Unit

Another type of fan switch is heated by a separate carbon resistor, Fig. 39. This is the 'fan delay unit' which consists of a micro switch operated by a bimetal strip which is heated by a resistor. The micro switch is in the mains voltage supply to the fan and the resistor is in series with the roomstat, in the low voltage circuit, Fig. 40.

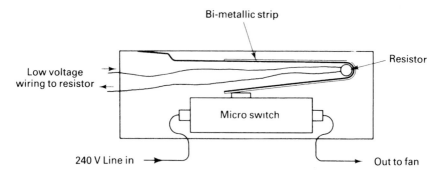

Fig. 39   Fan delay unit

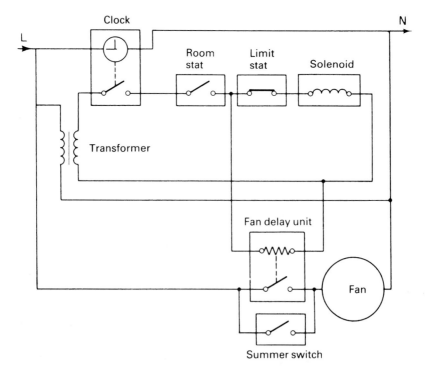

Fig. 40   Wiring to fan delay unit

When the clock is on and the roomstat calls for heat the solenoid valve is energised and the fan delay unit resistor starts to heat up.

After about 90 seconds, the bimetal strip operates the microswitch and switches on the fan. When the room reaches the required temperature the roomstat opens and the supply to the resistor is cut off. The fan then continues to run for a further 120 seconds while the resistor cools down.

The resistor has a value chosen by the manufacturer and is not adjustable. It will operate satisfactorily if the gas rate to the heater is correct.

Summer/Winter Switch

This is a simple, manually operated switch, fitted in the main voltage supply to the fan. It overrides the fan switch or fan delay unit. In the 'winter' position the switch is open and the fan switch controls the fan. If the user wishes to run the fan in the summer when the heating unit is not operating, the switch is changed to the 'summer' position. In this way the fan may be used to circulate cool air through the rooms.

Fan Speed Controls

On some belt-driven models the fan speed may be varied by altering the adjustable pulley, as described in Chapter 11. On most heaters the speed control is via a variable resistance or a transformer. These may vary the supply voltage and so control the speed.

Variable Resistance

This may be either a straight slide resistor or a small circular rheostat. The slide resistor consists of a wire-wound resistance with a sliding contact which can be clamped in any position, Fig. 41. One end of the coil is attached to the input terminal and the slide is moved away from the input end to increase the resistance and reduce the voltage. So the fan speed is reduced as the length of wire used is increased. Slide resistors can only be adjusted when the power is off.

The circular variable resistor, or rheostat, has a carbon brush on an arm which is moved round over a wire coil, Fig. 42. The moving arm carries current to one of two fixed connections. The other connection is attached to one end of the wire coil. Moving the arm clockwise increases the length of wire in circuit and so increases the resistance. This reduces the voltage and the fan speed. The arm may be moved by a setting screw which can be adjusted while the power is on and the fan is running. It is, therefore, much easier to get an accurate speed setting with this device.

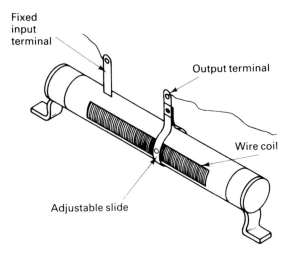

Fig. 41 Adjustable slide resistor

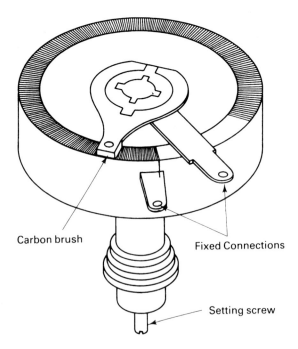

Fig. 42   Rheostat

Transformer

Voltage may be reduced by a step-down transformer. The tappings would be taken from the secondary winding as shown in Fig. 43.

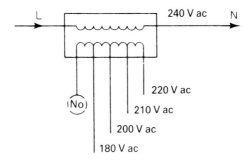

Fig. 43 Step-down transformer

More frequently the lower voltage tappings are taken from an auto-transformer, Fig. 44, or from a step-down transformer primary winding, Fig. 45. This allows the transformer to have its secondary winding supplying the lower voltage control circuit at the same time.

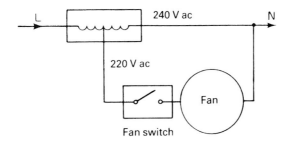

Fig. 44 Auto transformer

The voltage required may be selected by a rotary switch or by altering the position of a two-pin plug.

Modulating Control System

An electronic control system has been produced which can eliminate the fluctuations in room temperature experienced with the ordinary room thermostat.

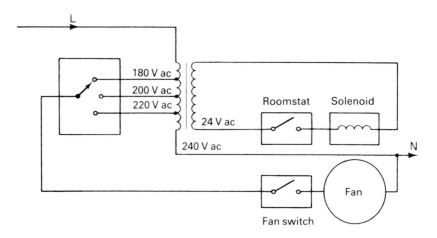

Fig. 45   Tappings on primary windings

This is achieved by modulating the fan speed to maintain air delivery temperatures at the required comfort level, within 2° C. The fan is controlled by a thermistor, which replaces the fan switch on the heat exchanger. At loads below 25% the fan cycles on and off so that the speed is never less than the 400 r.p.m. necessary for effective lubrication.

To maintain combustion efficiency, the gas burner is either full on or off. At loads below 95% the burner operates in 2 minute cycles with the 'on' period proportional to the load. So, at 50% load, the burner is on for 1 minute and off for 1 minute. Control is effected by a thermistor-based room thermostat linked to the control box on the heater.

One such system, marketed by Johnson and Starley, is the Modair-flow System, the diagram Fig. 46, shows how the system operates.

Fault Diagnosis and Remedy

Faults in central heating systems can be divided into three categories:

- gas faults
- electrical faults
- system faults.

A number of these faults have already been dealt with at the appropriate stages of the Manual and electrical work will be covered in Chapter 13. However, the following notes on the diagnosis and remedy of faults specifically on heating sytems may be of use.

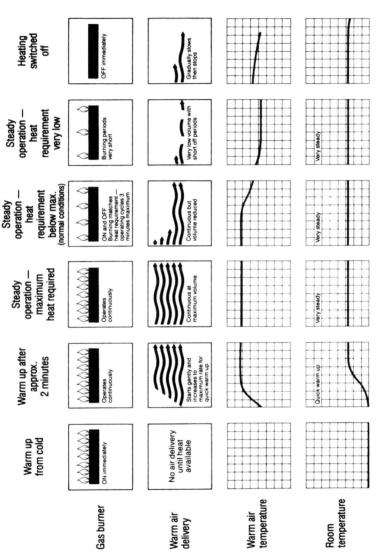

Fig. 46   Operation of Modairflow System (Johnson & Starley)

Gas faults or system faults in wet or warm air systems with electrical controls and a multifunctional control may be as follows:

Symptom	Action
No pilot light	Check that gas is turned on and present at appliance tap If a separate pilot supply, check at pilot tap Check for blockage of pilot supply or filter Check for blockage in pilot burner port Check pilot adjustment
Pilot ignites but does not stay alight	Check correct size of flame Check correct pilot position on thermocouple Check thermocouple – clean and tight connections – correct voltage output Check sealing on balanced flue heater or excessive draughts
Pilot correct no mains gas	Check supply from main gas solenoid to burner, burner cock on, injector clear Ensure all components controlling the valve are switched on and operating Check valve is not jammed closed

WET CENTRAL HEATING SYSTEMS

Main gas does not shut off	Check boilerstat phial correctly located and in good condition Check that boilerstat is operative Check main gas solenoid is not jammed open Examine valve and seating
Pump does not operate	Ensure that all components controlling the pump are switched on and operative Check whether pump impeller is jammed

Symptom	Action
Gas valve and pump work normally but no heat to radiators	*Check all isolating valves* *Check radiator valves, zoning valves diverter valve* *Check all air vents and bleed air* *Check water level and ball valve in feed and expansion cistern* *Check pump is fitted right way round*
Gas valve and pump work normally but no domestic hot water	*Check cylinder thermostat operative* *Check zone or diverter valve* *Check air vents on primary flow* *Check water level and ball valve in feed and expansion cistern*
Rusty or contaminated water at draw-off taps from single feed cylinder	*Check boilerstat for overheating* *Drain and refill system; if discoloration persists, cylinder is faulty*

WARM AIR SYSTEMS

Symptom	Action
Main gas does not shut off, fan continues running	*Ensure that all components controlling the solenoid valve are switched off* *Check that gas solenoid is not jammed open* *Examine valve and seating*
Main gas does not re-ignite	*Check all components controlling the solenoid valve are switched on* *Check overheat control, manually reset if necessary* *Check for cause of overheating, filter, registers, fan*
Gas valve opens, fan operates inter-mittently	*Check and adjust gas rate* *Check and adjust fan switch* *Check and adjust fan speed*
Fan runs too long after valve has closed, giving cool air	*Check and adjust fan switch* *Check fan delay unit cooling down*

Symptom	Action
Fan runs normally, gas burner operates intermittently	*Check all outlet registers open* *Check filter for blockage* *Check and adjust gas rate* *Check and adjust fan speed* *Check calibration of limit and overheat controls* *Check calibration and siting of roomstat*
Unit operates normally but gives too little heat	*Check filter for blockage* *Ensure all outlet registers fully open* *Check and adjust gas rate* *Check and adjust fan speed* *Check limit thermostat calibration* *Check roomstat calibration and siting*

CHAPTER 13

Electrical Work on Gas Appliances

Chapter 13 is based on an original draft by J. Norwebb

Introduction

Basic electrical theory was covered in Vol. 1, Chapter 9. Subsequent chapters have described a number of electrical instruments and control devices. It remains for this chapter to deal with the practical details of the installation and servicing of the electrical equipment associated with gas appliances.

Installation

The gas service engineer is not an electrician. So he will not normally install electrical circuits or socket outlets in a house. He may, however, connect gas appliances to the existing electrical supplies. And he should know how to test those supplies to ensure that they have the right polarity and have been properly and safely installed.

Flexes and Cables

Most gas appliances with electrical components are supplied with a 2 m length of flexible cord or 'flex', already attached by the manufacturer. If there is no plug fitted, the service engineer may have to fit one.

There are various types of 3-core flexible cords and the type selected must be suitable for the appliance and its situation. The flex commonly used is circular, with a PVC or vulcanised rubber sheath. This is suitable for internal or external use but it does not withstand high temperatures. With excessive heat PVC may melt and rubber will harden and crack.

When a heat-resistant flex is required, the sheathing may be of butyl rubber. For temperatures up to 200° C, glass fibre sheathing may be used, provided that it is not subjected to undue flexing, abrasion or dampness.

The size of the flex required depends on the current load of the appliance. Table 38 gives the size of the conductors and their current rating.

TABLE 38   Sizes and Ratings of Flexible Cords

| | Size | |
Current Rating (amps)	Cross-sectional area (mm^2)	Diameter (mm)
3	0.5	0.8
6	0.75	0.98
10	1.0	1.13
13	1.25	1.26
16	1.5	1.38
25	2.5	1.78
32	4.0	2.25

Some appliances are wired into a switched and fused spur. These may be fitted with a length of cable, rather than flex.

Cables generally have fewer, but heavier conductor wires and are consequently not fully flexible. Like flex, they are usually sheathed in PVC or rubber. In some cables the earth wire has no insulation and may be a single copper wire or about three wires twisted into a helix. Where there is no insulation on the earth wire the visible ends should be covered by a green/yellow sheath.

The colour coding of the wires in cables and flexes is:

- Brown (red)   = Live
- Blue (black)   = Neutral
- Green/Yellow (green) = Earth

(The old colours are shown in brackets)

Methods of Jointing

There are four principal methods of connecting wires:

- screwed terminal
  – wires secured beneath a screw head
  – wires in a sleeve, secured by a set screw
- push-on connector
- plug and socket
- soldered connections.

Secured by Screwhead

Where the wire is secured under a screwhead, it should be covered by a plate or washer. The end of the wire may be crimped or soldered on to a ring or spade end, Fig.1. 'Crimping' is carried out using the special tool shown in Fig. 2 and forms a strong joint on the wire.

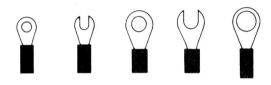

Fig. 1   Spade and ring ends

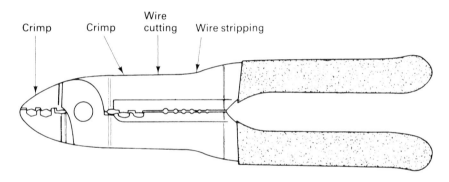

Fig. 2   Crimping tool

Alternatively, the end of the wire may be formed into a ring, Fig. 3. The procedure at (a) is as follows:

- strip off the insulation to give about 20 mm of bare wire
- divide the conductors into two and pull apart into a vee
- twist the two groups of wires round each other two or three times, then form a ring by placing a circular object, like a screwdriver blade, into the vee
- twist the ends of the wires three times, tighten the twist with pliers and cut off the surplus wire.

Another method shown at (b) has the wire twisted round itself after being formed into a ring round a screwdriver blade. With both these methods the blade used to form the ring should be slightly larger than the securing screw.

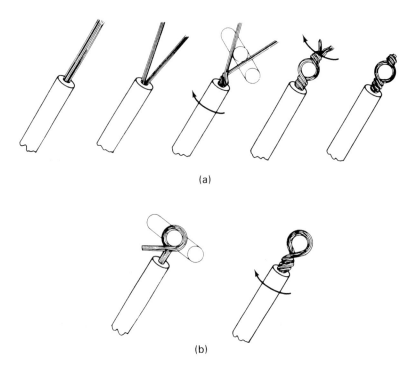

(a)

(b)

Fig. 3   Forming a ring end: (a) strands divided; (b) strands together

The joint can be made stronger mechanically by running solder into the ring and the twisted wire. In this case the wire should be cleaned and fluxed before the ring is formed.

Wire should never be just wrapped round the head of a screw. The insulation should be clear of the washer and there must be no stray ends of wire protruding.

Secured in a Sleeve

The type of terminal block in common use is the plastic bar containing a number of brass sleeves, each fitted with two set screws to secure the wires. These blocks are sometimes referred to as 'chocolate bars' because they are usually a dark brown colour and resemble a block of chocolate.

The conductor should fill the sleeve. In most cases, if a single wire is inserted, there is room to spare and it is usual to bend over the end of the conductor. This gives the screw plenty of wire to bite on and forms a mass behind the screw so that the wire cannot easily be pulled out. The preparation of the wire is shown in Fig. 4(a) and a completed joint in section at (b).

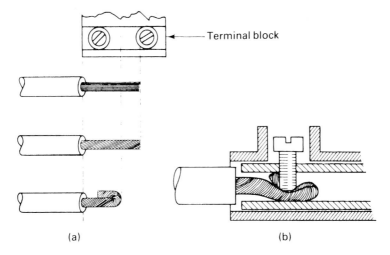

Fig. 4   Fitting wire in terminal blocks: (a) preparing the wire; (b) section through terminal block

If two wires are to be held by one screw the ends should be fanned out and then twisted together before inserting in the sleeve.

The insulation should be stripped so that it is well clear of the securing screw but there must be no uninsulated wire outside the terminal block.

With any screwed terminals, always use the correct screwdriver blade for the size of screw.

Push-on Connectors

These are made in a variety of forms, some of which are shown in Fig. 5. The ends are usually crimped on to the wire using the tool shown in Fig. 2.

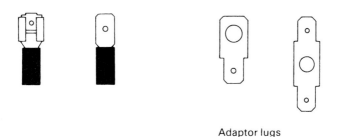

Adaptor lugs

Fig. 5   Push-on connectors

Some special tools can cut the wire, strip the insulation and compress the connector round the bare conductors. This forms a good electrical joint which is also mechanically strong.

Insulating sleeves are available which can be slipped over the connecting ends. When these are not fitted care must be taken not to touch any exposed live mains voltage connections.

Plug and Socket

As well as the mains plug and socket connection to the house wiring, many appliances have the individual electrical components connected by plug and socket connections.

These vary from a simple two-pin connection for a lamp, up to an eight-pin socket on a complex control system. The use of these connections simplifies the task of fault diagnosis by making it easy to isolate or remove a component.

Components may also be renewed with the minimum disruption to the wiring system.

Soldering

Soldering electrical connections is best done with an electric soldering iron. These are usually up to about 25 watt and may have interchangeable parts.

The wires must be cleaned and the ends twisted together so that there are no stray wires protruding. Usually components have soldering tags through which the wire is passed and then bent back on itself, Fig. 6. Resin cored solder is usually recommended for these joints.

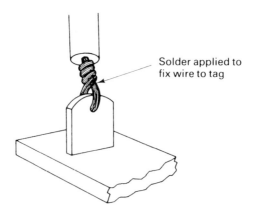

Solder applied to
fix wire to tag

Fig. 6   Soldered joint

A 'dry joint' is one which is not a good conductor of electricity. It may be due to dirt on the surface or lack of heat when making the joint.

Wires may be soldered directly together using the type of joint shown in Fig. 7. To make the joint carry out the following procedure:

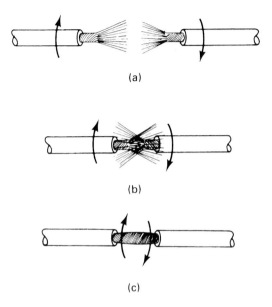

(a)

(b)

(c)

Fig. 7   Jointing two wires: (a) ends fanned and twisted; (b) wires married; (c) twisted ready for soldering

- slip an insulating sleeve over one wire and roll it clear of the end
- strip the insulation from both wires, fan out the conductors and twist them round for about $\frac{1}{3}$ of their length
- spread out the conductors in the shape of a cone
- bring the two ends of wire together so that the conductors interlace
- twist the conductors round so that each group marries with the twist of the opposite wire
- solder the joint
- roll the insulating sleeve over the bare section of wire.

Wiring Diagrams

There are different types of wiring diagrams of which the four most likely to be met are:

- ordinary electrical circuit diagram
- schematic wiring diagram
- functional flow diagram
- illustrated (pictorial) wiring diagram.

Circuit Diagram

An example is shown in Fig. 8 which is a diagram of the wiring for a central heating boiler and its controls. The purpose of the drawing is to show the path of the electric current to and through the system's components.

Although the actual connections to the main terminal block form the basis of the diagram, the rest of the wiring is theoretical rather

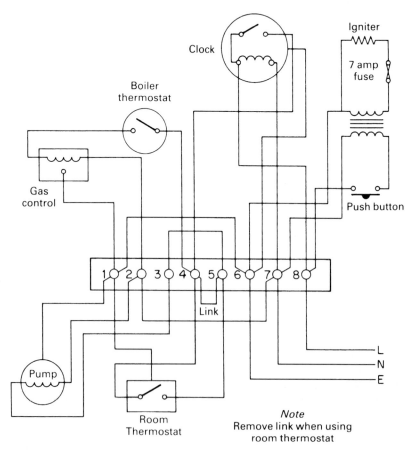

Fig. 8   Electrical circuit diagram

than actual. Each connection is shown as a separate wire whereas, in fact, they are grouped in two or three-core cables. Some diagrams may show common L or N connections to components whilst on the appliance itself are separate leads, or vice versa.

The layout of the diagram bears no relationship to the layout of the components and it is usually designed to facilitate drawing. Wires cross each other indiscriminately so that even a simple system like the one in Fig. 8 appears complicated.

BS 3939 specifies the symbols which should be used on wiring diagrams to indicate the various electrical components. Unfortunately many of the existing manufacturers' wiring diagrams use non-standard or non-preferred symbols.

Although the circuit diagram shows how the components are inter-connected, it does not clearly indicate the relationship of one device to another. Fault diagnosis is made easier by redrawing the wiring in a schematic or functional form.

Schematic Diagrams

Manufacturers have often produced schematic diagrams for inclusion in their installation and servicing instructions. These frequently dispense with the earth wires, which usually do not take any part in the operation of the components. So schematic diagrams are generally less complicated than circuit diagrams. Unfortunately they often employ non-standard symbols and they do not conform to any standard format or layout. Figure 9 is the wiring diagram of Fig. 8, redrawn as a schematic diagram. In this case it is fairly easy to see the relationship of the components but on more complex diagrams this is not always possible.

Schematic diagrams generally do not show the terminal blocks or terminal numbers so it is not always easy to relate them to the actual appliance.

Functional Flow Diagrams

This type of diagram was developed by British Gas. It was an attempt to produce a diagram which would be of the greatest use to the gas service engineer for fault diagnosis on electrical equipment associated with gas appliances.

Functional flow diagrams show clearly the dependence of one component upon another and the sequence of the control devices. The flow of electrical energy is shown in a similar way to the flow of gas in an all-gas control system. The position of terminals and their number is also shown so that the points from which voltage readings may be taken can be located easily.

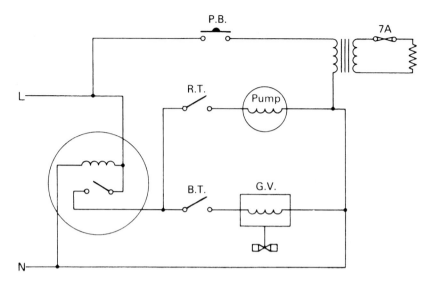

Fig. 9   Schematic wiring diagram

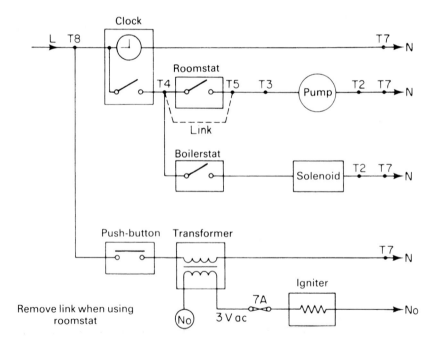

Fig. 10   Functional flow diagram

Figure 10 is a functional flow diagram of the system shown in Fig. 8. It conforms to the rules for drawing functional flow diagrams which are as follows:

- electrical energy is normally shown flowing from left to right and from the top of the page downwards
- the mains line connection is shown at the top left-hand side by an arrowhead and the letter L
- all the controls are shown in their correct sequence leading across to the 'working' components, that is, pumps, fans, solenoids, where the energy is made to do work
- all earth connections are ignored, unless they are part of an ignition or flame monitoring circuit
- neutral connections are taken across to the right-hand side of the diagram and end with an arrowhead and the letter N; the return to a mains double-pole switch is ignored
- the low-voltage output from a transformer is shown passing to the working component with the returning connection drawn as a neutral and marked No, as is the other secondary winding connection
- terminals are shown as a full dot on the line, as is a junction of two wires; the terminal number is also given, where possible
- wherever possible, wire colours are indicated by the following code:

b	– blue	Multicolours are shown:
bk	– black	g/y – green and yellow
br	– brown	bk/w – black and white
g	– green	and so on
gy	– grey	
or	– orange	
pk	– pink	
r	– red	
w	– white	
v	– violet	
y	– yellow	

- unless it is unavoidable, wires should not cross each other on the drawing
- components are shown enclosed in a rectangle or circle, whichever approximates to the actual shape of the component
- components are identified by their name beside or inside the shape
- where it may be necessary to fault-find in a component, its construction is indicated by the appropriate symbol; if the shape is left blank the component should simply be exchanged

- components which are linked mechanically are shown joined by a broken line
- switches are shown in their normal rest position.

Figure 10 gives an example of the application of these rules. The symbols in general use are listed in the later section on symbols.

Illustrated (Pictorial) Diagrams

These consist of drawings of the electrical components and wiring as they actually appear in their positions on the appliance, Fig. 11.

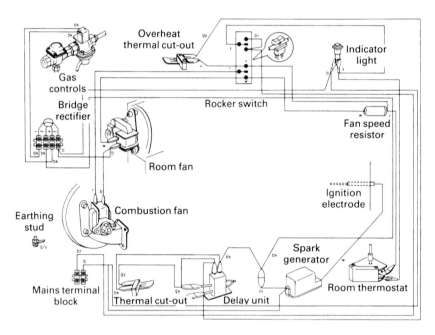

Fig. 11   Illustrated diagram (air heater)

The diagrams enable any component or cable to be immediately identified and its position located on an actual appliance.

Reading and Drawing Wiring Diagrams

Although at first glance Fig. 8 appears fairly complicated it is not difficult, with some practice, to follow the circuits in a logical sequence. To do this, consider one wire at a time and start with the mains line connection. If you follow the rules you can easily produce a functional flow diagram.

The points to remember are:

- locate the input terminals, L, N, E
- trace the path of the line wire first
- where it enters a terminal block and several wires leave the same terminal they are all line wires; trace one at a time
- trace each wire until it reaches a working component and note which of the switching components, that is, switches or thermostats, it goes through on the way
- each line wire ends at a working component and the wire which leaves the component is then neutral
- all neutral wires will go back to N on the input terminals
- in the case of low-voltage outputs from the secondary windings of a transformer they will leave from one tapping and pass through the switched components to the working component, returning to the other tapping and never to neutral, N
- the return tapping on a transformer is the one which is connected to the earth wire.

As an example of producing a functional flow diagram from a circuit diagram, compare Fig. 8 with Fig. 10 and follow the logic through.

On Fig. 8, the line input is at the lower right-hand side and the line wire goes to terminal 8 on the main block.

Here it splits into two paths. Take one vertically upwards.

This goes to the clock motor (a working component) and then returns through terminal 7 to N, neutral.

Within the clock it also goes through the switch and then to terminal 4 where it splits into three paths.

First, if there is no room thermostat, the link will be in position and it will go on to terminal 5.

If there is a roomstat, the link will have been removed and the line wire continues down to the roomstat and back to terminal 5.

From 5 it goes to terminal 3 and down to the pump.

The wire returning from the pump windings is a neutral and goes back to N via terminals 2 and 7.

The other switched line wire from terminal 4 goes up to the boilerstat and on to the solenoid valve.

Leaving the valve it becomes a neutral which goes back to N via terminals 2 and 7.

The only other line wire to be traced is that from terminal 8 which goes up to the right, through the ignition switch to the transformer.

Leaving the transformer it is a neutral returning to terminal 7 and N.

The low-voltage tapping supplies the glow-coil through a fuse and the circuit returns to the other transformer tapping, No.

Symbols

Although British Standard symbols have existed for a number of years, some appliance manufacturers appear to have disregarded BS 3939 and invented their own symbols. It is useful, therefore, to have some knowledge of the various symbols which may be met. Functional flow diagrams generally use either the BS symbol or a box labelled with the name of the component. The symbols for the principal electrical components are as follows:

Wiring and Connections (Fig. 12)

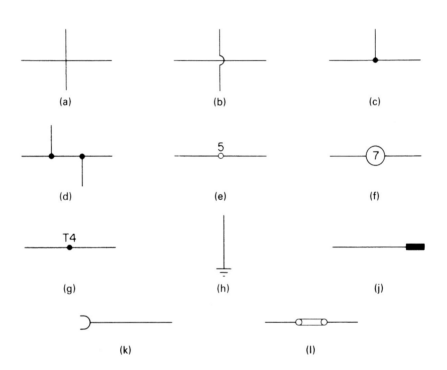

Fig. 12 Symbols, wiring and connections:

(a) wires crossing

(b) old, now incorrect symbol for wires crossing

(c) wires joined

(d) double junction

(e) terminal

(f) alternative terminal

(g) terminal, functional flow diagram

(h) earth

(j) plug

(k) socket

(l) link, normally closed, easily disconnected

Where conductors cross without making an electrical connection they should be shown as a straight crossover as at (a). Some manufacturers have used a semicircle as at (b) but this is not correct and tends to be confusing. Where wires are joined together the joint should be shown by a full dot on the line as at (c) and (d). Terminals may be shown as (e) or (f) and terminals on functional flow diagrams are shown at (g). Earth connections are shown at (h). Plug and socket connections are at (j) and (k) and a link which is normally closed by two plug-in contacts is shown at (l).

Switches (Fig. 13)

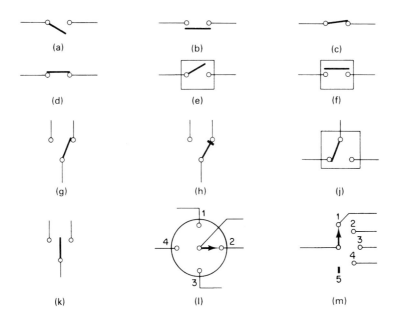

Fig. 13 *Symbols, switches:*

(a) *make contact*
(b) *make contact*
(c) *break contact*
(d) *break contact*
(e) *on/off switch, functional diagram*
(f) *push button or micro switch, functional flow diagram*
(g) *changeover, break before make*

(h) *changeover, make before break*
(j) *changeover, functional flow diagram*
(k) *two-way switch with off position*
(l) *rotary switch, functional flow diagram*
(m) *5-position switch with off position at 5*

Figures 13(a) and (b) show the normal type of switch with contacts that are usually open and which close when the switch is operated.

The switches in (c) and (d) have normally closed contacts which break when the switch is operated, (e) and (f) are symbols used in functional flow diagrams.

Changeover switches are shown at (g), (h) and (j), while (k) is a changeover switch with a neutral or off position when the contact arm is central. Figures 13(l) and (m) indicate rotary switches which might be used in programmers or as fan speed selectors.

All the contacts or switches illustrated are single-pole. Where it is necessary to indicate a double-pole switch this is done by showing two contacts with a mechanical link between them.

In functional flow diagrams, special types of switches, for example air pressure operated switches, are labelled clearly with their name.

Thermostats (Fig. 14)

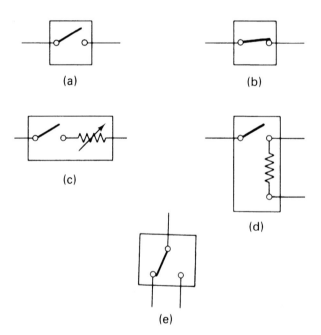

Fig. 14   *Symbols, thermostats: (a) simple thermostat; (b) limitstat or overheatstat; (c) roomstat with series accelerator, adjustable; (d) roomstat with shunt accelerator; (e) changeover thermostat for motorised valve control*

Thermostats are switches and so the same symbols can be used. Air or room thermostats, boiler, cylinder or frost thermostats, are generally shown as open contacts as at (a). Limitstats or overheat devices are usually shown closed as at (b).

Roomstats with accelerators are illustrated at (c) and (d). Some room or cylinder thermostats may be designed to operate zone valves and may incorporate changeover switching as shown at (e).

Clocks and Timers (Fig. 15)

The **BS** symbol for a clock is a circle with clock hands at 9 o'clock, Fig. 15(a). Where the clock controls single contacts, these are included in the circle, (b). When multiple contacts are controlled these may be shown outside the circle but connected to it by a broken line. Functional flow diagrams of three and four-terminal clocks are shown at (c) and (d) respectively. Where timers have multiple or changeover contacts these are shown within the enclosing rectangle.

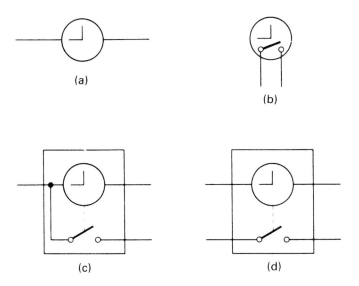

(a)

(b)

(c)

(d)

Fig. 15   Symbols, clocks and timers: (a) clock; (b) time switch, BS; (c) time switch, 3 pin, functional flow diagram; (d) time switch, 4 pin, functional flow diagram

Relays (Fig. 16)

In **BS** symbols, the mechanical construction of the unit may not be indicated. Make and break contact units are shown at (a) and (b).

The coil is shown as a rectangle with the connections to its longer sides. The resistance of the winding may be shown in the rectangle, the number indicating the number of ohms, Fig. 16(c).

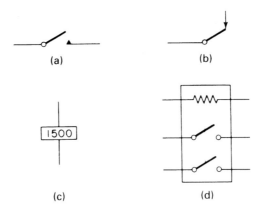

Fig. 16   *Symbols, relays: (a) make contact; (b) break contact; (c) relay coil, 1500 Ω; (d) relay, functional flow diagram*

In functional flow diagrams the relay is shown as a single, complete unit with the winding and the contacts contained in a rectangle, Fig. 16(d).

Fuses (Fig. 17)

The BS general symbols are shown at (a) and (b) with the alternative symbol at (c). This latter is used in functional flow diagrams and the value of the fuse rating is normally indicated as at (d).

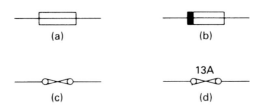

Fig. 17   *Symbols, fuses: (a) fuse, general symbol; (b) supply side indicated by black bar; (c) alternative symbol, functional flow diagram; (d) 13 amp fuse*

Resistors (Fig. 18)

The general symbol for a fixed resistor is shown at (a). Its alternative, at (b), is used in functional flow diagrams, usually to indicate glow coils or heater elements. The number of vertices may be varied as required.

Variable resistances are shown at (c) and (d). The alternative symbol is used in functional flow diagrams.

When a variable resistor is pre-set, it is shown as at (e).

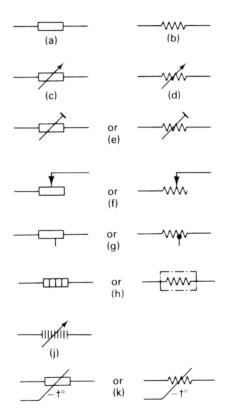

Fig. 18   Symbols, resistors:
(a) fixed resistor
(b) alternative symbol,
    functional flow diagram
(c) variable resistor
(d) alternative symbol
(e) resistor with pre-set
    adjustment
(f) resistor with moving contact
(g) resistor or voltage divider
    with fixed tapping
(h) heater
(j) carbon pile resistor
(k) thermistor

Resistors with moving contacts may be indicated by (f).

Voltage dividers, or resistors which have a tapping taken from them are shown at (g). If the tapping is by means of a moving contact, it is indicated by an arrowhead as used in (f).

Resistors which are used as heating elements are indicated by (h) and a carbon-pile resistor is shown at (j).

A thermistor, that is, a resistor with a resistance which decreases as its temperature rises, is shown at (k).

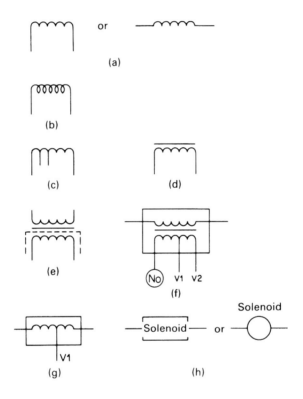

Fig. 19   *Symbols, windings and coils:*

(a) winding of a coil, solenoid, transformer, choke coil or inductor, functional flow diagram

(b) alternative symbol, not recommended

(c) winding with tappings

(d) winding with a core

(e) transformer with core and screen

(f) transformer with two voltage tappings, V1, V2, functional flow diagram

(g) auto-transformer, functional flow diagram

(h) solenoid valves, functional flow diagram

Windings and Coils (Fig. 19)

The general symbol for a winding or coil is shown at (a). This symbol is also used on functional flow diagrams.

The old symbol which shows a spiral coil, as at (b), is no longer recommended and is being superseded by symbol (a).

Figure 19(c) shows a winding with two tappings such as might be used on a transformer.

Where the winding has a core it is indicated by a line, as at (d). The core is magnetic iron unless otherwise indicated.

A transformer with an iron core and screen between the primary and secondary windings is shown at (e). Functional flow diagrams of transformers have the secondary neutral marked No and include the voltage of the various tappings, Fig. 19(f). An auto-transformer is shown at (g).

On functional flow diagrams the gas solenoid valves are shown by rectangles or circles with the name inside or adjacent, Fig.19 (h).

Capacitors (Fig. 20)

The **BS** general symbols for ordinary and variable capacitors are shown at (a) and (b) respectively. When a capacitor is polarised this is denoted by a + sign against the positive plate. In functional flow diagrams the capacitor general symbol is used, enclosed within a rectangle.

(a)   (b)

Fig. 20   Symbols, capacitors: (a) capacitor; (b) variable capacitor

Pumps and Fans (Fig. 21)

The actual windings of pump or fan motors are rarely shown. Most appliance manufacturers use a rectangle or, more commonly, a circle such as is used in the functional flow diagrams at (a) and (b). When the motor incorporates a capacitor which can easily be renewed this is indicated by the symbol touching the circle, Fig.21 (c).

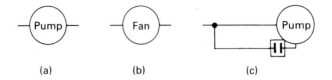

Fig. 21   Symbols, pumps and fans: (a) pump; (b) fan; (c) pump with capacitor

Lamps (Fig. 22)

Illuminating lamps are shown at (a) and (b), the first being an ordinary filament lamp and the second a neon. When a lamp (of whatever design) is used to indicate that a device is operating it should be represented by symbol (c).

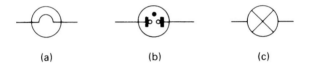

Fig. 22   Symbols, lamps: (a) illuminating lamp, filament type; (b) illuminating lamp, neon type; (c) indicating lamp, all types

Rectifiers (Fig. 23)

A single diode is shown at (a) and a full bridge circuit at (b). The functional symbol for a bridge rectifier is a rectangle or a circle with the input a.c. connections marked with ~ and the load shown as + and −.

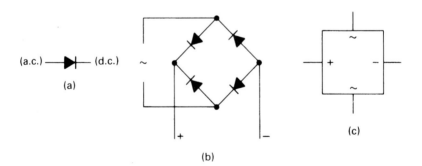

Fig. 23   Symbols, rectifiers: (a) single diode; (b) bridge rectifier; (c) bridge rectifier, functional flow diagram

Batteries (Fig. 24)

A single primary or secondary cell is shown by a long thin line representing the positive pole and a short thick line representing the negative pole. Batteries or cells are shown by alternate long and short lines as at (b) or by two cells joined by a broken line, Fig.24 (c). The voltage of the battery may be written along the broken line.

(a)   (b)   (c)

Fig. 24   Symbols, cells and batteries: (a) single cell; (b) battery of three cells; (c) alternative symbol

Instruments (Fig. 25)

Instruments for measuring current, voltage or power are indicated by circles or rectangles containing the initial letter of the units being measured.

(a)   (b)   (c)   (d)   (e)

Fig. 25   Symbols, indicating or measuring instruments: (a) galvanometer; (b) ammeter; (c) voltmeter; (d) wattmeter; (e) watt-hour meter

Time Delay Units (Fig. 26)

The heater is usually on a low-voltage supply although it may, on some heaters, operate on mains voltage. The symbol shown is used on functional flow diagrams and represents a unit which closes on heating. Some units open the switch when heated and are shown with closed contacts.

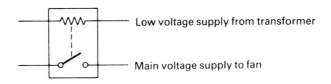

Low voltage supply from transformer

Main voltage supply to fan

Fig. 26   Symbols, time delay unit

Ignition and Flame Electrodes (Fig. 27)

On functional flow diagrams the spark electrodes of ignition devices and the flame electrodes of flame monitoring devices use a similar symbol. The difference is that the ignition electrode has a zigzag line, representing a spark, between the contacts.

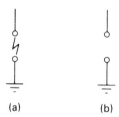

(a)   (b)

Fig. 27   Symbols, electrodes: (a) spark electrode; (b) flame electrode

Miscellaneous Functional Symbols

There are a number of other electrical or electronic devices for which symbols have not been included. This is because the functional flow symbols for them are easy to recognise and self-explanatory.

In the case of programmers and electronic full sequenced control boxes, a rectangle with the numbered terminal positions is adequate. Other controls such as heat motors, motorised valves and zone or diverter valves are indicated by rectangles or circles with the name of the control in, or adjacent to, the symbol.

FAULT DIAGNOSIS

Preliminary Tests

Before attempting to find any reported fault on an electrical system there are a number of tests which must be carried out to ensure that the equipment is safe to work on. These are:

- check that the appliance is properly earthed
- check that there is mains voltage at the appliance terminal block
- if not, check the inlet fuse
- if the fuse has blown, check for a short circuit and remedy any fault before replacing the fuse
- if the fuse has not blown there is a fault on the inlet wiring which must be remedied before continuing

- when the mains voltage is correct check the polarity of the terminals and remedy any fault
- when the polarity is correct ensure that the live supply is satisfactorily insulated by checking the resistance between line and earth connections
- when the resistance is satisfactory, the electrical supply should be correct and may be operated safely.

The tests should be carried out in the order in which they are listed. They should go before any other fault finding procedure. If remedying a fault requires the breaking and remaking of electrical connections, then the checks for earth continuity, polarity and resistance to earth must be repeated when the work is completed.

Details of the tests are as follows:

1. Earth Continuity Check

Isolate the appliance by removing the plug from the socket outlet.

Using a multimeter set to ohms × 1 scale, test between any appliance earth point and the earth pin on the plug. The resistance should be less than 1. If the resistance is greater than 1 check all earth wires for continuity and ensure that all connections are clean and tight.

If the resistance is still greater than 1 the fault still exists and it could be dangerous to work with the appliance live. Either report the fault or call in an electrician.

2. Mains Voltage and Polarity Check

The first operation of the polarity check tests the mains voltage. So the checks can follow consecutively if mains voltage is correct.

After the earth continuity check, reconnect the appliance to the mains supply.

Set the meter to the 300 V a.c. scale and test at the appliance terminal block.

Test at terminals:

- L and N — meter should read approximately 240 V a.c.
- L and E — meter should read approximately 240 V a.c.
- N and E — meter should read from 0 to 15 V a.c.

These readings indicate that the L or line terminal is live. If the voltage (0 to 15 V a.c.) reading occurs between any terminals other than Neutral and Earth, there is an electrical fault.

The fault should be dealt with as follows:

All terminals read zero. Check whether the main fuse has blown, if so check for a short circuit. Remedy the fault and replace the fuse.

If the fuse has not blown, check the terminal at the appliance plug and, if necessary, at the socket outlet. If the fault is still present it is on the house wiring and must be dealt with by an electrician.

N to E reads 240 V a.c. L to N will also read 240 V a.c. and L to E 0 to 15 V a.c. This indicates the wrong polarity and the fault must be remedied before the appliance may be used.

As with the previous fault, repeat the test at the appliance plug to the appliance. If necessary check the socket outlet or spur box.

If the fault still occurs at this point it is on the house wiring system and should be dealt with by an electrician. The customer should be warned not to use the appliance until the polarity has been corrected.

3. Short Circuit Check

Isolate the appliance, set the clock and all switches on and all thermostats calling for heat.

Set the meter on ohms × 1 scale.

Test from L to N on appliance terminal block – if meter reads zero there is a short circuit.

Set the meter on ohms × 100 scale.

Test from L to E on appliance terminal block – if meter reads less than infinity there is a fault.

On some occasions it may be found that a fuse has failed but the fault is not apparent. Check for traces of burning or arcing and, if necessary, carry out a continuity check on each component separately.

4. Resistance to Earth Check

This is the same as the second part of the short circuit test.

Isolate the appliance, set all switches on the thermostats calling for heat.

Set the meter to ohms × 100 scale.

Test between L and E on appliance terminal block – if meter reads less than infinity there is a fault.

Isolate the faulty section by turning off the switches or thermostats in turn and carry out a continuity check to trace the faulty component.

Fault Finding, Basic Points

The basic elements of fault finding were discussed in Vol. 1, Chapter 11. The main points to remember are:

- find out as much as you can from the user, take time to get the whole story before taking any action
- always refer to the manufacturer's instruction sheets where provided
- study the wiring diagram to discover the relationship between switching and working components; the component which is not operating may not be the one which is faulty
- if the wiring diagram appears complicated, try redrawing it in a functional flow form, tracing one wire at a time
- when the sequence of the components and their method of connection is understood it is often possible to check faulty devices directly from the main appliance terminal block
- if you have an intrinsically safe electrical test meter, like the BGC multimeter, the simplest way of checking components and circuits is by checking the voltage through the system, in the same way that you would check gas pressure
- always ensure that the appliance is properly earthed before testing mains voltage connections
- if you have some other type of meter you may have to resort to continuity and resistance testing which is often inconclusive; a number of components are wound to very wide tolerances and it is not easy to judge whether a resistance reading indicates a fault condition or not
- when carrying out continuity checks ensure that the appliance is electrically isolated
- it is often possible to discover the location of a fault by operating switches or thermostats to check whether other circuits are working and precisely which components are affected
- always check for signs of obvious damage including
  – loose wires
  – scorch marks
  – overheated components
  – arcing
- it is not enough to remedy a fault, you have to remove the cause of the fault, otherwise the fault will recur
- treat your electrical test meter with respect and carry out any daily checks to ensure that it is reading correctly
- when using a multirange instrument to read voltage or current, always start with it set to the highest scale and then move down through the ranges until a reasonable scale deflection is obtained
- examine components to discover their operating voltage before taking readings.

CHECKING COMPONENTS

The tests which follow are designed to be carried out using a suitable Multimeter. It could be dangerous to attempt them with any other instrument.

Solenoid Valves

Solenoids may operate at input voltages from 240 V to 20 m V and be either a.c. or d.c. If the voltage is not known start with the meter on its highest range and work down.

Check:

- input connections clean, tight and correctly positioned
- nominal voltage at input connections is correct.

If the input voltage is incorrect, there is a fault elsewhere in the system.

If the valve does not operate with the correct voltage input, isolate and exchange the coil or the valve as appropriate.

Repeat the check.

Transformers

Transformers may operate at the following voltages:

Input voltage 240 V a.c. mains

Output voltages, typical values 200 to 150 V a.c.

24 ''
12 ''
8.5 ''
7.5 ''
6.5 ''
3 ''
1.5 ''

Check:

- input connections clean, tight and correctly positioned
- correct nominal voltage at input connections
- output voltage is correct value.

If the input voltage is incorrect, there is a fault elsewhere in the system.

If the output voltage is incorrect, isolate and exchange the transformer.

Repeat the check.

Rectifiers

Diodes

Isolate the component, set the meter to ohms × 100 scale.
Test across the rectifier:
1. from left to right
2. from right to left
One reading should be infinity.
The other should be a low ohms reading.
If both readings are infinity or if both readings are low the rectifier is faulty and should be exchanged.
Repeat the check.

Bridge Rectifier

With the electrical supply switched on, check:

● correct nominal a.c. voltage at input terminals
● correct d.c. voltage at output terminals

If the input voltage is incorrect there is a fault elsewhere in the system.
If the output voltage is incorrect, isolate and exchange the rectifier.
Repeat the check.

Capacitors

Isolate the capacitor connections, set the meter to ohms × 100 scale.
Connect the meter across the capacitor connections, the pointer should 'kick'.
If the pointer does not kick, the capacitor is faulty and should be exchanged.
If the pointer kicks and then it shows any reading other than infinity, the capacitor is faulty.
If the pointer kicks and then reads infinity, set the meter to the 30 mA scale and connect it across the capacitor terminals in the opposite direction to the first test. The pointer should kick.
If the pointer does not kick the capacitor is faulty and should be exchanged.
If the meter pointer kicks, the capacitor is satisfactory.
When exchanging a capacitor, check the replacement component.

Capacitor Faults

A capacitor is open circuit when:

- the pointer fails to kick on the first test and then gives an ohms reading above zero.

A capacitor is short circuit when:

- on the first test, whether the pointer kicks or not, it reads zero ohms.

A capacitor is leaking when:

- after giving a kick on the first test it shows an ohms reading between zero and infinity
- the pointer fails to kick on the second test.

Pumps and Fans

With the electrical supply switched on, check the nominal voltage at the input terminals.

If the input voltage is incorrect, check that the connections are clean and tight and wiring is satisfactory.

If the input voltage is still incorrect, there is a fault elsewhere in the system.

If the input voltage is satisfactory, switch off the supply and check whether the rotor is free to rotate.

If it is not free, this must be remedied before continuing with the test.

If the rotor is free, check any capacitor which may be fitted and exchange if faulty.

If the pump or fan motor still does not work it is faulty and must be exchanged.

After renewing, adjust pump head or fan speed.

To Free a Pump

If the pump has a manual clutch device, operate this by means of a screwdriver after first switching off the electricity and closing the isolating valves. Turn the spindle about six times to free any sediment. If may be necessary to press in against a spring to engage the clutch. Open the valves and vent the pump before switching on.

If it is necessary to strip the pump down, continue as follows:

Isolate the pump, remove it from the pipework and dismantle. Clean out any sludge from the casing and impeller. Clean the rotor shaft, the outer surface of the rotor and the inside of the rotor housing.

Check the bearings. If the bearings are worn, change the pump. Do not oil the bearings, they are water-lubricated.

Do not run a new pump without water.

To Free a Fan

Isolate the fan, check the impeller and the bearings. Lubricate if necessary in accordance with the manufacturer's instructions. Clean the impeller if necessary. If the bearings are worn or seized, exchange the fan, rotor or bearings as appropriate.

Switches

Check:

- input connection clean, tight and correctly positioned
- correct nominal voltage from L to E terminals (this confirms that the meter is reading correctly)
- switch in 'off' position, correct nominal voltage between input and output terminals
- with switch in 'on' position, meter reads zero volts between input and output terminals.

If the voltage is incorrect there is a fault elsewhere in the system.

If the voltage does not fall to zero when the switch is turned on, the contacts are dirty or faulty.

If the switch is a changeover type test the contacts in each of the two 'on' positions.

Thermostats

Thermostats may operate at the following voltages:

Roomstats	240 V a.c.	24 V d.c./a.c.	12 V d.c./a.c.
Boilerstats	240 V a.c.	24 V d.c./a.c.	
Cylinderstats	240 V a.c.	24 V d.c./a.c.	
Froststats	240 V a.c.	24 V d.c./a.c.	
Fanstats	240 V a.c.		
Limitstats	240 V a.c.	30 mV d.c.	

Thermostats are basically switches and should be tested in a similar manner.

Generally, boilerstats, fanstats, and some roomstats are simple switches. Cylinderstats may be changeover switches. Some roomstats may have accelerator heaters wired in series or in parallel to the switch. To check the heater, isolate and test for continuity across the connections. If the heater is adjustable, check that it is set in accordance with the manufacturer's instructions.

If the thermostat may be adjusted over a range of temperature settings, turn the setting knob from the maximum to the minimum setting and back again to maximum.

Check that the contacts make and break satisfactorily.

If the thermostat is faulty it should be exchanged and the replacement should be checked.

Always ensure that the user understands the operation of the thermostat.

Clocks and Timers

These usually operate on 240 or 24 V a.c. They have either three or four terminal connections.

On three-terminal clocks, both the clock motor and the switched output line are at the same voltage. On four-terminal clocks the voltages may differ as follows:

- clock and switched output line both 240 V a.c.
- clock and switched output line both 24 V a.c.
- clock 240 V a.c., switched output line 24 V a.c.
- clock 24 V a.c., switched output line 240 V a.c.

Before working on any timer ensure that the isolating switch has isolated both the clock motor and the switched supplies.

To check the timer proceed as follows.

With the electrical supply switched on, check the nominal voltage at the input terminals.

If the voltage is incorrect, check that the connections are clean and tight and the wiring is satisfactory.

If the input voltage is still incorrect, there is a fault elsewhere in the system.

If the input voltage is satisfactory, check that the clock motor is operating.

If not, isolate the timer and check the motor and the fuse for continuity. Replace the fuse or the motor as necessary.

Turn the dial so that the tappets revolve and check that the contacts operate and switch the outlet line. If correct, set the dial to the correct time of day.

Always ensure that the user understands how to operate the timer.

If the contact mechanism is faulty, replace the timer.

When replacing a timer it is usually essential that it is exchanged for one of exactly the same type. However, under some circumstances a three-terminal clock may be used to replace a four-terminal type provided that the voltages of the clock and the switched output line are the same.

Figure 28 shows the usual wiring for a four-terminal clock with the control system fitted on the output switched line. Figure 29 shows a three-terminal clock with a similar wiring system. If this three-terminal was replacing an existing four-terminal type it would only require for the inlet wiring to the contact terminal, shown as a broken line, to be removed. The clock would then operate satisfactorily.

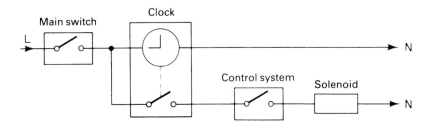

Fig. 28   Four-terminal clock with control system on switched line

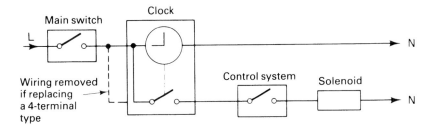

Fig. 29   Three-terminal clock with control system on switched line

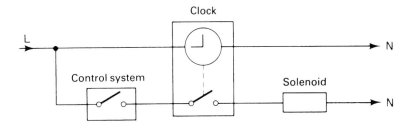

Fig. 30   Four-terminal clock with control system on input line

Similarly, a four-terminal clock could be used to replace a three-terminal type, if the additional link between input line and the input contact terminal was fitted externally.

However, when the control system is fitted on the input line to a four-terminal clock, Fig. 30, the clock cannot be replaced by a three-terminal type under any circumstances. If the exchange was to be made, the internal link between input line and the contact would bypass the control system and render it inoperative.

Programmers

The programmer should first be checked as a timer. If the timing mechanism is correct, set all switches and stats to 'on' and operate the programming switches.

Check that power is switched to all output switched lines.

If faulty, exchange the programmer for one of an appropriate type.

Thermocouples

Faults on thermoelectric devices and thermocouples are dealt with in Vol. 1, Chapter 11.

The output voltages of thermocouples, both on and off load may be checked by means of a multimeter and thermocouple interrupter as shown in Fig. 31.

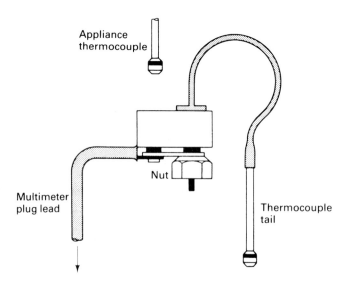

Fig. 31   Thermocouple interrupter

If the pilot can be lit, the flame is the right size and correctly positioned, the thermocouple is clean, uncorroded and properly connected, but the flame goes out when the button is released, either the thermocouple or the solenoid valve is faulty.

The solenoid can be checked using the previously described procedure. The thermocouple can be tested as follows.

Disconnect the appliance thermocouple from the control device.

Connect the appliance thermocouple to the interrupter by the existing nut or the M10 split nut if the existing nut is not threaded M10.

TABLE 39   Typical Thermocouple Voltages

Open Circuit Voltage (mV)	Expected On-load Voltage Range (mV)	
	Thermocouple only	Thermocouple + switch
15	3 – 8	6 – 12
16	3 – 9	7 – 12
17	4 – 9	7 – 13
18	4 – 10	8 – 13
19	5 – 10	8 – 14
20	5 – 11	9 – 15
21	5 – 12	10 – 16
22	6 – 12	10 – 17
23	6 – 13	11 – 17
24	7 – 13	11 – 18
25	7 – 14	12 – 19
26	7 – 15	13 – 20
27	8 – 15	13 – 21
28	8 – 16	14 – 21
29	9 – 16	14 – 22
30	9 – 17	15 – 23

Connect the thermocouple tail to the control device using the appropriate split nut.

Set the meter to the 30 mV d.c. scale and connect the interrupter lead to the meter socket.

With the nut unscrewed the meter will read the open circuit voltage.

With the nut screwed down tightly on to the connecting bar the meter will read the on-load voltage. Typical voltages are given in Table 39.

If the open circuit voltage is less than 15 mV for thermocouple only or 16 mV for thermocouple and switch and the solenoid is satisfactory, renew the thermocouple.

If the on-load voltage is greater than the expected value, the solenoid is probably faulty.

If the on-load voltage is less than 8 mV for thermocouples or 12 mV for thermocouple and switch and the solenoid is satisfactory, renew the thermocouple.

With all low-voltage circuits and particularly with thermocouples it is essential that all contacts are clean and tight.

Ignition Systems

Battery Filament Systems

Check that the pilot can be lit manually and correct any faults.

Remove the batteries and check that the housing contacts are clean and tight.

Check voltage output of the batteries. This should be at least 1.5V d.c. per cell on open circuit. New batteries should always give an open circuit reading above the nominal voltage.

If the output voltage is less than 1.5 V d.c. exchange the batteries. Check the operation of the switch as previously described.

With the switch 'on', check:

● minimum voltage at batteries, 2.4 V d.c.
● minimum voltage at igniter head, 2.2 V d.c.

If the outputs are below these values, check the wiring, clean and tighten the contacts. Check the igniter head and exchange if necessary.

Repeat the check.

If the ignition system still fails to work there is a fault in the gas system or in the orientation of the igniter head.

Mains Filament Systems

Check that the pilot can be lit manually and correct any faults.

Check input and output voltages from the transformer as previously described. Note the open circuit output voltage.

Check the switch and the fuse, replace fuse if necessary.

With the switch 'on', check:

● minimum voltage at transformer output terminals
● minimum voltage at igniter head.

Minimum values are given in Table 40.

TABLE 40   Typical Transformer Voltages

Open Circuit Output at Transformer Output Terminals (V a.c.)	Expected Minimum On-load Voltages (V a.c.)	
	At transformer output terminals	*At igniter head*
3.0	2.4	2.2
6.3	5.2	4.8
7.5	6.0	5.5
8.5	6.8	6.2
12.0	9.6	8.8

If the on-load voltages are below the expected values check the wiring, clean and tighten the contacts.

Check the igniter head and exchange if necessary.

Repeat the check.

If the ignition system still fails to work there is a fault in the gas system or in the orientation of the igniter head.

Battery Spark Systems

Check, as for battery filament systems:

- pilot can be lit
- battery housing contacts are satisfactory
- output voltage of batteries is at least 1.5 V d.c. per cell
- switch operates correctly.

Attempt to light the burner. If there is no spark carry out the following procedure.

Check the input voltage to the spark generator, this is usually 1.5 or 3 V d.c.

Check connections are clean and tight and wiring is satisfactory.

Check continuity of the high tension (HT) wire from generator to spark electrode.

Check visually for sparks from HT lead or electrode to adjacent metal component, or for cracks in the electrode insulator.

Set the meter to ohms × 100 scale, disconnect HT lead from spark generator and test from the disconnected lead to any earth point. The meter should read infinity.

Check continuity from earth electrode to generator earth.

Check the width of the spark gap and the alignment of the electrodes. Adjust if necessary.

Re-attempt to light the burner. If there is still no spark, exchange the generator unit.

If there are sparks but the burner does not light there is a fault in the gas system.

Mains Spark Systems

These systems operate at the following voltages:

240 V a.c.

3 V a.c.

1.5 V a.c.

On some models the transformer and switch may be integral with the spark generator unit.

Check the components of the system as previously described:

- pilot can be lit
- transformer
  - input voltage correct
  - output voltage correct
  - properly earthed
- switch operates satisfactorily
- fuse is correct
- generator input voltage is correct
- connections clean and tight, wiring satisfactory
- HT lead
  - continuity satisfactory
  - no spark leakage
  - resistance is infinity
- earth continuity from electrode satisfactory
- spark gap and electrodes correctly adjusted.

When the igniter unit works satisfactorily, always ensure that the user understands how to operate it.

Index